Biomathematics

Volume 21

Thomas Nagylaki

Introduction to
Theoretical
Population Genetics

With 55 Figures

Springer-Verlag

Berlin Heidelberg New York
London Paris Tokyo
Hong Kong Barcelona
Budapest

Thomas Nagylaki
Department of Ecology and Evolution
University of Chicago
Chicago, IL 60637, USA

ISBN-13: 978-3-642-76216-1   e-ISBN-13: 978-3-642-76214-7
DOI: 10.1007/978-3-642-76214-7

Library of Congress Cataloging-in-Publication Data
Nagylaki, Thomas, 1944-. Introduction to theoretical population genetics /
Thomas Nagylaki. p. cm.
(Biomathematics; v. 21)
Includes bibliographical references and indexes.
ISBN-13: 978-3-642-76216-1
1. Population genetics – Mathematical models.   I. Title   II. Series.
QH455.N32   1992   575.1'5'011–dc20   91-18943   CIP

© Springer-Verlag Berlin Heidelberg 1992

Softcover reprint of the Hardcover 1st edition 1992

Typesetting: Springer T$_E$X in-house system

41/3140-543210   Printed on acid-free paper

Preface

This book covers those areas of theoretical population genetics that can be investigated rigorously by elementary mathematical methods. I have tried to formulate the various models fairly generally and to state the biological assumptions quite explicitly. I hope the choice and treatment of topics will enable the reader to understand and evaluate detailed analyses of many specific models and applications in the literature.

Models in population genetics are highly idealized, often even over-idealized, and their connection with observation is frequently remote. Furthermore, it is not practicable to measure the parameters and variables in these models with high accuracy. These regrettable circumstances amply justify the use of appropriate, lucid, and rigorous approximations in the analysis of our models, and such approximations are often illuminating even when exact solutions are available. However, our empirical and theoretical limitations justify neither opaque, incomplete formulations nor unconvincing, inadequate analyses, for these may produce uninterpretable, misleading, or erroneous results. Intuition is a principal source of ideas for the construction and investigation of models, but it can replace neither clear formulation nor careful analysis. Fisher (1930; 1958, pp. x, 23–24, 38) not only espoused similar ideas, but he recognized also that our concepts of intuition and rigor must evolve in time.

The book is neither a review of the literature nor a compendium of results. The material is almost entirely self-contained. The first eight chapters are a thoroughly revised and greatly extended version of my published lecture notes (Nagylaki, 1977a).

The little basic genetics needed for understanding this book is presented at certain points. Although neither experimental population genetics nor data analysis are discussed, typical parameter values are given in order to motivate approximations. The demands on the reader's mathematical ability and industry exceed those on his formal preparation. Calculus and linear algebra are used freely. Although other mathematical techniques are developed as required, previous exposure to elementary probability theory would be helpful. Some knowledge of elementary differential equations would ease the reader's way through the three sections on continuous selective models.

The aim of the brief, qualitative introductory chapter is to orient the reader by describing the evolutionary forces and the three most influential theories of evolution, due to Fisher, Wright, and Kimura. The next chapter concerns asexual haploid populations. Its purpose is to formulate and analyze in a simple setting some of the problems studied in this book. After obtaining this overview of ideas, methods, and results, the reader will be better prepared for the more complex formulations and analyses that include mating, Mendelian segregation, and recombination.

In Chap. 3, we shall examine the structure of a randomly mating Mendelian population in the absence of evolutionary forces. The approaches and results presented here are useful in human genetics and are required for the rest of the book. The basic theory of selection at an autosomal locus, and some of its variants and extensions, are developed in Chap. 4. This fundamental theory is needed in parts of all the succeeding chapters.

Chapter 5 illustrates the great variety of models that have been proposed for nonrandom mating and exhibits their diverse dynamics. The theory of migration and selection in subdivided populations is the subject of Chap. 6. The next two chapters extend the study of natural selection to an X-linked locus and to two autosomal loci, respectively. In Chap. 9, we shall explore the intimately related processes of inbreeding and random genetic drift. Chapter 10 is an introduction to quantitative genetics.

Chapters 5 to 9 are independent of each other; the theory of inbreeding is used in Chap. 10.

I have formulated all the models for multiple alleles because that is not only biologically desirable, but also more concise and transparent. Some of the analyses and many of the examples, however, are diallelic.

The major mathematical theme of the book is the dynamical analysis of recursion relations. We shall study many exact and approximate techniques for investigating linear and nonlinear recursion relations in one dimension and in several dimensions. Most of these methods are introduced in the text and problems of Chaps. 4, 5, and 6.

The problems are an essential part of this book. Although some of them ask the reader merely to supply details of derivations in the text, many illuminate, rigorize, or extend these derivations. Some of the results in the problems are useful and important. The problems are intended to deepen and augment the comprehension of the diligent, careful, self-critical student and to exhibit the looseness and inadequacy of the understanding acquired by the casual reader. The problems vary in difficulty; the harder ones offer a challenge even to good students.

If all the material is covered carefully, it suffices for a one-year course. The reader who already has a general knowledge of population genetics and a thorough understanding of Mendelism (including sex-linkage and recombination) could start with Chap. 4. Chapters 4 and 9 comprise the core of the book, and a fairly elementary one-quarter course, confined to discrete models, could consist of Chap. 1; Sects. 3.1 to 3.3; Sects. 4.1, 4.2, 4.3, and 4.9; and most of Chap. 9. In such a course, I simplify the formulation of selection (Sect. 4.1) by assuming that there are no fertility differences, cover frequency-dependent selection (Problem 4.15), and summarize basic quantitative genetics (Chap. 10).

The restriction to elementary mathematics unfortunately excludes many important classical and modern topics whose study requires advanced analytical or probabilistic methods. Chapter 9 provides only a brief introduction to the study of random genetic drift; the reader may consult Moran (1962), Ewens (1969, 1979, 1990), Crow and Kimura (1970), Kingman (1980), Nagylaki (1990), and Svirezhev and Passekov (1990) for more far-reaching investigations. Theoretical work on geographical variation is reviewed in Nagylaki (1989b, c). Felsenstein's (1981) *Bibliography of Theoretical Population Genetics* is indispensable.

I am very grateful to Wayne Boucher, James F. Crow, Carter Denniston, Russell Lande, and Michael Moody for helpful comments; to Mitzi Nakatsuka for beautifully and carefully typing parts of the manuscript, being a conscientious second proofreader, and serving as an expert consultant on TeX; and to the National Science Foundation for its support.

Chicago, Illinois   Thomas Nagylaki
June 1991

Contents

1. Introduction 1

2. Asexual Haploid Populations 5
   2.1 Selection 5
   2.2 Mutation and Selection 9
   2.3 Migration and Selection 13
   2.4 Continuous Model with Overlapping Generations 13
   2.5 Random Drift 24
   2.6 Problems 25

3. Panmictic Populations 28
   3.1 The Hardy-Weinberg Law 28
   3.2 X-Linkage 35
   3.3 Two Loci 37
   3.4 Population Subdivision 40
   3.5 Genotypic Frequencies in a Finite Population 40
   3.6 Problems 42

4. Selection at an Autosomal Locus 47
   4.1 Formulation for Multiple Alleles 47
   4.2 Dynamics with Two Alleles 52
   4.3 Dynamics with Multiple Alleles 56
   4.4 Two Alleles with Inbreeding 63
   4.5 Variable Environments 65
   4.6 Intra-Family Selection 71
   4.7 Maternal Inheritance 74
   4.8 Meiotic Drive 75
   4.9 Mutation and Selection 75
   4.10 Continuous Model with Overlapping Generations 79
   4.11 Density and Frequency Dependence 91
   4.12 Problems 95

5. Nonrandom Mating 102

    5.1 Selfing with Selection 103
    5.2 Assortative Mating with Multiple Alleles and
        Distinguishable Genotypes 107
    5.3 Assortative Mating with Two Alleles and Complete
        Dominance 108
    5.4 Random Mating with Differential Fertility 112
    5.5 Self-Incompatibility Alleles 116
    5.6 Pollen and Zygote Elimination 119
    5.7 Problems 125

6. Migration and Selection 128

    6.1 The Island Model 128
    6.2 General Analysis 132
    6.3 The Levene Model 144
    6.4 Two Diallelic Niches 148
    6.5 Problems 151

7. X-Linkage 153

    7.1 Formulation for Multiallelic Selection and Mutation 153
    7.2 Selection with Two Alleles 159
    7.3 Mutation-Selection Balance 162
    7.4 Weak Selection 164
    7.5 Problems 170

8. Two Loci 174

    8.1 General Formulation for Multiple Loci 174
    8.2 Analysis for Two Multiallelic Loci 176
    8.3 Two Diallelic Loci 185
    8.4 Continuous Model with Overlapping Generations 189
    8.5 Problems 195

9. Inbreeding and Random Drift 200

    9.1 The Inbreeding Coefficient 201
    9.2 Calculation of the Inbreeding Coefficient from Pedigrees .. 203
    9.3 Identity Relations Between Relatives 210
    9.4 Phenotypic Effects of Inbreeding 219
    9.5 Regular Systems of Inbreeding 222
    9.6 Panmixia in Finite Populations 236
    9.7 Heterozygosity under Mutation and Random Drift 242
    9.8 The Inbreeding Effective Population Number 243
    9.9 The Model for Random Drift of Gene Frequencies 248
    9.10 Random Drift of Gene Frequencies 253

9.11 Gene-Frequency Change due to Mutation and Random
     Drift 257
9.12 The Island Model 259
9.13 The Variance Effective Population Number 261
9.14 Problems 267

10. Quantitative Genetics 279

10.1 The Decomposition of the Variance with Panmixia 280
10.2 The Correlation Between Relatives with Panmixia 284
10.3 The Change in Variance due to Assortative Mating 290
10.4 The Correlation Between Relatives with Assortative
     Mating 302
10.5 Selection 311
10.6 Mutation-Selection Balance 321
10.7 Problems 329

References 340

Author Index 356

Subject Index 360

1. Introduction

We begin this brief introductory chapter with a minimum of very much simplified genetic background. More information will be introduced when it is required. Cavalli-Sforza and Bodmer (1971) present a good summary of the pertinent genetics. Crow's (1983) lucid book is more detailed, but still quite concise.

The genetic material, deoxyribonucleic acid (DNA), consists of four bases: adenine (A), guanine (G), thymine (T), and cytosine (C). Each base is linked to a sugar and a phosphate and forms a *nucleotide*. These nucleotides are arranged in a double helix, the only normal pairings being A-T and G-C. Therefore, the information is in the sequence along a single helix. Three bases code for an amino acid, of which there are 20. A polypeptide chain is a folded string of (typically a few hundred) amino acids, and a protein comprises at least one polypeptide chain. Roughly, the region of DNA that determines a polypeptide chain is a *gene*, its position is a *locus*, and a particular sequence there is an *allele*. In population genetics, however, "gene" is sometimes used in the sense of "allele", as defined above: e.g., the proportion of a particular allele in the population may be called a gene frequency rather than an allelic frequency. A *chromosome* is a single thread-like molecule of DNA, along which genes are arranged linearly, surrounded by various proteins. The number of different chromosomes is characteristic of each species. If the chromosomes in a set are single, the organism is called *haploid*; if they are doubled, the organism is *diploid*. Generally, bacteria and many algae, mosses, and fungi are haploid organisms, whereas higher plants and animals are diploid. We shall not consider polyploids, in which chromosomes are at least tripled. The *genotype* of an individual is the set of his genes at the loci under consideration; his *phenotype* is the observable expression of his genotype. A diploid individual is *homozygous* (*heterozygous*) at a particular locus if his two genes at that locus are the same allele (different alleles). *Recessive* alleles are expressed only in homozygotes. Very crudely, asexual haploids just duplicate themselves. In diploids, at *meiosis* the chromosomes separate, and each *gamete* (sperm or egg) carries a single set of chromosomes. The sets from a sperm and egg unite at *fertilization* to from a diploid *zygote*, from which the individual develops. In plants, pollen fertilize ova to form seeds.

Population genetics concerns the genetic structure and evolution of populations. It has important interfaces with ecology, demography, and epidemiology; its principal applications are to human genetics and to animal and plant breeding. The genetic composition of a population is usually described by genotypic or allelic frequencies in deterministic models and by the probability distribution of these frequencies (or functionals of such distributions) in stochastic models. These variables may depend on space and time. Their dynamics is governed by a few elementary genetic principles and the evolutionary forces described below. The major theme of theoretical population genetics is the study of the amount and pattern of genetic variability under sundry combinations of these evolutionary forces. Observation of such patterns may yield inferences about the relative significance and domain of action of evolutionary forces in natural populations.

The *mating system* can itself have significant evolutionary consequences, and it strongly affects the action of the other evolutionary forces. If mating occurs without regard to ancestry and the genotypes under consideration, we say it is *random*. This is the simplest situation and, at least approximately, appears to occur frequently in nature. Its most important aspect is the conservation of allelic frequencies, and therefore of genetic diversity. It leads also to the random combination of the genes at each locus, and therefore (after one generation) to the conservation of genotypic frequencies.

We say there is *inbreeding* if related individuals are more likely to mate than randomly chosen ones. This does not change the allelic frequencies, but raises the homozygosity (the frequency of homozygous individuals), thereby exhibiting the phenotypic effects – usually harmful – of recessive alleles.

Assortative mating refers to the tendency of individuals who resemble each other with respect to the trait in question to mate with each other. *Disassortative mating* means that phenotypically dissimilar individuals mate more often than those chosen at random. Assortative and disassortative mating may change allelic frequencies if some genotypes have a higher probability of mating than others. They always change the genotypic frequencies. Assortative mating has its most important influence on quantitative characters (i.e., traits that vary continuously or almost continuously): by increasing their variance, it can greatly raise the proportion of individuals with extreme trait values. Incompatibility systems in plants, which preserve or augment genetic diversity, exemplify disassortative mating.

Recombination is the formation of a nonparental gamete from the maternal alleles at a set of loci and the paternal alleles at the remaining loci. It changes gametic frequencies, but not allelic frequencies. In the absence of other evolutionary forces, it leads to the random combination of genes within gametes. It is an essential feature of multilocus systems.

Various genotypes may have different probabilities of surviving to adulthood and may reproduce at different rates. Differential mortality and fertility are the components of *selection*. Unless the population is in equilibrium, selection will change the genotypic and allelic frequencies in accordance with the expected number of progeny, called fitness, of the various genotypes. Natural selection has been recognized since Darwin as the directive force of adaptive evolution.

In each generation, selection reduces the frequency of every deleterious allele in a population by a fraction proportional to its selection coefficient, or relative fitness defect. Selection can also maintain a stable polymorphism (i.e., a stable equilibrium with at least two alleles present) by various mechanisms, such as overdominance (superior fitness of heterozygotes) and frequency dependence that favors rare genotypes. If lethality and sterility are excluded, selection coefficients seldom exceed a few percent.

Mutation designates the change from one allelic form to another. Clearly, it directly alters allelic frequencies. Although some other molecular processes, such as gene conversion, can also produce new alleles, mutation is the only force that does so in classical evolutionary models. Furthermore, mutation keeps deleterious alleles in the population at low frequencies. The rate of gene frequency change due to mutation is of the same order as the mutation rate per locus, about 10^{-6} per generation.

Since many, perhaps most, natural populations are divided into subpopulations that mate at random only within groups, *migration* must often be taken into account. Geographically distributed populations provide the most important example. Migration can affect not only the pattern of geographical differentiation in the population, but the amount of genetic variability as well.

Unless some of the parameters, such as the selection coefficients and mutation rates, required to specify the above evolutionary forces fluctuate at random, these forces will be deterministic. In a finite population, however, allelic frequencies will vary probabilistically because of the random sampling of genes from one generation to the next. This process is called *random genetic drift*, *random sampling drift*, or simply *random drift*. Its two causes are (nonselective) random variation of the number of progeny of different individuals and the stochastic nature of gamete production (Mendel's Law of Segregation). In natural populations, the former generally contributes somewhat more to the random fluctuation of allelic frequencies than the latter. These fluctuations lead to the decay of genetic variability and the eventual loss or fixation of alleles. The rate of decrease of genetic variability and the rate of increase of the variance of each allelic frequency per generation are both of the order of the reciprocal of the (variance effective) population number. Even in infinite populations, however, the fate of rare alleles still depends strongly on random sampling.

The three most influential theories of evolution are those of Fisher, Wright, and Kimura. The primary significance of these theories is that they stimulate, focus, and put into perspective empirical and theoretical research. Each of them surely describes and illuminates many particular cases. But characters, organisms, and species are too complex and diverse for these theories to provide an accurate, general mathematical description of the evolutionary process.

All the above theories identify mutation as the source of raw material for evolutionary change, but they differ in the emphasis placed on the other factors.

The main underpinning of Fisher's theory is his Fundamental Theorem of Natural Selection – that the rate of change of the mean fitness of a population is equal to the additive component of its genetic variance in fitness (Fisher, 1930). Fisher held that evolution occurs primarily by the deterministic increase in fitness of large, randomly mating populations under the action of natural selection. We may imagine the mean fitness as a surface in the space of suitable dynamical variables, such as allelic frequencies, and consequently envisage that the population is climbing a hill on this surface. In Fisher's picture, random drift is responsible only for small chance fluctuations of the trajectory of the population. On account of mutation and environmental fluctuations, the fitness surface is constantly changing, so a population always has the opportunity to climb higher.

In his "shifting-balance" theory, Wright (1931; 1970; 1977, Chap. 13) stressed that, due to multiple effects (pleiotropy) and interactions (epistasis) of loci in the determination of characters and to selection for an intermediate phenotypic optimum for each character (stabilizing selection), the fitness surface has many selective peaks. Small populations can "test" this surface by random drift, sometimes crossing a saddle from a lower selective hill to a higher one. He pointed out that if a species is divided into many small subpopulations that exchange relatively few migrants per generation, then more rapid growth of fitter subpopulations and dispersion of selectively favored individuals may enable it to reach the highest peak on the surface. A comprehensive analytical or numerical treatment of this complex theory may not be feasible.

The neutral theory of Kimura (1968, 1983) and of King and Jukes (1969) ascribes much of evolution, especially at the molecular level, to mutation and random drift. In its strongest form, this theory attributes a large proportion of the variation even in morphological characters to these two forces (Nei, 1975, pp. 251–253). This is still consistent with the fact that natural selection determines the nature of adaptation and with the view that most mutants are harmful. The weakest form of the theory holds only that most polymorphisms and substitutions of amino acids and, to an even greater degree, of nucleotides are neutral. The neutral theory has been extended to incorporate slightly deleterious alleles (Ohta, 1976) and stabilizing selection at the phenotypic level (Kimura, 1981). Although the neutral theory is more recent than the schemes of Fisher and Wright, its mathematical and statistical development are more advanced (Ewens, 1979, 1989).

The reader may wish to consult Kimura (1983), Nei (1987), and Hartl and Clark (1989) for much more detailed biological discussion.

2. Asexual Haploid Populations

The study of haploid populations in this chapter will enable us to formulate and analyze without the additional complication of mating and Mendelian segregation and recombination many of the problems that concern us. If alleles are interpreted as genotypes and mutation rates refer to zygotes rather than alleles, the formalism applies, regardless of ploidy, also to asexual species, such as the dandelion. As explained in Chap. 3, it applies also to a Y-linked locus in sexually reproducing diploids. We shall expound the basic selection model with discrete nonoverlapping generations in Sect. 2.1, include mutation and migration in Sects. 2.2 and 2.3, and treat overlapping generations in continuous time in Sect. 2.4. We shall discuss random drift briefly in Sect. 2.5.

2.1 Selection

We consider a single locus with alleles A_i, where $i = 1, 2, ..., k$, and assume that generations are discrete and nonoverlapping. Thus, the adults are replaced by their (juvenile) offspring in each generation. Although this assumption will hold for some laboratory populations, for natural populations of haploids it should be viewed as a simple approximation. Let the number of offspring that carry A_i in generation t, where $t = 0, 1, 2, ...$, be $n_i(t)$. The total number of offspring,

$$N = \sum_i n_i, \tag{2.1}$$

must be sufficiently large to allow us to neglect random drift.

Let v_i designate the probability that an A_i offspring survives to reproductive age. The average number of progeny of an A_i adult is f_i. The *viabilities* v_i and *fertilities* f_i may be functions of the time t and the vector of population numbers, denoted by $\mathbf{n} = \mathbf{n}(t)$. The product $w_i = v_i f_i$ represents the *fitness* of an A_i individual. The $v_i n_i$ A_i adults in generation t contribute $f_i(v_i n_i) = w_i n_i$ A_i offspring to generation $t+1$. Therefore, the basic recursion relations

$$n_i(t+1) = w_i(t, \mathbf{n})n_i(t) \tag{2.2}$$

depend only on the fitnesses, and not on the viabilities and fertilities separately. Since w_i is the expected number of progeny of an A_i juvenile, this is not surprising. The fundamental difference equations (2.2) determine $\mathbf{n}(t)$ iteratively in terms of $\mathbf{n}(0)$, provided the fitnesses $w_i(t, \mathbf{n})$ are specified.

If A_i is lethal, $v_i = 0$, or causes sterility, $f_i = 0$, then $w_i = 0$. Otherwise, the w_i will usually not differ from each other by more than a few percent. If the population size is approximately constant, the average of the w_i will be close to unity. The numbers $w_i - 1$ are called *selection coefficients*.

The proportion, or *frequency*, of the allele A_i among offspring is

$$p_i = \frac{n_i}{N}. \tag{2.3}$$

Unless stated otherwise, a prime will always signify the next generation. Then (2.1), (2.2), and (2.3) yield

$$N' = \sum_i n_i' = \sum_i w_i \left(\frac{n_i}{N}\right) N = \overline{w} N, \tag{2.4}$$

where

$$\overline{w} = \sum_i w_i p_i \tag{2.5}$$

is the *mean fitness* of the population and gives its rate of growth. The mean fitness is of great conceptual and analytical importance in the theory of selection.

The gene frequencies satisfy the recursion relations

$$p_i' = \frac{n_i'}{N'} = \frac{w_i n_i}{\overline{w} N} = p_i \left(\frac{w_i}{\overline{w}}\right). \tag{2.6}$$

We see at once from (2.5) and (2.6) that the relation

$$\sum_i p_i = 1 \tag{2.7}$$

holds for all time if initially true, as it must be. It is also apparent from (2.5) and (2.6) that the gene frequencies depend only on ratios of the fitnesses. All the w_i may be multiplied by the same constant without altering the evolution of the allelic frequencies p_i. Exploiting this scale invariance often simplifies the algebra. We shall always employ a Δ to indicate the change in one generation. Thus, (2.6) gives

$$\Delta p_i = p_i' - p_i = \frac{p_i(w_i - \overline{w})}{\overline{w}}. \tag{2.8}$$

If the fitnesses are functions only of time, it is useful to iterate (2.2):

$$n_i(t) = n_i(0) \prod_{\tau=0}^{t-1} w_i(\tau). \tag{2.9}$$

Then the gene frequencies read

$$p_i(t) = \frac{p_i(0) \prod_{\tau=0}^{t-1} w_i(\tau)}{\sum_j p_j(0) \prod_{\tau=0}^{t-1} w_j(\tau)} \ . \tag{2.10}$$

It is often assumed that the fitnesses are constant, meaning, really, that they vary much more slowly than the other pertinent evolutionary parameters. In that case, (2.9) and (2.10) reduce to

$$n_i(t) = n_i(0)w_i^t \ , \tag{2.11}$$

$$p_i(t) = \frac{p_i(0)w_i^t}{\sum_j p_j(0)w_j^t} \ . \tag{2.12}$$

Suppose A_1 is the fittest allele: $w_1 > w_i$ for $i > 1$. Equation (2.12) informs us immediately that $p_1(t) \to 1$ as $t \to \infty$. Assuming that the population size remains finite (as can be arranged by regulating it in each generation without changing the gene frequencies), this means that the fittest gene will be ultimately *fixed*, the others being *lost*. Of course, all statements of this sort presuppose that the allele under consideration is initially present in the population. For instance, here we posit $p_1(0) > 0$.

It will be helpful to distinguish three levels of description of the evolution of a population. We shall refer to the specification of the variables of interest as functions of time as a *complete solution*. Equation (2.12) is an example. Often, even though one cannot obtain a complete solution, one can determine the fate of the population for all initial conditions. We may call this a *complete*, or *global, analysis*. The statement that with constant fitnesses the fittest allele is fixed, falls into this category. If we cannot carry out a complete analysis, we may still obtain some information of evolutionary interest by locating all the equilibria of the system and investigating its behavior in the neighborhood of these stationary states. In the problem treated above, a part of such a *local analysis* would be to observe that $p_1 = 1$ is an equilibrium, and to show that if $p_1(0)$ is sufficiently close to 1, then $p_1(t) \to 1$ as $t \to \infty$.

We define the local stability of an equilibrium as follows: If, in terms of a suitable metric for the variables of the problem, the population will remain within an arbitrarily small, preassigned distance of the equilibrium, provided it starts sufficiently close to the equilibrium, we say the equilibrium is *stable*. Otherwise, it is *unstable*. An equilibrium is *asymptotically stable* if it is stable and a population that starts sufficiently close to the equilibrium converges to it. The gene-frequency equilibrium $p_1 = 1$ is globally asymptotically stable, whereas the equilibrium $p_2 = 1$ is unstable. If $w_i = 1$ for all i, every point is a stable equilibrium, but the stability is not asymptotic. The word "asymptotic" is frequently omitted in population genetics.

We proceed now to study the change in *mean fitness*. From (2.5) we have

$$\Delta \overline{w} = \sum_i (p_i' w_i' - p_i w_i)$$

$$= \sum_i [p_i'(\Delta w_i + w_i) - p_i w_i]$$

$$= \overline{\Delta w} + \sum_i w_i \Delta p_i \ , \tag{2.13}$$

where

$$\overline{\Delta w} = \sum_i p_i' \Delta w_i \tag{2.14}$$

is the mean of the fitness changes over the next generation. Substituting (2.8) into (2.13), we find

$$\Delta \overline{w} = \overline{\Delta w} + \overline{w}^{-1} \sum_i p_i w_i (w_i - \overline{w}) \ . \tag{2.15}$$

But (2.5) informs us that

$$\sum_i p_i (w_i - \overline{w}) = 0 \ ,$$

so we may subtract $\overline{w}$ from the first w_i in (2.15) to obtain

$$\Delta \overline{w} = \overline{\Delta w} + \overline{w}^{-1} V \ , \tag{2.16}$$

where

$$V = \sum_i p_i (w_i - \overline{w})^2 \tag{2.17}$$

is the genic variance in fitness. The simple steps in this derivation are often useful.

Equation (2.16) is a simple case of Fisher's Fundamental Theorem of Natural Selection (Fisher, 1930). With constant fitnesses, $\overline{\Delta w} = 0$, so $\Delta \overline{w} = \overline{w}^{-1} V \geq 0$, i.e., the mean fitness is nondecreasing. Since $V = 0$ if and only if $p_i = 0$ or $w_i = \overline{w}$ for all i, (2.6) implies that $\Delta \overline{w} = 0$ only at equilibrium. Thus, selection increases the mean fitness, using up the genic variance. In view of (2.4), this implies that the relative rate of population growth increases.

As mentioned below (2.7), all the w_i may be multiplied by the same genotype-independent quantity without affecting the dynamics of the gene frequencies. Such scaling converts *absolute* fitnesses, which equal the expected numbers of progeny for the various genotypes, to *relative* ones. Since resource limitations preclude indefinite growth of natural populations, absolute fitnesses cannot long remain constant. It is, therefore, more reasonable to posit constant relative fitnesses. For instance, we might have $w_i = z_i g(N)$, with z_i constant and $w_i < 1$ for sufficiently large N for all i. Then it is the mean relative fitness, $\overline{z}$, rather than the mean absolute fitness $\overline{w}$, that will be maximized by natural selection.

The biological significance of the maximization of the mean relative fitness is the following. Let A_1 be the fittest genotype: $w_1 > w_i$ for $i > 1$. Then natural selection minimizes $(w_1 - \overline{w})/w_1 = (z_1 - \overline{z})/z_1$, the relative reproductive excess of the optimal genotype. This reproductive excess, usually called the *genetic load*, is discussed in detail by Crow (1970). Note that with pure viability selection it is simply the relative amount of selective mortality.

If the selection coefficients are small, we may choose all the relative fitnesses to be close to 1. Then the mean fitness will be approximately unity, and its rate of change will be roughly equal to the genic variance.

With only two alleles, it is customary to put $p = p_1$ and $q = p_2 = 1 - p$. From (2.8) we deduce

$$\Delta p = pq(w_1 - w_2)/\overline{w} \ , \tag{2.18}$$

and the variance in fitness reduces to

$$V = pq(w_1 - w_2)^2 \ . \tag{2.19}$$

2.2 Mutation and Selection

Let us consider first *mutation without selection*. We designate the probability that an A_i individual has an A_j offspring for $i \neq j$ by the mutation rate u_{ij}. It will be convenient to use the convention $u_{ii} = 0$ for all i. Generally, mutation rates are quite small; 10^{-6} is a representative value. Mutation rates at the nucleotide level are on the order of 10^{-9}. The gene-frequency change in one generation clearly reads

$$\Delta p_i = \sum_j p_j u_{ji} - p_i \sum_j u_{ij} \ . \tag{2.20}$$

By interchanging dummy variables in one of the sums – often a useful device – we observe directly that

$$\sum_i \Delta p_i = 0 \ ,$$

so that (2.7) is preserved. Since the total mutation rate must not exceed unity, therefore $\Delta p_i \geq -p_i$, whence $p_i' \geq 0$, as required.

For two alleles, one commonly writes $u = u_{12}$ and $v = u_{21}$. With $p = p_1$, as above, (2.20) becomes

$$\Delta p = v - (u + v)p \ . \tag{2.21}$$

At equilibrium, $\Delta p = 0$, so the frequency of A_1 is

$$\hat{p} = \frac{v}{u + v} \ . \tag{2.22}$$

We shall follow convention and indicate equilibrium values by a caret. Equilibria such as (2.22), with more than one allele present, are called *polymorphic*.

As expected, if mutation is irreversible, i.e., $u = 0$ or $v = 0$, the allele whose frequency is decreasing is absent at equilibrium.

It is frequently convenient to study the deviation from equilibrium. Substituting $x = p - \hat{p}$ into (2.21), we find

$$x' = (1 - u - v)x \ ,$$

with the solution

$$x(t) = x(0)(1 - u - v)^t \ . \tag{2.23}$$

Therefore, there is global convergence to (2.22) at the geometric rate $1 - u - v$. Since (2.23) yields

$$t = \frac{\ln[x(t)/x(0)]}{\ln(1 - u - v)} \approx \frac{\ln[x(0)/x(t)]}{u + v} \ ,$$

we see that the characteristic time for gene-frequency change (i.e., for reduction of $x(t)$ by a factor e) is very long, typically about 10^6 generations.

It is often useful to approximate powers like the one in (2.23) by exponentials. For $|\varepsilon| \ll 1$ and $\varepsilon^2 t \ll 1$, $(1 - \varepsilon)^t = \exp[t \ln(1 - \varepsilon)] = \exp[-t(\varepsilon + \frac{1}{2}\varepsilon^2 + \dots)] \approx e^{-\varepsilon t}$. Thus, we may rewrite (2.23) as

$$x(t) \approx x(0)e^{-(u+v)t} \ . \tag{2.24}$$

The fact that (2.24) becomes inaccurate as $(u + v)^2 t$ approaches 1 is irrelevant because by that time $x(t)$ is extremely close to 0.

To *include selection*, we set up the formal scheme

Offspring $\longrightarrow$ Adult $\longrightarrow$ Offspring $\longrightarrow$ Offspring

viability   fertility   mutation

p_i $p_i^\star$ p_i'

with the indicated gene frequencies. Let R_{ij} be the probability that a gamete from a juvenile A_i offspring carries A_j. Recalling (2.6), we have

$$p_i^\star = p_i \left(\frac{w_i}{\overline{w}} \right) \ , \tag{2.25a}$$

$$p_i' = \sum_j p_j^\star R_{ji} \ , \tag{2.25b}$$

where $\overline{w}$ is still given by (2.5). It is important to note that (2.25) correctly describes the biological situation that, while selection acts on the phenotype, which develops from the offspring genotype, the germ cells mutate with no phenotypic effect at rates u_{ij}, related to R_{ij} by

$$R_{ij} = \delta_{ij} \left(1 - \sum_k u_{ik} \right) + u_{ij} \ . \tag{2.26}$$

The Kronecker delta, δ_{ij}, is defined by $\delta_{ij} = 1$ if $i = j$ and $\delta_{ij} = 0$ if $i \neq j$.
From (2.25b) and (2.26) we derive

$$\Delta p_i = p_i' - p_i = p_i^* - p_i + \sum_j p_j^* u_{ji} - p_i^* \sum_j u_{ij} \ .$$

If selection is weak, since $u_{ij} \ll 1$, we may neglect $(p_i^* - p_i)u_{ij}$ for all i and j to obtain

$$\Delta p_i \approx \Delta p_i(\text{selection}) + \Delta p_i(\text{mutation}) \ ,$$

where

$$\Delta p_i(\text{selection}) = p_i^* - p_i \ ,$$

$$\Delta p_i(\text{mutation}) = \sum_j p_j u_{ji} \quad p_i \sum_j u_{ij} \ .$$

Let us analyze the *diallelic case*. Choosing, without loss of generality, $w_1 = 1$ and $w_2 = 1 + \dot{s}$, where $s > 0$, we easily reduce (2.25) to the *linear fractional transformation*

$$p' = \frac{\alpha + \beta p}{\gamma + \delta p} \ , \tag{2.27}$$

with $\alpha = v(1+s)$, $\beta = 1 - sv - u - v$, $\gamma = 1 + s$, and $\delta = -s$. Since (2.27) occurs in several models, we shall discuss it for arbitrary values of its parameters. The two solutions of $p' = p$ are

$$p_\pm = (2\delta)^{-1}(\beta - \gamma \pm Q^{1/2}) \ , \tag{2.28}$$

where

$$Q = (\beta - \gamma)^2 + 4\alpha\delta \ . \tag{2.29}$$

The trivial case $\delta = 0$ corresponds to (2.21), so we suppose $\delta \neq 0$. We assume also that $\alpha\delta \neq \beta\gamma$, for otherwise (2.27) shows that $p' = \alpha/\gamma$.

If $Q \neq 0$, then $p_+ \neq p_-$, and hence

$$y = \frac{p - p_+}{p - p_-} \tag{2.30}$$

satisfies $y' = \lambda y$, where

$$\lambda = \frac{\beta + \gamma - Q^{1/2}}{\beta + \gamma + Q^{1/2}} \ . \tag{2.31}$$

Therefore,

$$y(t) = y(0)\lambda^t, \tag{2.32}$$

and (2.27) has the solution

$$p(t) = \frac{p_- y(t) - p_+}{y(t) - 1} \ . \tag{2.33}$$

Of course, $y(0)$ is evaluated from (2.30). There are two cases with $Q \neq 0$.

1. $Q < 0$: Here p_+ and p_- are complex. Since $|\lambda| = 1$, (2.32) has the form

$$y(t) = y(0)e^{-i\theta t} , \qquad (2.34)$$

where

$$\theta = 2\tan^{-1}\left[\frac{(-Q)^{1/2}}{\beta + \gamma}\right]. \qquad (2.35)$$

2. $Q > 0$: Now p_+ and p_- are real, and there are three subcases.
   (a) $\beta + \gamma > 0$: From (2.31) we see that $|\lambda| < 1$, whence $y(t) \to 0$ as $t \to \infty$. Therefore, (2.33) implies that $p(t) \to p_+$ as $t \to \infty$.
   (b) $\beta + \gamma < 0$: Since $|\lambda| > 1$, therefore, $y(t) \to \infty$, and consequently $p(t) \to p_-$ as $t \to \infty$.
   (c) $\beta + \gamma = 0$: Here $\lambda = -1$, and hence $y(t) = y(0)(-1)^t$. Therefore, $p(t)$ alternates between $p(0)$ and $p(1)$. This holds also for $Q < 0$, for then (2.35) yields $\theta = \pi$.

It remains to analyze the case with $Q = 0$.

3. $Q = 0$: Equation (2.28) reveals that there is a single equilibrium

$$\hat{p} = (2\delta)^{-1}(\beta - \gamma) . \qquad (2.36)$$

We find that for $\beta + \gamma \neq 0$

$$z = (p - \hat{p})^{-1} \qquad (2.37)$$

satisfies

$$z' = z + \frac{2\delta}{\beta + \gamma} , \qquad (2.38)$$

with the obvious solution

$$z(t) = z(0) + \frac{2\delta t}{\beta + \gamma} . \qquad (2.39)$$

Equations (2.37) and (2.39) inform us that

$$p(t) \approx \hat{p} + \frac{\beta + \gamma}{2\delta t} \qquad (2.40)$$

as $t \to \infty$. Thus, the ultimate rate of convergence to $\hat{p}$ is algebraic. If $\beta + \gamma = 0$, (2.29) tells us that $\alpha\delta = \beta\gamma$, so that $p' = \alpha/\gamma$.

In genetic problems, we shall be concerned with the mapping of some interval $I : [a, b]$ into itself. Then we can restrict the possible equilibrium structures for any continuous map $p' = f(p)$. Suppose that the number of equilibria is finite and a and b are not equilibria: $f(a) \neq a$ and $f(b) \neq b$. Hence, $f(a) > a$ and $f(b) < b$, so $g(p) = f(p) - p$ must change sign an odd number of times in I. Since f maps I into itself, therefore $g(p)$ is finite for p in I, and hence, counting multiplicity, has an odd number of zeroes in I. Thus, still counting multiplicity, f has an odd number of fixed points in I. For

(2.27), provided $p_\pm \neq 0, 1$, this means Cases 1 and 3 may be excluded, and in Case 2 either p_+ or p_- is in I, but not both.

Let us apply the theory just developed to our mutation-selection problem. It is easy to see that $Q \geq 0$, where equality holds if and only if $u = 0$ and $s = v(1 + s)$. Since usually $s \gg v$, we assume $Q > 0$. Furthermore, it is trivial to verify that $\beta + \gamma \geq 0$, with equality only in the unbiological situation $u = v = 1$. Therefore, $p(t)$ converges to p_+ globally. It follows that $0 \leq p_+ \leq 1$. From the formula

$$p_\pm = (2s)^{-1}\{s + sv + u + v \mp [(s + sv + u + v)^2 - 4sv(1 + s)]^{1/2}\}, \quad (2.41)$$

with a bit of algebra one can show that $p_- \geq 1$. With the biologically trivial assumption $u + v \leq 1$, one can prove from (2.31) that $\lambda \geq 0$, so that we conclude from (2.30) that the convergence is without oscillation. If $v = 0$, both selection and mutation decrease p, so we expect $p_+ = 0$. Indeed, (2.41) yields $p_+ = 0$ and $p_- = 1 + us^{-1}$. With $u = 0$, (2.41) reduces to $p_+ = v(1 + s)/s$, $p_- = 1$, provided $v \leq s(1 + s)$. In the unlikely event that $v > s(1 + s)$, we have $p_+ = 1$, $p_- = v(1 + s)/s$. In the biologically important case $u, v \ll 1, s$, linearizing (2.41) in u and v yields $p_+ \approx v(1 + s)/s$ and $p_- \approx 1 + us^{-1}$. As expected, the equilibrium frequency $p_+ \ll 1$.

2.3 Migration and Selection

We assume that a proportion m of the population is replaced each generation by migrants with fixed gene frequencies $\bar{p}_i$. More complicated migration-selection schemes than this island model will be discussed for diploids in Chap. 6. To write our recursion relations, we replace mutation by migration in the formal mutation-selection scheme of Sect. 2.2. Then (2.25b), which we may rewrite as

$$p_i' = p_i^* + \sum_j p_j^* u_{ji} - p_i^* \sum_j u_{ij} , \quad (2.42)$$

becomes

$$p_i' = p_i^* + m(\bar{p}_i - p_i^*) . \quad (2.43)$$

But the substitution $u_{ij} = m\bar{p}_j$, $i \neq j$, reduces (2.42) to (2.43), which shows that migration is a special case of mutation.

2.4 Continuous Model with Overlapping Generations

Our formulation will be based on that of Cornette (1975) for diploids. Time, measured in arbitrary units, flows continuously. Let $\nu_i(t, x)\Delta x$ be the number of A_i individuals between the ages of x (measured in the same units as t) and $x + \Delta x$ at time t. The total number of A_i individuals at time t is

$$n_i(t) = \int_0^\infty \nu_i(t,x)dx \ . \tag{2.44}$$

If no individual survives beyond age X, then $\nu_i(t,x) = 0$ for $x > X$. The total population size is

$$N(t) = \sum_i n_i(t) \ . \tag{2.45}$$

We set up equations for our fundamental variables, $\nu_i(t,x)$, with the aid of the *life table*, $l_i(t,x)$. We start observing the population at time $t = 0$. For $t \geq x$, $l_i(t,x)$ represents the probability that an A_i individual born at time $t - x$ survives to age x. If $t < x$, $l_i(t,x)$ designates the probability that an A_i individual aged $x - t$ at time 0 survives to age x. By definition,

$$l_i(t,0) = 1 \quad \text{and} \quad l_i(0,x) = 1 \ . \tag{2.46}$$

The number of A_i individuals born from time $t - \Delta x$ to time t is the same as the number of A_i individuals between ages 0 and Δx at time t, $\nu_i(t,0)\Delta x$. This assertion neglects the mortality of the newborn, since that is proportional to $(\Delta x)^2$. It follows that the number of A_i individuals born per unit time at time t is $\nu_i(t,0)$. For $t \geq x$, we equate the number of A_i individuals aged x at time t to the number of births at time $t - x$ times the proportion surviving to age x. For $t < x$, we replace the number of births at time $t - x$ by the number aged $x - t$ at time 0. We conclude

$$\nu_i(t,x) = \begin{cases} \nu_i(t-x,0)\,l_i(t,x), & t \geq x, \\[2mm] \nu_i(0,x-t)\,l_i(t,x), & t < x. \end{cases} \tag{2.47}$$

Thus, we have expressed the age distribution in terms of its initial value, the birth rate, and the life table. In principle, the birth rate, $\nu_i(t,0)$, may be calculated from the integral equation given in Problem 2.8.

Let us demonstrate next that the life table is completely determined by the mortality. We designate the probability that an A_i individual aged x at time t dies between ages x and $x + \Delta x$ by $d_i(t,x)\Delta x$. Then

$$l_i(t,x) - l_i(t+\Delta x, x+\Delta x) = l_i(t,x)d_i(t,x)\Delta x + O[(\Delta x)^2], \tag{2.48}$$

where $O[(\Delta x)^2]$ refers to terms proportional to $(\Delta x)^2$, or smaller, as $\Delta x \to 0$. We rewrite (2.48) as

$$\frac{l_i(t+\Delta x, x+\Delta x) - l_i(t, x+\Delta x)}{\Delta x} + \frac{l_i(t, x+\Delta x) - l_i(t,x)}{\Delta x}$$
$$= -d_i(t,x)l_i(t,x) + O(\Delta x)$$

and let $\Delta x \to 0$ to obtain

$$\frac{\partial l_i}{\partial t} + \frac{\partial l_i}{\partial x} = -d_i(t,x)l_i(t,x) \ . \tag{2.49}$$

To solve this partial differential equation, we put $y = \frac{1}{2}(t+x)$, $z = \frac{1}{2}(t-x)$, $D_i(y,z) = d_i(t,x)$, and $L_i(y,z) = \ln l_i(t,x)$. From (2.46) and (2.49) we deduce the problem

$$\frac{\partial L_i}{\partial y} = -D_i(y,z), \quad L_i(y,y) = 0, \quad L_i(y,-y) = 0 \ . \tag{2.50}$$

Figure 2.1 shows that the solution of (2.50) in the first quadrant of the xt-plane reads

$$L_i(y,z) = \begin{cases} -\int_z^y D_i(y',z)dy', & z \geq 0 \ , \\[2mm] -\int_{-z}^y D_i(y',z)dy', & z < 0 \ . \end{cases} \tag{2.51}$$

Substituting $\xi = y' - z$ into (2.51) and reverting to the original variables leads to

$$l_i(t,x) = \begin{cases} \exp\left[-\int_0^x d_i(\xi+t-x,\xi)d\xi\right], & t \geq x \ , \\[2mm] \exp\left[-\int_{x-t}^x d_i(\xi+t-x,\xi)d\xi\right], & t < x \ . \end{cases} \tag{2.52}$$

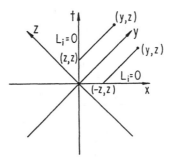

Fig. 2.1. The solution of (2.50).

Returning to the age distribution, we observe that (2.48) holds with l_i replaced by ν_i, and hence so does (2.49):

$$\frac{\partial \nu_i}{\partial t} + \frac{\partial \nu_i}{\partial x} = -d_i(t,x)\nu_i(t,x) \ . \tag{2.53}$$

Using (2.44) and (2.53) and noting that $\nu_i(t,\infty) = 0$, we calculate the time derivative

$$\dot{n}_i(t) = \int_0^\infty \frac{\partial \nu_i}{\partial t} dx$$
$$= -\int_0^\infty \left[\frac{\partial \nu_i}{\partial x} + d_i(t,x)\nu_i(t,x)\right] dx$$
$$= \nu_i(t,0) - \int_0^\infty d_i(t,x)\nu_i(t,x)dx \ . \tag{2.54}$$

We define $b_i(t, x)$ as the rate at which A_i individuals of age x at time t give birth. Then

$$\nu_i(t, 0) = \int_0^\infty b_i(t, x)\nu_i(t, x)dx \ .$$

(2.55)

Substituting (2.55) into (2.54) gives

$$\dot{n}_i(t) = \int_0^\infty m_i(t, x)\nu_i(t, x)dx \ ,$$

(2.56)

where

$$m_i(t, x) = b_i(t, x) - d_i(t, x) \ .$$

(2.57)

The usual *model without age structure* is obtained by assuming that $m_i(t, x) = m_i(t)$, independent of age. This will usually not happen unless the birth and death rates are both age independent: $b_i(t, x) = b_i(t)$ and $d_i(t, x) = d_i(t)$. Note that with age-independent mortalities, setting $\tau = \xi + t - x$ simplifies (2.52) to

$$l_i(t, x) = \begin{cases} \exp\left[-\int_{t-x}^t d_i(\tau)d\tau\right], & t \geq x \ , \\ \exp\left[-\int_0^t d_i(\tau)d\tau\right], & t < x \ . \end{cases}$$

(2.58)

Since the age at $t = 0$ is now irrelevant, $l_i(t, x)$ is independent of x for $t < x$. If mortality is also time independent, $d_i(t) = d_i$, (2.58) yields an exponential life table that depends on the time interval from the first observation of the individual to time t:

$$l_i(t, x) = \begin{cases} e^{-d_i x}, & t \geq x \ , \\ e^{-d_i t}, & t < x \ . \end{cases}$$

(2.59)

Let us suppose that $m_i(t, x)$ is independent of x. This is exact for some populations, such as bacteria, but must be viewed as an approximation that complements the model with discrete, nonoverlapping generations for others. Then (2.56) shows that the Malthusian parameters $m_i(t, \mathbf{n})$ determine the population numbers:

$$\dot{n}_i(t) = m_i(t, \mathbf{n})n_i(t) \ .$$

(2.60)

Owing to the restrictive nature of the age-independence condition, it is important to observe that (2.60) holds also if the population has reached a *stable age distribution*. This may often be approximately true in natural populations. Let $\lambda_i(x)\Delta x$, independent of time by assumption, be the proportion of A_i individuals between ages x and $x + \Delta x$. Then $\nu_i(t, x) = n_i(t)\lambda_i(x)$,

$$\int_0^\infty \lambda_i(x)dx = 1 \ ,$$

and (2.56) yields

$$\dot{n}_i(t) = n_i(t) \int_0^\infty m_i(t,x)\lambda_i(x)dx \ .$$

With the expected identification of the Malthusian parameters, this has the form (2.60).

Equation (2.60) is the analogue of (2.2), and our exposition will follow Sect. 2.1. The gene frequencies $p_i(t)$ are still given by (2.3). From (2.3) and (2.60) we derive

$$\dot{N} = \overline{m}N \ , \tag{2.61}$$

where

$$\overline{m} = \sum_i m_i p_i \tag{2.62}$$

is the mean fitness in continuous time. Again using (2.3) and (2.60), we find that the gene frequencies satisfy

$$\dot{p}_i = p_i(m_i - \overline{m}) \ . \tag{2.63}$$

As expected, $\dot{p}_i = 0$ if $p_i = 0$, and

$$\frac{d}{dt} \sum_i p_i = 0 \ ,$$

whence $p_i(0) \geq 0$ and $\sum_i p_i(0) = 1$ imply that $p_i(t) \geq 0$ and $\sum_i p_i(t) = 1$ for $t \geq 0$.

If $m_i(t,\mathbf{n}) = m_i(t)$, (2.60) gives

$$n_i(t) = n_i(0)\exp\left[\int_0^t m_i(\tau)d\tau\right]. \tag{2.64}$$

Hence,

$$p_i(t) = \frac{p_i(0)\exp\left[\int_0^t m_i(\tau)d\tau\right]}{\sum_j p_j(0)\exp\left[\int_0^t m_j(\tau)d\tau\right]} \ . \tag{2.65}$$

With constant fitnesses m_i, (2.64) and (2.65) reduce to

$$n_i(t) = n_i(0)e^{m_i t} \ , \tag{2.66}$$

$$p_i(t) = \frac{p_i(0)e^{m_i t}}{\sum_j p_j(0)e^{m_j t}} \ . \tag{2.67}$$

In generations, (2.66) agrees with (2.11) if $w_i = e^{m_i}$. If A_1 is the fittest allele, $m_1 > m_i$ for $i > 1$, then (2.67) informs us that $p_1(t) \to 1$ as $t \to \infty$.

We shall now examine the rate of change of the *mean fitness*, $\overline{m}$. From (2.62) and (2.63) we obtain

$$\dot{\overline{m}} = \sum_i (\dot{m}_i p_i + m_i \dot{p}_i)$$

$$= \dot{\overline{m}} + \sum_i p_i m_i (m_i - \overline{m})$$

$$= \dot{\overline{m}} + V, \tag{2.68}$$

where

$$\dot{\overline{m}} = \sum_i \dot{m}_i p_i \tag{2.69}$$

is the mean rate of change of the fitnesses, and

$$V = \sum_i p_i (m_i - \overline{m})^2 \tag{2.70}$$

represents the genic variance in fitness.

Equation (2.68) is the simplest continuous case of Fisher's Fundamental Theorem of Natural Selection (Fisher, 1930). With constant Malthusian parameters, $\dot{\overline{m}} = 0$, whence $\dot{\overline{m}} = V \geq 0$, with equality only at equilibrium. So, as for discrete nonoverlapping generations, selection increases the mean fitness at a rate equal to the genic variance, reducing the latter to zero in the process. Hence, (2.61) shows that the relative rate of population growth increases.

The interpretation of (2.68) is similar to that of (2.16). Since natural populations cannot increase indefinitely, the absolute Malthusian parameters cannot remain constant for very many generations. However, (2.63) informs us that conversion of the absolute Malthusian parameters into relative ones by subtraction of the same genotype-independent quantity from each does not affect the gene frequencies, and we may often reasonably postulate constant relative fitnesses. For instance, if $m_i = r_i + g(N)$, with r_i constant and $m_i < 0$ for sufficiently large N for all i, then natural selection will maximize the mean relative fitness $\overline{r}$, rather than the mean absolute fitness $\overline{m}$. If A_1 is the optimal genotype, $m_1 > m_i$ for $i > 1$, then its reproductive excess relative to the population average, $m_1 - \overline{m} = r_1 - \overline{r}$, is minimized by natural selection.

With only two alleles, (2.63) and (2.70) reduce to

$$\dot{p} = pq(m_1 - m_2) , \tag{2.71}$$
$$V = pq(m_1 - m_2)^2 . \tag{2.72}$$

As an example of a deteriorating environment, we take $m_i = r_i - at$, with r_i and a constant and $a > 0$. Since (2.63) yields $\dot{p}_i = p_i(r_i - \overline{r})$, the gene frequencies evolve as with $a = 0$, but (2.68) becomes $\dot{\overline{m}} = V - a$, where $V = \sum_i p_i(r_i - \overline{r})^2$. Therefore, $\dot{\overline{m}} < 0$ for sufficiently large a. Since we have already seen, and will find repeatedly, that the mean fitness is a most useful and illuminating function, this example leads us to seek a suitable generalization.

Suppose that the right-hand side of (2.63) depends only on $\mathbf{p}$ for every i. If there exists a continuous function $F(\mathbf{p})$ such that its time derivative along trajectories of the population is nonnegative,

$$\dot{F}(\mathbf{p}) = \sum_i \frac{\partial F}{\partial p_i} \dot{p}_i = \sum_i \frac{\partial F}{\partial p_i} p_i(m_i - \overline{m}) \geq 0 \ ,$$

then $\mathbf{p}(t)$ converges as $t \to \infty$ to the set $\Gamma = \{\mathbf{p}|\dot{F}(\mathbf{p}) = 0\}$ (LaSalle, 1977). Since $\dot{F} = 0$ if $\dot{\mathbf{p}} = \mathbf{0}$, the set $\Omega = \{\mathbf{p}|\dot{\mathbf{p}} = \mathbf{0}\}$ is a subset of Γ. If F is such that $\Gamma = \Omega$, then $\mathbf{p}(t)$ converges as $t \to \infty$ to Ω, i.e., to gene-frequency equilibrium. In particular, if Γ is a set of isolated equilibrium points, then $\mathbf{p}(t)$ converges to one of those points. From (2.63) and (2.70) we see that $V(\mathbf{p}) = 0$ if and only if $\dot{\mathbf{p}} = \mathbf{0}$, so if we can find $F(\mathbf{p})$ such that $\dot{F}(\mathbf{p}) = V(\mathbf{p})$, then indeed $\Gamma = \Omega$.

For instance, let us posit that

$$m_i = f_i(p_i) + g(\mathbf{n}, t) \ . \tag{2.73}$$

From (2.69) we get

$$\overline{\dot{m}} = \sum_i p_i \left(\frac{df_i}{dp_i} \dot{p}_i + \frac{dg}{dt} \right) \ .$$

We define

$$h(\mathbf{p}) = \sum_i \int_0^{p_i} f_i'(x) x \, dx \ ;$$

now the prime indicates differentiation. Therefore,

$$\overline{\dot{m}} = \frac{d}{dt}[h(\mathbf{p}) + g(\mathbf{n}, t)] \ ,$$

whence (2.68) yields $\dot{F} = V(\mathbf{p})$, with

$$F(\mathbf{p}) = \overline{m} - h(\mathbf{p}) - g(\mathbf{n}, t)$$
$$= \overline{f}(\mathbf{p}) - h(\mathbf{p}) \ . \tag{2.74}$$

Clearly, (2.73) subsumes constant fitnesses $(F = \overline{m})$ and $m_i = r_i - at$ $(F = \overline{r})$. We consider two less trivial examples.

1. Frequency Dependence

Rare alleles may be favored because of genotype-dependent resource requirements, special skills, etc. The simplest hypothesis is $m_i = r_i - a_i p_i$, where r_i and a_i are constant and $a_i > 0$. From (2.74) we calculate $F = \overline{r} - \frac{1}{2}\sum_i a_i p_i^2$. The diallelic case is easy to analyze. From (2.71) we have

$$\dot{p} = pq[(r_1 - r_2 + a_2) - (a_1 + a_2)p] \ . \tag{2.75}$$

Since $p(t)$ changes continuously and time does not appear explicitly in (2.75), all the global results follow from the sign of $\dot{p}$. We find:

(a) $r_1 - r_2 + a_2 \leq 0$: Evidently, $\dot{p} < 0$ for $0 < p < 1$. Therefore, $p(t) \to 0$ as $t \to \infty$.

(b) $r_1 - r_2 + a_2 > 0$: We separate two subcases.

(i) $r_1 - r_2 \geq a_1$: Now $\dot{p} > 0$ for $0 < p < 1$. Hence, $p(t) \to 1$.

(ii) $r_1 - r_2 < a_1$: This is the case of strong frequency dependence, i.e., large a_i. There is a polymorphic equilibrium

$$\hat{p} = \frac{r_1 - r_2 + a_2}{a_1 + a_2} \quad,$$

and $\operatorname{sgn} \dot{p} = \operatorname{sgn}(\hat{p} - p)$, where $\operatorname{sgn} x = \pm 1$ for $x \gtrless 0$. Consequently, $p(t) \to \hat{p}$.

It is, of course, trivial to integrate (2.75) directly if more detailed information is required.

2. Population Regulation

We consider $m_i = r_i + g(\mathbf{n})$, where the r_i are positive constants and there exists N_0 such that $m_i < 0$ for $N > N_0$. According to (2.74), $F = \bar{r}$, and (2.63) yields $\dot{p}_i = p_i(r_i - \bar{r})$. Thus, this simple form of population regulation has no effect on gene frequencies. In the special case $g(\mathbf{n}) = -\bar{r}N/K$, where K is a positive constant called the carrying capacity, from (2.61) we obtain

$$\dot{N} = \bar{r}N \left(1 - \frac{N}{K}\right) \quad, \tag{2.76}$$

which tells us that $N(t) \to K$ as $t \to \infty$. If the population is monomorphic, $\bar{r}$ is replaced by the constant r in (2.76); this gives the logistic equation, which has the solution

$$N(t) = K \left\{ 1 + \left[\frac{K - N(0)}{N(0)} \right] e^{-rt} \right\}^{-1} \quad,$$

sketched in Fig. 2.2.

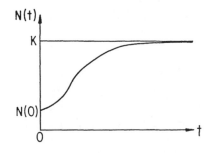

Fig. 2.2. The logistic curve.

As a final example, we shall discuss a model of *competitive selection* motivated by work of Kimura (1958), who analyzed the triallelic model with equal selection coefficients, and Nei (1971), who treated two alleles in discrete time.

We assume that the genotypes compete in pairs, c_{ij} being the probability that A_i defeats A_j. Hence, $c_{ij} = 1 - c_{ji}$. We suppose that a victorious individual gives birth and a defeated one dies. Measuring time in generations, we posit Δt combats for each individual in time Δt. Therefore, the fertilities and mortalities are the probabilities of victory and defeat. Assuming pairings at random and putting for convenience $c_{ij} = \frac{1}{2}(1 + s_{ij})$, where $s_{ij} = -s_{ji}$, we have

$$b_i = \sum_j c_{ij} p_j = \frac{1}{2} + \frac{1}{2} \sum_j s_{ij} p_j \ ,$$

$$d_i = \sum_j c_{ji} p_j = \frac{1}{2} - \frac{1}{2} \sum_j s_{ij} p_j \ .$$

Therefore, we obtain the frequency-dependent selection pattern

$$m_i = b_i - d_i = \sum_j s_{ij} p_j \ ,$$

from which it follows immediately that $\overline{m} = 0$. Hence, the population size is constant. The allelic frequencies satisfy

$$\dot{p}_i = p_i \sum_j s_{ij} p_j \ . \qquad (2.77)$$

The diallelic case is trivial: $\dot{p} = s_{12} pq$, so the frequency of the advantageous gene tends to unity.

For three alleles, we let $s_{12} = a$, $s_{23} = b$, and $s_{31} = c$ and write out (2.77):

$$\dot{p}_1 = p_1(ap_2 - cp_3) \ , \qquad (2.78a)$$

$$\dot{p}_2 = p_2(-ap_1 + bp_3) \ , \qquad (2.78b)$$

$$\dot{p}_3 = p_3(cp_1 - bp_2) \ . \qquad (2.78c)$$

We exclude the trivial case $a = b = c = 0$, for then $p_i(t) = p_i(0)$. There are always three trivial equilibria, $(1,0,0)$, $(0,1,0)$, and $(0,0,1)$, corresponding to the fixation of A_1, A_2, and A_3, respectively. If at least two of a, b, and c are nonzero and the nonzero selection coefficients have the same sign, then there exists the polymorphism $\hat{\mathbf{p}}$:

$$\left(\frac{b}{a+b+c}, \frac{c}{a+b+c}, \frac{a}{a+b+c} \right) \ .$$

The crucial element of the analysis is the observation that the function

$$G(\mathbf{p}) = b \ln p_1 + c \ln p_2 + a \ln p_3 \qquad (2.79)$$

is constant on orbits. Indeed, (2.78) and (2.79) yield

$$\dot{G} = b\dot{p}_1 p_1^{-1} + c\dot{p}_2 p_2^{-1} + a\dot{p}_3 p_3^{-1}$$

$$= b(ap_2 - cp_3) + c(-ap_1 + bp_3) + a(cp_1 - bp_2) = 0 \ .$$

[G is slightly more convenient than $e^G = p_1{}^b p_2{}^c p_3{}^a$, the natural generalization of Kimura's (1958) $p_1 p_2 p_3$ for $a = b = c = 1$.] In view of the symmetry of the problem, there are five distinct cases. The orbits in the $p_1 p_2$-plane sketched in Fig. 2.3 are given by $G(p_1, p_2, 1 - p_1 - p_2) = $ const., and the signs of $\dot{p}_i$ by (2.78).

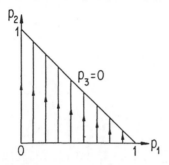

Fig. 2.3a. The triallelic competition model with $a = c = 0$ and $b > 0$. Here A_1 does not fight and its frequency is constant. The genotype A_2 is superior to A_3, which disappears.

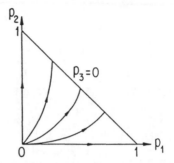

Fig. 2.3b. The triallelic competition model with $a = 0$, $b > 0$, and $c < 0$. Here A_3, being inferior to both A_1 and A_2, disappears. Figure 2.3b was sketched for $b > -c$. If $b = -c$, the lines emanating from the origin are straight; if $b < -c$, they curve downward.

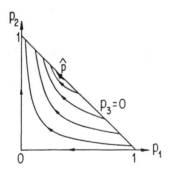

Fig. 2.3c. The triallelic competition model with $a = 0$ and $b, c > 0$. The equally fit genotypes A_1 and A_2 remain in the population, whereas A_3 is lost.

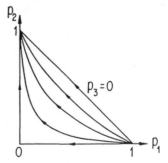

Fig. 2.3d. The triallelic competition model with $a < 0$ and $b, c > 0$. Here A_2, which defeats both A_1 and A_3, is ultimately fixed.

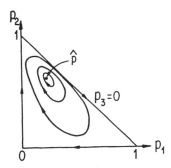

Fig. 2.3e. The triallelic competition model with $a, b, c > 0$. The population cycles around $\hat{\mathbf{p}}$. The equilibrium $\hat{\mathbf{p}}$ is stable, but not asymptotically stable.

Before ending this section, we shall briefly consider *mutation*. Let u_{ij} be the probability per unit time that A_i mutates to A_j for $i \neq j$. As in Sect. 2.2, $u_{ii} = 0$. Instead of (2.20), we now have

$$\dot{p}_i = \sum_j p_j u_{ji} - p_i \sum_j u_{ij} \; . \tag{2.80}$$

To include selection, we let selection act continuously and suppose mutation occurs every Δt time units. Separating these effects as in (2.25), from (2.63) and (2.80) we obtain

$$p_i^\star(t + \Delta t) = p_i(t) + p_i(t)(m_i - \overline{m})\Delta t + O[(\Delta t)^2] \; , \tag{2.81a}$$

$$p_i(t + \Delta t) = p_i^\star(t + \Delta t) + \left[\sum_j p_j^\star(t + \Delta t) u_{ji} \right.$$

$$\left. - p_i^\star(t + \Delta t) \sum_j u_{ij} \right] \Delta t + O[(\Delta t)^2] \; . \tag{2.81b}$$

Substituting (2.81a) into (2.81b) and letting $\Delta t \to 0$, we find

$$\dot{p}_i = p_i(m_i - \overline{m}) + \sum_j p_j u_{ji} - p_i \sum_j u_{ij} \; . \tag{2.82}$$

Thus, the approximate additivity of effects for weak selection and mutation in the discrete model of Sect. 2.2 becomes exact in continuous time.

We leave the analysis of (2.80) and (2.82) with two alleles for the problems.

In most organisms, it is necessary to distinguish zygotic and germ-line genotypic frequencies, and to take into account that selection acts on the phenotype, which develops from the offspring genotype, and germ cells mutate without phenotypic effect (Nagylaki, 1974). For the usual low mutation rates, however, (2.82) is a good approximation.

2.5 Random Drift

We suppose that generations are discrete and nonoverlapping, the population size being N in each generation. Selection and mutation are absent. The N offspring are sampled from the N members of the parental generation with replacement: if we imagine that N choices are made in sequence, each of the N parents has the same probability, $1/N$, of reproducing at each choice. Equivalently, we may envisage that each individual produces the same extremely large number of gametes and from these N are chosen at random to survive. Thus, sampling with replacement describes most accurately species, such as many plants, insects, and fish, that produce gametes greatly in excess of the number of surviving individuals. Even in this case, however, owing to sundry random factors, some individuals may make a decreased (or increased) contribution to the gametic pool, thereby increasing the sampling variation in the gene frequencies.

To study the evolution of the population, we employ the probability, f_t, that two randomly chosen distinct individuals in generation t carry alleles of the same type at the locus under consideration. Although f_t does not specify completely the genetic state of the population, it is highly informative and simple to analyze. We may express f_t explicitly as the sum of the probabilities that both genes are A_i, where $i = 1, 2, \ldots, k$, but this is not useful here. The probability, $h_t = 1 - f_t$, that the two homologous genes are different is clearly a measure of genetic variability in the population.

Consider two randomly chosen homologous genes in generation $t+1$. They may be identical for two mutually exclusive reasons: descent from the same gene or descent from distinct identical genes in generation t. Descent from the same gene evidently has probability N^{-1}. Hence, the two genes are descended from distinct genes in the previous generation with probability $1 - N^{-1}$, in which case they are identical if and only if the parental genes are identical, and this occurs with probability f_t. Therefore

$$f_{t+1} = N^{-1} + (1 - N^{-1})f_t \ , \tag{2.83}$$

whence

$$h_{t+1} = (1 - N^{-1})h_t \ . \tag{2.84}$$

We could have written (2.84) directly by noting that nonidentity is possible only by descent from distinct nonidentical genes. In more complicated situations, however, it is easier to start with the probability of identity.

The solution of (2.84) reads

$$h_t = h_0(1 - N^{-1})^t \ , \tag{2.85}$$

which can be approximated as

$$h_t \approx h_0 e^{-t/N} \tag{2.86}$$

in the biologically important case $N \gg 1$. Thus, $h_t \to 0$ as $t \to \infty$, demonstrating that ultimately only one allele remains in the population. The characteristic time for the decay of genetic variability is quite long, about N generations. It was essentially for this reason that Fisher (1930), who held that randomly mating natural populations were usually large, ascribed little evolutionary importance to random drift. In contrast, Wright (1931, 1977, 1978) asserted that many species are subdivided into small, randomly mating subpopulations that exchange migrants at a low rate, thereby allowing significant gene-frequency change by random drift. Refer to Chap. 1 for a fuller discussion of these classical viewpoints.

2.6 Problems

2.1 Choose $w_1/w_2 = 1 - s$, where s $(0 < s < 1)$ is a constant, in (2.18). Show that

$$t = \ln \left[\frac{q(0)p(t)}{p(0)q(t)} \right] \Big/ \ln (1 - s) \ .$$

Observe that the times required to reach given frequencies for weak selection $(s \ll 1)$ are inversely proportional to the selection intensity and for $p(t) \ll p(0), q(0)$

$$t \approx -s^{-1} \ln p(t) \ .$$

2.2 Consider two frequencies of A_1, $p = \pi_1, \pi_2$, with $\pi_1 > \pi_2$, in (2.18). If w_1 and w_2 are constant, we expect $\pi_1' > \pi_2'$. Prove this by deducing that

$$\frac{dp'}{dp} = \frac{w_1 w_2}{\overline{w}^2} > 0 \ .$$

2.3 Analyze completely (2.18) with $w_1 = 1 + s - ap$ and $w_2 = 1$, where $s > -1$ and $0 < a < 1 + s$.

2.4 Analyze completely (2.18) with $w_1 = 1 - s + ap$ and $w_2 = 1$, where $0 < s < 1$ and $a > 0$. What is the asymptotic rate of convergence to $p = 0$?

2.5 Calculate the ultimate rate of convergence for (2.18) with $w_1 = 1 - sp$ and $w_2 = 1$, where $0 < s < 1$.

2.6 Consider the multiallelic discrete model with the life cycle

Offspring	$\longrightarrow$	Adult	$\longrightarrow$	Adult	$\longrightarrow$	Offspring
	viability		mutation		fertility	
p_i		$\overline{p}_i$		$\tilde{p}_i$		p_i'

(a) Write the recursion relations.

(b) Show that if there is no differential viability ($v_i = 1$) or no differential fertility ($f_i = 1$), this system is equivalent to (2.25).

(c) Prove that if selection and mutation are weak, so that the equations may be expanded to first order, the change in the gene frequency reduces to the sum of changes due to the three evolutionary forces operating, where each change is calculated as if the corresponding force were the only one present.

(d) Demonstrate that to first order the model is the same as (2.25).

2.7 Show directly from (2.41) that $0 \le p_+ \le 1 \le p_-$, and if $u + v \le 1$, then $\lambda \ge 0$.

2.8 Demonstrate that the total birth rate, $\nu_i(t, 0)$, satisfies the integral equation

$$\nu_i(t,0) = \int_t^\infty \nu_i(0, x-t) l_i(t, x) b_i(t, x) dx + \int_0^t \nu_i(t-x, 0) l_i(t, x) b_i(t, x) dx.$$

Given the initial age distribution, the first integral, in principle, is known. If this integral equation is solved, then (2.47) determines $\nu_i(t, x)$ as a function of $\nu_i(0, x)$, $b_i(t, x)$, and $l_i(t, x)$.

2.9 Choose $m_2 - m_1 = s > 0$ in (2.71) and show that the time required to reach a specified allelic frequency $p(t)$ is

$$t = \frac{1}{s} \ln \left[\frac{p(0)q(t)}{q(0)p(t)} \right] .$$

Note that this is the same as the weak-selection approximation for discrete time, and $t \approx -s^{-1} \ln p(t)$ for $p(t) \ll p(0)$, $q(0)$.

2.10 For the gene-frequency equation $\dot{p} = f(p)$, find a function whose derivative is positive except at equilibrium, where it is zero.

2.11 Analyze completely (2.71) with $m_i = r_i(1 + a_i p_i)^{-1}$, where $a_i, r_i > 0$ for $i = 1, 2$.

2.12 Taking $u = u_{12}$ and $v = u_{21}$ for two alleles in (2.80), find $p(t) = p_1(t)$.

2.13 Add selection with $m_2 - m_1 = s > 0$ to Problem 2.12. Give a complete analysis. Show that if $u + v \ll s$, the equilibrium is the same as that for the discrete case with weak selection and mutation.

2.14 For weak selection, the discrete model (2.8) is well approximated by the continuous model (2.63). Suppose that $w_i = 1 + sr_i$, where r_i may depend on the gene frequencies, but not on time or the selection intensity s ($s > 0$). Rescale time according to $\tau = st$ and set $\pi_i(\tau) = p_i(t)$, $t = 0, 1, \ldots$. Deduce from (2.8) that in the limit $s \to 0$

$$\frac{d\pi_i}{d\tau} = \pi_i(r_i - \bar{r}) \ , \quad \bar{r} = \sum_i r_i \pi_i \ .$$

3. Panmictic Populations

Hereafter we shall be concerned with diploid populations. We begin by investigating in this chapter the genetic structure of a randomly mating population in the absence of selection, mutation, and random drift. This part of population genetics was the first to be understood, and a thorough grasp of its principles is required for the formulation and interpretation of most evolutionary models. To ensure the desired comprehension, some fairly detailed examples and problems, of a type that has proved useful in human genetics, are presented. In Sect. 3.1 we shall derive the Hardy-Weinberg Law. We shall proceed to sex-linkage and two loci in Sects. 3.2 and 3.3 and discuss the effect of population subdivision in Sect. 3.4. In Sect. 3.5 we investigate deviations from the Hardy-Weinberg Law in a finite population.

The material in Sects. 3.1 to 3.4 is based entirely on a few elementary genetic principles and the following simple rules of probability. Let $\mathcal{P}(A)$ and $\mathcal{P}(B)$ designate the probabilities of events A and B. We denote the probability of either A or B or both by $\mathcal{P}(A+B)$. If $\mathcal{P}(AB)$ represents the probability of both A and B, called the *joint* probability of A and B, then we have

$$\mathcal{P}(A+B) = \mathcal{P}(A) + \mathcal{P}(B) - \mathcal{P}(AB) \ .$$

For *mutually exclusive* events, $\mathcal{P}(AB) = 0$. We shall also need *conditional* probabilities: the probability of A given B is $\mathcal{P}(A|B)$. Clearly,

$$\mathcal{P}(AB) = \mathcal{P}(A|B)\mathcal{P}(B) \ .$$

If A and B are *independent*, $\mathcal{P}(A|B) = \mathcal{P}(A)$, so $\mathcal{P}(AB) = \mathcal{P}(A)\mathcal{P}(B)$.

3.1 The Hardy-Weinberg Law

We assume that generations are discrete and nonoverlapping, as is true, for instance, for annual plants and many insects. If the population is *monoecious*, i.e., every individual has both male and female sexual organs, as in most plants and some animals, it can be described by one set of genotypic proportions. It will be clear that this may also be done if the initial genotypic frequencies are identical in the two sexes of a *dioecious* population. After treating the monoecious case, we shall show that if the initial genotypic frequencies are different in the two sexes, the attainment of equilibrium is merely delayed by one generation.

We denote the alleles at the locus under consideration by A_i; the diploid individuals in the population will have genotypes $A_i A_j$. Mendel's Law of Segregation tells us that a gamete from an $A_i A_j$ individual is equally likely to carry A_i and A_j. Unless otherwise stated, we shall always employ ordered genotypic frequencies. Thus, $P_{ij}\,(= P_{ji})$ designates the frequency of ordered $A_i A_j$ genotypes, i.e., P_{ii} is the frequency of $A_i A_i$ *homozygotes*, and $2P_{ij}$, where $i \neq j$, is the frequency of unordered $A_i A_j$ *heterozygotes*. The frequency of A_i in the population, p_i, reads

$$p_i = \sum_j P_{ij} \ . \tag{3.1}$$

By Mendel's Law of Segregation, p_i is the frequency of A_i in the gametic output of the population.

If mating occurs without regard to the genotype at the A-locus, random union of gametes yields the genotypic proportions

$$P'_{ij} = p_i p_j \tag{3.2}$$

in the next generation. Therefore, the gene frequencies do not change,

$$p'_i = \sum_j P'_{ij} = p_i \ , \tag{3.3}$$

and Hardy-Weinberg proportions,

$$P'_{ij} = p'_i p'_j \ , \tag{3.4}$$

are attained in a single generation. For two alleles, various simple numerical cases were worked out first, then Weinberg (1908) obtained the result assuming the initial absence of heterozygotes, and, simultaneously and independently, Hardy (1908) analyzed the most general situation. Weinberg (1909) extended the principle to multiple alleles.

The most important aspect of the Hardy-Weinberg Law is the constancy of the allelic frequencies. This means that if no evolutionary forces are acting, under random mating genetic variability is conserved. The feature that equilibrium is reached in a single generation and thereafter the genotypic frequencies are uniquely determined by the gene frequencies is also of great significance. It allows one to analyze the evolution of a population with k alleles in terms of $k-1$ independent allelic frequencies, rather than $\frac{1}{2}k(k+1)-1$ independent genotypic frequencies. This is a considerable simplification. Note that the frequencies of the genotypes $A_i A_j$ are the terms in the expansion of $(\sum_i p_i)^2$. For k alleles, we have

$$\sum_{i=1}^{k} p_i^2 \geq \frac{1}{k} \ .$$

Therefore, the total frequency of homozygotes is at least $1/k$, with equality if and only if $p_i = 1/k$ for all i.

The very brief derivation of (3.4) presented above is directly applicable if large numbers of sperm and eggs are produced and some of these fuse at random to form zygotes, as exemplified by some marine organisms. For most species, however, explicitly considering matings approximates the biological situation more closely. This approach has the advantage of generalizing readily to nonrandom mating and differential fertility.

Let the frequency of ordered $A_i A_j \times A_k A_l$ matings be $X_{ij,kl}$ (e.g., the total proportion of $A_1 A_2 \times A_1 A_1$ matings, without regard to allelic and genotypic order, is $4X_{12,11}$). We shall impose the restriction of panmixia only when necessary. The mating frequencies satisfy the symmetry conditions $X_{ij,kl} = X_{ji,kl} = X_{kl,ij}$ and the normalization

$$\sum_{ijkl} X_{ij,kl} = 1 \ . \tag{3.5}$$

To calculate the genotypic proportions in the next generation, we distinguish homozygotes from heterozygotes throughout. We must include, owing to ordering, a factor of 2 for all heterozygotes and for all matings between distinct genotypes, and, owing to segregation, one of $\frac{1}{2}$ for all heterozygotes. In addition, we need to observe that the cross $A_i A_j \times A_i A_j$, where $i \neq j$, can produce $A_i A_j$ in two ways. Hence, we obtain

$$P'_{ii} = X_{ii,ii} + \sum_{k \neq i} 2(2)\tfrac{1}{2} X_{ik,ii} + \sum_{k \neq i} 2(2)(\tfrac{1}{2})\tfrac{1}{2} X_{ik,ki}$$

$$+ \sum_{k \neq i} \sum_{l \neq i, l > k} 2(2)(2)(\tfrac{1}{2})\tfrac{1}{2} X_{ik,li} \ ,$$

$$2P'_{ij} = 2X_{ii,jj} + \sum_{k \neq i} 2(2)\tfrac{1}{2} X_{ik,jj} + \sum_{l \neq j} 2(2)\tfrac{1}{2} X_{ii,lj}$$

$$+ 2(2)(2)(\tfrac{1}{2})\tfrac{1}{2} X_{ij,ij} + \sum_{\substack{k \neq i}} \sum_{\substack{l \neq j \\ (k,l) \neq (j,i)}} 2(2)(2)(\tfrac{1}{2})\tfrac{1}{2} X_{ik,lj} \ , \qquad i \neq j \ .$$

Removing the restrictions from the sums and combining terms yields

$$P'_{ij} = \sum_{kl} X_{ik,lj} \quad \text{for all } i, j \ . \tag{3.6}$$

At this point, we posit random mating:

$$X_{ik,lj} = P_{ik} P_{lj} \ . \tag{3.7}$$

Observe that (3.7) satisfies (3.5). Substituting (3.7) into (3.6) immediately gives (3.2).

We proceed to generalize to *separate sexes* with different initial genotypic frequencies. Let P_{ij} and Q_{ij} be the ordered frequencies of A_iA_j males and females, respectively. The gene frequencies in the two sexes are

$$p_i = \sum_j P_{ij} \ , \quad q_i = \sum_j Q_{ij} \ . \tag{3.8}$$

With random union of gametes, we have the same genotypic frequencies in the two sexes in the next generation:

$$P'_{ij} = Q'_{ij} = \tfrac{1}{2}(p_iq_j + p_jq_i) \ . \tag{3.9}$$

Therefore, the allelic frequencies in each sex are just the unweighted average allelic frequencies in the previous generation:

$$p'_i = q'_i = \tfrac{1}{2}(p_i + q_i) \ . \tag{3.10}$$

Thus, another generation of random union of gametes produces Hardy-Weinberg proportions

$$P''_{ij} = Q''_{ij} = p'_ip'_j \ . \tag{3.11}$$

The sex ratio does not influence (3.10), because males and females contribute equally to the next generation: (3.10) gives the allelic frequencies in successful gametes.

To derive (3.9) from the mating structure, we now define $X_{ij,kl}$ as the ordered frequency of matings between A_iA_j males and A_kA_l females. We still have $X_{ij,kl} = X_{ji,kl} = X_{ij,lk}$, but reciprocal crosses are not necessarily equally frequent: $X_{ij,kl} \neq X_{kl,ij}$. Equation (3.5) still applies, but now we must generalize (3.6) to

$$P'_{ij} = Q'_{ij} = \tfrac{1}{2} \sum_{kl}(X_{ik,lj} + X_{lj,ik}) \ , \tag{3.12}$$

and random mating now means

$$X_{ij,kl} = P_{ij}Q_{kl} \ . \tag{3.13}$$

Note, that (3.13) is consistent with (3.5). Inserting (3.13) into (3.12) and recalling (3.8) at once produces (3.9).

It will be convenient to refer to (3.9) and to the corresponding relation for a sex-linked locus as generalized Hardy-Weinberg proportions.

We end this section with some examples.

1. The *MN* Blood Group

There are two alleles, M and N, the three genotypes being easily distinguishable. This is the case of *no dominance*. The frequency, p_1, of M can be obtained from the measured genotypic proportions, P_{11}, $2P_{12}$, and P_{22}, of *MM*, *MN*, and *NN* using $p_1 = P_{11} + P_{12}$, and Hardy-Weinberg can be tested with one degree of freedom (see, e.g., Li, 1955; Crow and Kimura, 1970).

2. Phenylthiocarbamide (PTC)

The genotypes A_1A_1 and A_1A_2 can taste PTC strongly, whereas A_2A_2 individuals either cannot or can do so only very faintly. In general, if A_1A_2 has the same *phenotype* (i.e., expressed character) as A_1A_1, we say A_1 is *dominant* (to A_2) and A_2 is *recessive* (to A_1). The frequency, q, of A_2 is given by $q = \sqrt{P_{22}}$, but the genetic mechanism for a dominant gene cannot be tested without using some family data. If homozygous recessives are affected by some disease, e.g., phenylketonuria (PKU), then usually $q \ll 1$. This has the important implication that the frequency of carriers of the recessive allele, $2pq \approx 2q$, is much greater than the proportion affected, q^2. In applications to diseases, care must be exercised: owing to selection, the recessive frequency will usually be less in adults than in zygotes. Since A_2A_2 is extremely rare, however, the error is generally quite small.

3. The *ABO* Blood Group

For our purposes, it will suffice to consider three alleles, A, B, and O, with A and B dominant to O (for more detail, see Cavalli-Sforza and Bodmer, 1971), as shown in Table 3.1.

Table 3.1. The *ABO* blood group

Blood type	Genotypes	Frequency
A	AA, AO	$p_A^2 + 2p_A p_O$
B .	BB, BO	$p_B^2 + 2p_B p_O$
AB	AB	$2p_A p_B$
O	OO	p_O^2

If R_A, R_B, R_{AB}, and R_O are the observed frequencies of the blood types, we have $p_O = \sqrt{R_O}$ and $(p_A + p_O)^2 = R_A + R_O$, whence $p_A = \sqrt{R_A + R_O} - \sqrt{R_O}$. As for two alleles without dominance, we can test Hardy-Weinberg with one degree of freedom (Li, 1955; Crow and Kimura, 1970).

4. The Parentage of Homozygous Recessives

Here and in the following two examples, let A be dominant to the other allele, a. Homozygous recessives all come from the matings $Aa \times Aa$, $Aa \times aa$, $aa \times aa$. We wish to compute the proportion of aa individuals originating in each of the three matings. The calculation set out in Table 3.2 shows that if the recessive is rare, $q \ll 1$, almost all affected individuals come from unions of two carriers. Selection against homozygous recessives will again produce a small error here.

Table 3.2. The parentage of homozygous recessives

Mating	Freq. of mating	Freq. of aa progeny	Freq. among aa progeny
$Aa \times Aa$	$(2pq)^2$	$\frac{1}{4}(4p^2q^2)$	p^2
$Aa \times aa$	$2(2pq)q^2$	$\frac{1}{2}(4pq^3)$	$2pq$
$aa \times aa$	$(q^2)^2$	q^4	q^2

5. Parent-Offspring Relations

It will often be convenient to abbreviate the dominant and recessive phenotypes by D and R.

(a) Breeding ratios: These are the conditional probabilities of the birth of a child of phenotype i $(= D, R)$ in a $j \times k$ mating, denoted by $P_{i,jk}$. Obviously, $P_{R,RR} = 1$. We need not resort to a mating table to calculate Snyder's (1932) ratios, $P_{R,DR}$ and $P_{R,DD}$. When applicable, analysis in terms of gametes is much quicker. Let x be the probability of drawing a gamete that carries a from a dominant phenotype. Then $P_{R,DR} = x$ and $P_{R,DD} = x^2$, where

$$x = \frac{(1/2)2pq}{p^2 + 2pq} = \frac{q}{1 + q} \ . \tag{3.14}$$

As a check, note that recessives have been correctly allocated to the three types of matings:

$$(q^2)^2 P_{R,RR} + 2q^2(1 - q^2)P_{R,DR} + (1 - q^2)^2 P_{R,DD} = q^2 \ .$$

Parent-offspring resemblance is confirmed by the inequalities $P_{R,RR} > P_{R,DR} > P_{R,DD}$. The conditional probabilities of dominants are given by $P_{D,ij} = 1 - P_{R,ij}$. Family frequencies are calculated by multiplying the conditional probabilities by the appropriate mating frequencies.

(b) Mother-child probabilities: We desire to compute the probability that a mother (or father) of phenotype j has a child of phenotype i, $P_{i,j}$. This can be done, of course, from the breeding ratios:

$$P_{i,j} = q^2 P_{i,jR} + (1 - q^2)P_{i,jD} \ .$$

But an even faster method is to note that a randomly chosen gamete carries A and a with respective probabilities p and q, so, with panmixia, $P_{R,R} = q$ and $P_{R,D} = xq$. Mother and child resemble each other, for $P_{R,R} > P_{R,D}$. As a check, observe here that $q^2 P_{R,R} + (1 - q^2)P_{R,D} = q^2$.

6. A Sibling Distribution

We wish to derive the probability, F_{sn}, that a sibship of size s has exactly n $(= 0, 1, \ldots, s)$ dominant individuals (Cotterman, 1937). We must separate the six genotypically distinct matings, as done in Table 3.3.

Table 3.3. The sibling distribution for a dominant gene

Mating	Freq. of mating	Conditional probability of sibship		
		$n = s$	$0 < n < s$	$n = 0$
$AA \times AA$	$(p^2)^2$	1	0	0
$AA \times Aa$	$2p^2(2pq)$	1	0	0
$AA \times aa$	$2p^2q^2$	1	0	0
$Aa \times Aa$	$(2pq)^2$	$\left(\frac{3}{4}\right)^s$	$\binom{s}{n}\left(\frac{3}{4}\right)^n\left(\frac{1}{4}\right)^{s-n}$	$\left(\frac{1}{4}\right)^s$
$Aa \times aa$	$2(2pq)q^2$	$\left(\frac{1}{2}\right)^s$	$\binom{s}{n}\left(\frac{1}{2}\right)^n\left(\frac{1}{2}\right)^{s-n}$	$\left(\frac{1}{2}\right)^s$
$aa \times aa$	$(q^2)^2$	0	0	1

We obtain $F_{sn}(q)$ from the sums of the unconditional probabilities in the columns of Table. 3.3. With some simplifications, we find

$$F_{ss} = p\left\{p(2 - p^2) + q^2\left[4\left(\tfrac{3}{4}\right)^s p + \left(\tfrac{1}{2}\right)^{s-2}q\right]\right\} \ , \tag{3.15a}$$

$$F_{s0} = q^2\left[\left(\tfrac{1}{2}\right)^{s-1}p + q\right]^2 \ , \tag{3.15b}$$

$$F_{sn} = \left(\tfrac{1}{2}\right)^{s-2}\binom{s}{n}pq^2\left[3^n\left(\tfrac{1}{2}\right)^s p + q\right] \ , \quad 0 < n < s \ . \tag{3.15c}$$

As partial checks, we observe that (3.15) gives correctly the trivial results $F_{11} = 1 - q^2$ and $F_{10} = q^2$, and with a little algebra we may verify that

$$\sum_{n=0}^{s} F_{sn} = 1 \ . \tag{3.16}$$

For two siblings, (3.15) yields

$$F_{22} = p\left[1 + pq\left(1 + \tfrac{1}{4}q\right)\right] \ ,$$
$$F_{20} = \tfrac{1}{4}q^2(1 + q)^2 \ ,$$
$$F_{21} = \tfrac{1}{2}pq^2(3 + q) \ .$$

The probability that there are precisely n dominants among s randomly chosen individuals is

$$f_{sn} = \binom{s}{n}(1 - q^2)^n(q^2)^{s-n} \ ,$$

whence

$$f_{22} = p^2(1 + q)^2 \ , \quad f_{20} = q^4 \ , \quad f_{21} = 2pq^2(1 + q) \ .$$

It is easy to see that concordant sib pairs are more and discordant ones less frequent than the corresponding sib pairs among randomly chosen individuals:

$$F_{22} > f_{22} \ , \quad F_{20} > f_{20} \ , \quad F_{21} < f_{21} \ . \tag{3.17}$$

3.2 *X*-Linkage

In most higher animals, such as mammals, sex is determined by a single chromosomal difference between males and females. One sex has a pair of *sex chromosomes* of the same type, XX, and the other has two different ones, denoted XY. Although in some taxa (e.g., birds) XX is male and XY is female, we shall always refer to XX as female and XY as male, this situation being far more common (e.g., mammals and *Drosophila*). Chromosomes that are not sex chromosomes are called *autosomes*, and the genes on them are *autosomal*. Genes on the sex chromosomes are said to be *sex-linked*; the terms *X-linked* and *Y-linked* distinguish genes on the X and Y chromosomes. Since there are far fewer Y-linked loci with known phenotypic effects than X-linked ones, sex-linkage and X-linkage are often used synonymously. It is apparent that if we interpret all variables as referring only to males, Chap. 2 applies to the dynamics of Y-linked genes. Since X-linked recessives are expressed in males, X-linked loci are of particular interest in human, mouse, and *Drosophila* genetics. Furthermore, in species with *arrhenotokous* parthenogenesis, all males are haploid because they develop from unfertilized eggs of the mother, whereas females are diploid. Arrhenotoky occurs in the insect orders Hymenoptera, Coleoptera, Thysanoptera, and Homoptera, the arachnid order Acarina, and the rotifer order Monogononta (Hartl, 1971). In such animals, all loci are effectively X-linked.

Let the frequency of A_i be p_i in males and q_i in females. We denote the proportion of ordered $A_i A_j$ females by Q_{ij}. Random union of gametes informs us that

$$p_i' = q_i \ , \tag{3.18a}$$

$$Q_{ij}' = \tfrac{1}{2}(p_i q_j + p_j q_i) \ . \tag{3.18b}$$

To deduce (3.18) from explicit consideration of matings, let $X_{i,jk}$ designate the frequency of unions between A_i males and ordered $A_j A_k$ females. Thus,

$$\sum_{ijk} X_{i,jk} = 1 \ . \tag{3.19}$$

Then

$$p_i' = \sum_{jk} X_{j,ik} \ , \tag{3.20a}$$

$$Q_{ij}' = \tfrac{1}{2} \sum_{k}(X_{i,kj} + X_{j,ik}) \ . \tag{3.20b}$$

With random mating,

$$X_{i,jk} = p_i Q_{jk} \ ; \tag{3.21}$$

noting that

$$q_i \doteq \sum_{j} Q_{ij} \tag{3.22}$$

reduces (3.20) to (3.18).

From (3.18b) we find the expected relation

$$q_i' = \tfrac{1}{2}(p_i + q_i) \ . \tag{3.23}$$

Using (3.18a) and (3.23), we see that the average frequency of A_i in the entire population, $x_i = \tfrac{1}{3}(p_i + 2q_i)$, is conserved: $x_i' = x_i$, whence $x_i(t) = x_i(0)$. The male-female gene-frequency difference, $y_i = p_i - q_i$, changes sign and is reduced by $\tfrac{1}{2}$ every generation: $y_i' = -\tfrac{1}{2}y_i$, whence $y_i(t) = y_i(0)(-\tfrac{1}{2})^t$. Therefore,

$$p_i(t) = x_i(0) + \tfrac{2}{3}(-\tfrac{1}{2})^t y_i(0) \ , \tag{3.24a}$$

$$q_i(t) = x_i(0) - \tfrac{1}{3}(-\tfrac{1}{2})^t y_i(0) \ . \tag{3.24b}$$

Thus, the Hardy-Weinberg equilibrium

$$p_i = x_i(0), \quad Q_{ij} = x_i(0)x_j(0) \tag{3.25}$$

is approached in a rapid, oscillatory manner, but in contrast to the autosomal dioecious case, it is not attained in two generations. Notice that, as in the autosomal case, the average frequency is defined without reference to the sex ratio: the weighting is by the number of X chromosomes carried by an individual, because regardless of the sex ratio, each mating involves one male and one female.

For two alleles, numerical examples of the behavior (3.24) were first given by Jennings (1916); the analytic solution was deduced by Robbins (1918).

An important consequence of (3.25) is that if q is the frequency of a rare recessive allele, the corresponding trait (various types of color blindness are examples) is expressed with a much higher frequency, q, in males than in females, in whom the frequency is q^2.

As an *example*, suppose A, with frequency p, is completely dominant to a and the population is in equilibrium. What are the parent-offspring probabilities, $P_{i,j}$, of Example 5(b) of Sect. 3.1? We shall use an asterisk to refer to females.

For father-son, clearly, $P_{R,R} = P_{R,D} = q$, indicating that the father's genotype is irrelevant, since he contributes a Y chromosome to his son. Father-daughter is equally simple : $P_{R*,R} = q$ and $P_{R*,D} = 0$. For mother-son, we evidently have $P_{R,R*} = 1$, and since $x = q/(1+q)$ is the probability of drawing a from a dominant mother, we see that $P_{R,D*} = x$. The mother-daughter relation is the same as for an autosome: $P_{R*,R*} = q$ and $P_{R*,D*} = xq$. The following relations are satisfied, as they must be:

$$qP_{R,R} + pP_{R,D} = q \ ,$$

$$qP_{R*,R} + pP_{R*,D} = q^2 \ ,$$

$$q^2 P_{R,R*} + (1-q^2)P_{R,D*} = q \ ,$$

$$q^2 P_{R*,R*} + (1-q^2)P_{R*,D*} = q^2 \ .$$

3.3 Two Loci

We shall suppose that the sexes need not be distinguished: the dioecious case is treated in Problem 3.10. Let the alleles at the A- and B-loci, A_i and B_i, have frequencies p_i and q_i, the number of alleles at each locus being arbitrary. We denote the frequency of $A_i B_j$ gametes in the gametic output of generation t by $P_{ij}(t)$. The gametic frequencies are our basic variables; the gene frequencies

$$p_i = \sum_j P_{ij}, \quad q_j = \sum_i P_{ij} \tag{3.26}$$

no longer suffice to describe the population.

Gametes $A_i B_j$ and $A_k B_l$ unite to form an individual $A_i B_j / A_k B_l$. A proportion $1 - c$ of the gametes produced by this individual are parental (or nonrecombinant) gametes, i.e., $\frac{1}{2}(1 - c)$ are $A_i B_j$ and $\frac{1}{2}(1 - c)$ are $A_k B_l$, and a fraction c are nonparental (or recombinant), i.e., $\frac{1}{2}c$ $A_i B_l$ and $\frac{1}{2}c$ $A_k B_j$. If the two loci are on the same chromosome, the *recombination* (or *cross-over*) frequency c satisfies $0 \le c \le \frac{1}{2}$. The numerical value of c depends on the structure of the chromosome and the position of the two loci. If we measure distance, y, along the chromosome from the A-locus in a particular direction, the recombination fraction between A and a locus at y is a monotone increasing function of y. We shall assume $c > 0$, because it is manifest that for our purposes $c = 0$ corresponds to a single locus. For loci on different chromosomes, almost invariably $c = \frac{1}{2}$, as expected. Loci with $c = \frac{1}{2}$ are called *unlinked* (or, less accurately, *independent*).

A proportion $1 - c$ of the gametes in generation $t+1$ are produced without recombination. Clearly, the contribution to P'_{ij} from these events is $(1-c)P_{ij}$. With random mating, the contribution of recombinant events is $cp_i q_j$. Therefore,

$$P'_{ij} = (1 - c)P_{ij} + cp_i q_j \ . \tag{3.27}$$

It follows immediately from (3.26) and (3.27) that the gene frequencies are conserved, as they must be. We define the *linkage disequilibria*

$$D_{ij} = P_{ij} - p_i q_j \tag{3.28}$$

in order to have measures of the departure from random combination of alleles within gametes. From (3.27) and (3.28) we deduce at once

$$D'_{ij} = (1 - c)D_{ij} \ , \tag{3.29}$$

whence

$$D_{ij}(t) = D_{ij}(0)(1 - c)^t \ . \tag{3.30}$$

We conclude that $D_{ij} \to 0$ at the rate $1 - c$. Thus, *linkage equilibrium* (or *gametic phase equilibrium*), $P_{ij} = p_i q_j$, is approached gradually and without oscillation. The larger c, the faster is the rate of convergence, the most rapid being $(\frac{1}{2})^t$ for unlinked loci.

For two alleles at each locus, numerical cases were worked out by Jennings (1917); the analytic solution was obtained by Robbins (1918a). Jennings and Robbins considered the gametic output of all the genotypes, as we shall do in Chap. 8; the elegant argument used here to deduce (3.27) is due to Malécot (1948). Geiringer (1944) showed that even with multiple loci and multiple alleles the gametic frequencies converge to products of the appropriate allelic frequencies. Bennett (1954) gave a simplified proof.

The ordered frequency of A_iB_j/A_kB_l is $P_{ij}P_{kl}$, provided there has been at least one generation of random mating. By summing $P_{ij}P_{kl}$ over j and l, or i and k, we recover the Hardy-Weinberg Law at each locus. In linkage equilibrium, $P_{ij}P_{kl}$ simplifies to $(p_ip_k)(q_jq_l)$, the product of the single-locus Hardy-Weinberg genotypic frequencies. This observation has the important consequence that since linkage has no effect on equilibrium genotypic frequencies, family data are required to detect linkage. Note also that in linkage equilibrium the *coupling* (A_iB_j/A_kB_l) and *repulsion* (A_iB_l/A_kB_j) phases are equally frequent. We may combine (3.27) and (3.28) to obtain

$$\Delta P_{ij} = -cD_{ij} \ . \tag{3.31}$$

For two alleles at each locus, the usual notation is $x_1 = P_{11}$, $x_2 = P_{12}$, $x_3 = P_{21}$, and $x_4 = P_{22}$. One finds easily from (3.28)

$$D_{11} = D_{22} = -D_{12} = -D_{21} = x_1x_4 - x_2x_3 \equiv D \ , \tag{3.32}$$

so that (3.31) becomes

$$\Delta x_i = -\varepsilon_i cD \ , \tag{3.33}$$

where $\varepsilon_1 = \varepsilon_4 = 1$ and $\varepsilon_2 = \varepsilon_3 = -1$.

As an example, let us consider *duplicate genes*. For any number of loci, let the alleles at locus i be $A^{(i)}$ and $a^{(i)}$, with frequencies $p^{(i)}$ and $q^{(i)}$. An individual has phenotype R (recessive) if he is homozygous for $a^{(i)}$ at every locus; otherwise his phenotype is D (dominant).

(a) Breeding ratios: We designate the probability of drawing an R gamete (i.e., one with $a^{(i)}$ at every locus) from a D individual by y. Evidently, $P_{R,RR} = 1$, $P_{R,DR} = y$, and $P_{R,DD} = y^2$, as for a single locus. If the population is in linkage equilibrium, then for any recombination frequencies, the single-locus result generalizes to $y = z/(1 + z)$, where $z = \Pi_i q^{(i)}$ is the proportion of R gametes. To see this explicitly for two diallelic loci, suppose the alleles $A, a; B, b$ have frequencies $p, q; u, v$. Then the gametic output of the genotypes $AaBb$, $Aabb$, and $aaBb$ yields

$$y = \frac{(1/4)4pquv + (1/2)2pqv^2 + (1/2)2q^2uv}{1 - q^2v^2} = \frac{qv}{1 + qv} \ .$$

(b) Sibling distribution: We shall calculate in the diallelic two-locus case the probability, Q_{sn}, that there are exactly n D individuals among s siblings. Although in linkage equilibrium the two types of double heterozygotes, AB/ab and Ab/aB, are equally frequent in the population, within a family one must take into account that these have different gametic outputs, which depend on c. The extreme case $c = 0$ is trivial: $Q_{sn} = F_{sn}(qv)$, F_{sn} being the single-locus distribution given by (3.15).

Let us derive Q_{sn} for unlinked loci, $c = \frac{1}{2}$. It is convenient to use d and r for single-locus phenotypes; thus, dr denotes an individual with the dominant phenotype at the A-locus (i.e., AA or Aa) and the recessive at the other (i.e., bb). Let the numbers of dd, dr, rd, and rr siblings be i, j, $n-i-j$, and $s-n$, respectively. For unlinked loci in linkage equilibrium, the probability of having $(i+j)$ $d-$ sibs and $(n-j)$ $-d$ ones is $F_{s,i+j}(q)F_{s,n-j}(v)$. Given that we have $(i+j)$ $d-$ sibs and $(n-j)$ $-d$ ones, we require the probability of choosing i dd sibs from $(i+j)$ $d-$ ones and $(n-i-j)$ rd sibs from $(s-i-j)$ $r-$ ones. Since we have conditioned on choosing $(n-j)$ $-d$ individuals, the desired probability is

$$\binom{i+j}{i}\binom{s-i-j}{n-i-j}\Big/\binom{s}{n-j} \ .$$

Hence,

$$Q_{sn} = \sum_{i=0}^{n}\sum_{j=0}^{n-i}\left[\binom{i+j}{i}\binom{s-i-j}{s-n}\Big/\binom{s}{n-j}\right]F_{s,i+j}(q)F_{s,n-j}(v)$$

$$= \sum_{k=0}^{n}\sum_{l=n-k}^{n}\left[\binom{k}{n-l}\binom{s-k}{s-n}\Big/\binom{s}{l}\right]F_{sk}(q)F_{sl}(v) \ , \tag{3.34}$$

where we changed variables to $k = i+j$ and $l = n-j$.

It is easy to check that (3.34) reduces sensibly for $s = 1, 2$. Let us verify that

$$\sum_{n=0}^{s} Q_{sn} = 1 \ .$$

We put $m = n - l$ to deduce

$$\sum_{n=0}^{s} Q_{sn} = \sum_{k=0}^{s} F_{sk}(q) \sum_{l=0}^{s}\left[F_{sl}(v)\Big/\binom{s}{l}\right]\sum_{m=m_1}^{m_2}\binom{k}{m}\binom{s-k}{s-l-m} \ ,$$

where $m_1 = \max(0, k-l)$ and $m_2 = \min(k, s-l)$. But writing

$$(1+x)^i(1+x)^j = (1+x)^{i+j} \ ,$$

expanding, and equating coefficients proves that

$$\sum_{j}\binom{i}{j}\binom{k-i}{l-j} = \binom{k}{l} \ .$$

Therefore, recalling (3.16), we get

$$\sum_{n=0}^{s} Q_{sn} = \sum_{k=0}^{s} F_{sk}(q) \sum_{l=0}^{s} F_{sl}(v) = 1 \ .$$

3.4 Population Subdivision

Many natural populations are divided into subpopulations. We shall show that if each subpopulation is panmictic, the effect of the subdivision is to raise the *homozygosity*, f, compared with that of a randomly mating population with the same allelic frequencies, f_r. Let c_j and $p_i^{(j)}$ represent the proportion of individuals and the frequency of A_i in subpopulation j. Using $\mathcal{E}$ (expectation) to designate averages over subpopulations, we may write the allelic frequencies in the population as

$$\overline{p_i} \equiv \mathcal{E}(p_i) = \sum_j c_j p_i^{(j)} \ .$$

The variance of p_i among subpopulations is

$$V_i = \mathcal{E}[(p_i - \overline{p_i})^2] = \overline{p_i^2} - \overline{p_i}^2 \ . \tag{3.35}$$

Therefore, the frequency of $A_i A_i$ in the population,

$$\overline{p_i^2} = \overline{p_i}^2 + V_i \ , \tag{3.36}$$

is higher than in a randomly mating population by the variance of p_i, which is zero if and only if the allelic frequencies in all the subpopulations are the same. Equation (3.36) is Wahlund's principle (Wahlund, 1928).

The proportion of homozygotes in the population reads

$$f = \sum_i \overline{p_i^2} = f_r + \sum_i V_i \ , \tag{3.37}$$

with

$$f_r = \sum_i \overline{p_i}^2 \ .$$

The *heterozygosity*, h, is decreased according to

$$h = 1 - f = h_r - \sum_i V_i \ , \tag{3.38}$$

where $h_r = 1 - f_r$, the proportion of heterozygotes with random mating. For two alleles, observing that $V_1 = V_2 \equiv V$, we can specialize (3.36) and (3.38) to

$$P_{11} = \overline{p}^2 + V \ , \qquad P_{12} = \overline{p}\,\overline{q} - V \ , \qquad P_{22} = \overline{q}^2 + V \ .$$

3.5 Genotypic Frequencies in a Finite Population

Hardy-Weinberg proportions do not hold exactly in a finite population: e.g., if an allele is represented only once, it obviously cannot appear in homozygotes. Given the vector of allelic frequencies, $\mathbf{p}$, the unordered genotypic frequencies, Q_{ij}, where $i \leq j$, are random variables. (We use unordered, rather than ordered, genotypic frequencies because independent random variation of the ordered frequencies P_{ij} and P_{ji} would destroy the symmetry $P_{ij} = P_{ji}$.)

Let N and q_{ij} denote the number of monoecious individuals and the probability that, given the allelic frequencies, an individual chosen at random has genotype $A_i A_j$. If the random variable $R_{ij}^{(k)}$ equals one or zero according as the kth individual is $A_i A_j$ or not, the number of $A_i A_j$ individuals is

$$NQ_{ij} = \sum_{k=1}^{N} R_{ij}^{(k)} \ .$$

The expectation of $R_{ij}^{(k)}$ reads

$$\mathcal{E}(R_{ij}^{(k)}|\mathbf{p}) = q_{ij}(1) + (1 - q_{ij})(0) = q_{ij} \ ;$$

substituting this into the expectation of the previous equation gives

$$\mathcal{E}(Q_{ij}|\mathbf{p}) = q_{ij} \ . \tag{3.39}$$

We employ an argument of Crow and Kimura (1970) to evaluate q_{ij}.

If gametes are combined at random, drawing an individual at random is equivalent to the successive random choice of two genes. The first gene is A_i with probability p_i; given that it is A_i, the second gene is A_i with probability $(2Np_i - 1)/(2N - 1)$ and A_j, where $j \neq i$, with probability $2Np_j/(2N - 1)$. Therefore, (3.39) yields

$$\mathcal{E}(Q_{ii}|\mathbf{p}) = p_i(2Np_i - 1)/(2N - 1) = p_i^2 - \alpha p_i(1 - p_i) \ , \tag{3.40a}$$
$$\mathcal{E}(Q_{ij}|\mathbf{p}) = 2p_i(2Np_j)/(2N - 1) = 2p_i p_j(1 + \alpha), \quad i \neq j \ , \tag{3.40b}$$

where $\alpha = 1/(2N - 1)$, and the first factor of 2 in (3.40b) takes into account the fact that drawing A_j before A_i also produces genotype $A_i A_j$.

Thus, the expected frequency of homozygotes is less, whereas that of heterozygotes is greater, than the Hardy-Weinberg result. The relative errors in the expected heterozygote frequencies are small for large populations, $N \gg 1$; for homozygote frequencies, they are small if each allele is represented many times, $Np_i \gg 1$, which is a much stronger condition than $N \gg 1$ for rare alleles.

Levene (1949) has obtained the conditional distribution and hence some of the moments of the genotypic frequencies. From his results one can infer that the absolute values of the deviations of the genotypic frequencies from their means have high probability of being much smaller than these means if $\sqrt{N}p_i \gg 1$ for all i, i.e., if the square root of the number of homozygotes is large for every allele. But then $Np_i \geq \sqrt{N}p_i \gg 1$, so the means are close to Hardy-Weinberg proportions; therefore, the Hardy-Weinberg Law is a good approximation.

If a sample of N individuals is drawn from an extremely large population, the allelic frequencies in the population must be estimated from those in the

sample. Levene's (1949) work may then be used to test whether the population
is in Hardy-Weinberg proportions.

3.6 Problems

3.1   The genotypic frequencies at a diallelic locus in a monoecious population
can be represented on a de Finetti (1926) diagram, sketched in Fig. 3.1.
With $D = P_{11}$, $H = 2P_{12}$, and $R = P_{22}$, we have $D + H + R = 1$,
so the distances from a point to the sides of an equilateral triangle of
unit altitude can be taken to correspond to D, H, and R. Prove that
the Hardy-Weinberg curve $H^2 = 4DR$ is a parabola and the vertical line
divides the base of the triangle in the ratio of $p : q$.

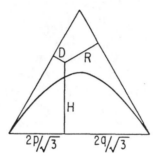

Fig. 3.1. The de Finetti diagram.

3.2   Consider an arbitrary diallelic monoecious population. Using $Q = P_{12} - p_1 p_2$ as abscissa and $p = p_1$ as ordinate, find the equations of the curves
bounding the permitted region in the Qp-plane and sketch this region.
The fact that Hardy-Weinberg populations fall on the p-axis makes this
plot convenient for displaying the evolution of populations with overlapping generations.

3.3   For a diallelic Hardy-Weinberg population, find the probabilities $P_{i,j}$ that
a mother of genotype j $(i, j = AA, Aa, aa)$ has a child of genotype i.

3.4   Prove (3.17).

3.5   Calculate the probability that in a diallelic Hardy-Weinberg population,
a sibship of size s has i AA, j Aa, and k aa individuals.

3.6   *Penetrance* is the extent to which a genotype is expressed as a phenotype.
The first model considered for the inheritance of human handedness was
diallelic autosomal, right being completely dominant to left, i.e., AA and

Aa had phenotype R, whereas aa was L. This was immediately seen to fail because about half the progeny of $L \times L$ matings are R. A more sophisticated model (Trankell, 1955), which still cannot account for the related phenotypic variation in cerebral organization (Levy and Nagylaki, 1972), postulates incomplete penetrance for a. Suppose AA and Aa are always R, but of aa genotypes only a fraction x among parents and y among children are L. Assume mating is random and A has frequency p in both generations. Notice that $x > y$ ($x < y$) implies a higher (lower) frequency of sinistrality among parents than children. For instance, $x < y$ might reflect a decrease in cultural pressure toward dextrality. Obtain the proportion of sinistral progeny in the three kinds of matings and the probability that there are exactly n sinistrals among s siblings.

3.7   Assuming equilibrium, derive for a diallelic X-linked trait with complete dominance the probability that in a sibship precisely i of m males and j of n females are dominant.

3.8   Two populations with respective genotypes $AABB$ and $aabb$ are crossed. If the A and B loci are unlinked, show that they are in linkage equilibrium after two generations. Why is (3.30) inapplicable?

3.9   A model for the inheritance of human handedness (Levy and Nagylaki, 1972; Nagylaki and Levy, 1973; Levy, 1976; Levy and Reid, 1976, 1978) posits the existence of two diallelic loci, one determining cerebral dominance and the other governing hand control. At the first locus, the allele L (frequency p) for left-hemisphere language is completely dominant to the allele l (freq. q) for right-hemisphere language. Thus, verbal and analytic functions reside primarily in the left hemisphere of LL and Ll, but in the right hemisphere of ll individuals. In each case, the other hemisphere performs mainly spatial and synthetic functions. At the second locus, the allele C (freq. u) for contralateral hand control is completely dominant to c (freq. v), which specifies ipsilateral hand control. Therefore, $LLCC$, $LLCc$, $LlCC$, $LlCc$, and $llcc$ are dextral (freq. γ); $LLcc$ and $Llcc$ are sinistral (freq. $\lambda = 1 - \gamma$) with ipsilateral hand control (freq. β); and $llCC$ and $llCc$ are sinistral with contralateral hand control (freq. $\alpha = \lambda - \beta$). Panmixia and linkage equilibrium are assumed.

(a) Calculate the allelic frequencies in terms of α and β. To obtain a unique solution, use the fact that the proportion of dextrals with right-hemisphere language is very small, i.e., $q^2 v^2 \ll 1$.

(b) Derive the proportion of sinistrals in the three types of matings by using the single-locus breeding ratios.

(c) Solve (b) directly by calculating the gametic output of the two phenotypes.

(d) For unlinked loci, prove that the probability of having exactly n sinistrals among s sibs reads

$$P_{sn} = \sum_{j=0}^{n} \sum_{i=0}^{s-n} A_{snij} F_{s,s-n-i+j}(q) F_{s,s-i-j}(v) \ ,$$

where

$$A_{snij} = \binom{s-n-i+j}{j} \binom{n+i-j}{i} \Big/ \binom{s}{i+j} \ .$$

(e) Demonstrate that

$$\sum_{n=0}^{s} P_{sn} = 1 \ .$$

3.10 To generalize Sect. 3.3 to separate sexes, suppose P_{ij} and $\tilde{P}_{ij}$, p_i and $\tilde{p}_i$, q_j and $\tilde{q}_j$, and c and $\tilde{c}$ are the frequencies of $A_j B_j$, A_i, B_j, and the recombination fractions in males and females, respectively.

(a) Show that with random mating the recursion relations are

$$P'_{ij} = (1-c)\tfrac{1}{2}(P_{ij} + \tilde{P}_{ij}) + c\tfrac{1}{2}(p_i \tilde{q}_j + \tilde{p}_i q_j) \ ,$$
$$\tilde{P}'_{ij} = (1-\tilde{c})\tfrac{1}{2}(P_{ij} + \tilde{P}_{ij}) + \tilde{c}\tfrac{1}{2}(p_i \tilde{q}_j + \tilde{p}_i q_j) \ .$$

(b) Deduce that after one generation the gene frequencies in the two sexes are constant and equal to each other, being then given by

$$\bar{p}_i = \tfrac{1}{2}[p_i(0) + \tilde{p}_i(0)] \ , \qquad \bar{q}_j = \tfrac{1}{2}[q_j(0) + \tilde{q}_j(0)] \ .$$

(c) Prove that after one generation the frequency of $A_i B_j$ in the population, $\overline{P}_{ij} = \tfrac{1}{2}(P_{ij} + \tilde{P}_{ij})$, satisfies the same difference equation as the gametic frequency in a monoecious population, provided the mean recombination fraction $\bar{c} = \tfrac{1}{2}(c + \tilde{c})$ is used:

$$\overline{P}'_{ij} = (1-\bar{c})\overline{P}_{ij} + \bar{c}\,\bar{p}_i \bar{q}_j \ .$$

(d) Demonstrate that

$$\overline{P}_{ij}(t) = \bar{p}_i \bar{q}_j + (1-\bar{c})^{t-1}[\overline{P}_{ij}(1) - \bar{p}_i \bar{q}_j] \ , \qquad t \geq 1 \ ,$$
$$P_{ij}(t) = \left(\frac{1-c}{1-\bar{c}}\right)\overline{P}_{ij}(t) + \tfrac{1}{2}\left(\frac{c-\tilde{c}}{1-\bar{c}}\right)\bar{p}_i \bar{q}_j \ , \qquad t \geq 2 \ ,$$
$$\tilde{P}_{ij}(t) = \left(\frac{1-\tilde{c}}{1-\bar{c}}\right)\overline{P}_{ij}(t) - \tfrac{1}{2}\left(\frac{c-\tilde{c}}{1-\bar{c}}\right)\bar{p}_i \bar{q}_j \ , \qquad t \geq 2 \ ,$$

whence $P_{ij}(t), \tilde{P}_{ij}(t) \to \bar{p}_i \bar{q}_j$ as $t \to \infty$.

For two alleles at each locus, these results are due to Robbins (1918a). The generalization to multiple loci and multiple alleles was presented by Geiringer (1948) and, more concisely, by Bennett (1954).

3.11 Assume that locus A is autosomal and locus B is X-linked. The alleles A_i have frequencies p_i and $\tilde{p}_i$ in sperm and eggs, whereas the corresponding frequencies of B_j are q_j and $\tilde{q}_j$. Generations are discrete and nonoverlapping, and no evolutionary forces are present.

(a) Show that the frequencies, P_{ij} and $\tilde{P}_{ij}$, of $A_i B_j$ in sperm and eggs satisfy

$$P'_{ij} = \tfrac{1}{2}(\tilde{P}_{ij} + p_i \tilde{q}_j) \ ,$$
$$\tilde{P}'_{ij} = \tfrac{1}{4}(P_{ij} + \tilde{P}_{ij} + p_i \tilde{q}_j + \tilde{p}_i q_j) \ .$$

(b) Deduce that the allelic frequencies have the usual single-locus behavior.

(c) Establish the recursion relations

$$D'_{ij} = \tfrac{1}{2}\tilde{D}_{ij} \ ,$$
$$\tilde{D}'_{ij} = \tfrac{1}{4}(D_{ij} + \tilde{D}_{ij}) \ ,$$

for the linkage disequilibria $D_{ij} = P_{ij} - p_i q_j$ and $\tilde{D}_{ij} = \tilde{P}_{ij} - \tilde{p}_i \tilde{q}_j$.

(d) Infer that

$$D_{ij}(t) = \tfrac{1}{3}[D_{ij}(0) + 2\tilde{D}_{ij}(0)](\tfrac{1}{2})^t + \tfrac{2}{3}[D_{ij}(0) - \tilde{D}_{ij}(0)](-\tfrac{1}{4})^t \ ,$$
$$\tilde{D}_{ij}(t) = \tfrac{1}{3}[D_{ij}(0) + 2\tilde{D}_{ij}(0)](\tfrac{1}{2})^t - \tfrac{1}{3}[D_{ij}(0) - \tilde{D}_{ij}(0)](-\tfrac{1}{4})^t \ .$$

Thus, the linkage disequilibria in sperm and eggs rapidly converge to equality, and, as Wright (1933) demonstrated, they tend to zero at the ultimate rate $(\tfrac{1}{2})^t$.

The reader may find some of the mathematical techniques developed in Sect. 5.1 helpful for this problem and the next one.

3.12 Suppose that loci A and B are both X-linked, with recombination frequency c in females (and zero in males). Retain the other assumptions and notation of Problem 3.11.

(a) Prove that

$$P'_{ij} = \tilde{P}_{ij} \ ,$$
$$\tilde{P}'_{ij} = \tfrac{1}{2}(1 - c)(P_{ij} + \tilde{P}_{ij}) + \tfrac{1}{2}c(p_i \tilde{q}_j + \tilde{p}_i q_j) \ .$$

(b) Verify that the allelic frequencies satisfy the usual recurrence relations for an X-linked locus.

(c) Derive the difference equations

$$D'_{ij} = \tilde{D}_{ij} \ , \tag{3.41}$$
$$\tilde{D}'_{ij} = \tfrac{1}{2}(1 - c)(D_{ij} + \tilde{D}_{ij}) + Q_{ij} \ ,$$

where

$$Q_{ij} = \tfrac{1}{4}(1 - 2c)(p_i q_j + \tilde{p}_i \tilde{q}_j - p_i \tilde{q}_j - \tilde{p}_i q_j) \ .$$

Observe that $Q_{ij} \to 0$ as $c \to \tfrac{1}{2}$.

(d) Demonstrate that

$$Q_{ij}(t) = Q_{ij}(0)(\tfrac{1}{4})^t$$

and

$$D''_{ij} - \tfrac{1}{2}(1-c)(D'_{ij} + D_{ij}) = Q_{ij} \ .$$

(e) Express the linkage disequilibrium in sperm as

$$D_{ij}(t) = \alpha_{ij}\lambda_+^t + \beta_{ij}\lambda_-^t - 16(9-10c)^{-1}Q_{ij}(0)(\tfrac{1}{4})^t \ ,$$

where

$$\lambda_\pm = \tfrac{1}{4}\{1 - c \pm [(1-c)(9-c)]^{1/2}\} \ ,$$
$$\alpha_{ij} = (\lambda_+ - \lambda_-)^{-1}[-\lambda_- D_{ij}(0) + \tilde{D}_{ij}(0)$$
$$- 4(9-10c)^{-1}(4\lambda_- - 1)Q_{ij}(0)] \ ,$$
$$\beta_{ij} = (\lambda_+ - \lambda_-)^{-1}[\lambda_+ D_{ij}(0) - \tilde{D}_{ij}(0)$$
$$+ 4(9-10c)^{-1}(4\lambda_+ - 1)Q_{ij}(0)] \ .$$

$\tilde{D}_{ij}(t)$ is given by (3.41).

(f) Conclude that the asymptotic rate of convergence is λ_+^t (Wright, 1933) and that λ_+ increases from approximately 0.640 to 1 as c decreases from $\tfrac{1}{2}$ to 0.

The complete solution of this problem (for two alleles at each locus) was first obtained in a different manner by Bennett (1963).

4. Selection at an Autosomal Locus

We owe the basic ideas and models of selection to Fisher (1922, 1930), Haldane (1924–34), and Wright (1931). After developing the multiallelic equations of selection in Sect. 4.1, we shall study their dynamics for two alleles in Sect. 4.2 and generalize to multiple alleles in Sect. 4.3. We shall analyze the diallelic case with a constant level of inbreeding in Sect. 4.4 and investigate variable environments in Sect. 4.5. Sections 4.6, 4.7, and 4.8 treat intra-family selection, maternal inheritance, and meiotic drive as examples of the many possible less common selection schemes that require modifications of the formulation of Sect. 4.1. We shall discuss multiallelic mutation-selection balance in Sect. 4.9 and consider overlapping generations in Sect. 4.10. A weak-selection analysis for density- and frequency-dependent genotypic fitnesses will be performed in Sect. 4.11.

Age structure is considered in this chapter only in the derivation of the continuous model in Sect. 4.10. Consult Charlesworth (1980) for an extensive treatment of evolution in age-structured populations.

4.1 Formulation for Multiple Alleles

We shall generalize now the fundamental model of Sect. 3.1 to incorporate selection. As before, we assume that generations are discrete and nonoverlapping. We suppose that the sexes need not be distinguished. This will be so for a monoecious population and for random mating in a dioecious one with the same viabilities in the two sexes and the same sex ratio in all matings. The second requirement is almost invariably satisfied. The general dioecious case is discussed in Problem 7.4.

We suppose that $A_i A_j$ zygotes have ordered frequency P_{ij}. Then the frequency of A_i in zygotes is

$$p_i = \sum_j P_{ij} \ . \tag{4.1}$$

Let v_{ij}, $a_{ij,kl}$, $X_{ij,kl}$ represent the viability of $A_i A_j$ individuals and the fertility and ordered frequency of $A_i A_j \times A_k A_l$ matings, respectively. These quantities must be specified as functions of the genotypic frequencies and time, on which they may depend. Notice that the fertilities $a_{ij,kl}$ may incorporate the effect of differential early survival due to variable parental care.

The frequency of A_iA_j adults is

$$\tilde{P}_{ij} = P_{ij}v_{ij}/\bar{v} \; , \tag{4.2}$$

where

$$\bar{v} = \sum_{ij} v_{ij}P_{ij} \tag{4.3}$$

denotes the mean viability of the population. Hence, the frequency of A_i in adults reads

$$\tilde{p}_i = \sum_j \tilde{P}_{ij} = p_iv_i/\bar{v} \; , \tag{4.4}$$

where we have introduced the viability of A_i- individuals, given by

$$p_iv_i = \sum_j P_{ij}v_{ij} \; . \tag{4.5}$$

It is important to note that the adults are generally not in Hardy-Weinberg proportions. According to Problem 4.1, provided the zygotes are in Hardy-Weinberg proportions, the adults will be in Hardy-Weinberg proportions if and only if the viabilities are multiplicative (i.e., there exist c_i such that $v_{ij} = c_ic_j$ for all i and j).

According to Sect. 3.1, with random mating and no fertility differences, the gene frequencies in zygotes in the next generation would be $\tilde{p}_i$ and the Hardy-Weinberg zygotic frequencies would be $\tilde{p}_i\tilde{p}_j$. This is the standard viability selection model. Our more general formulation will demonstrate that the form of the equation for gene frequency change is always the same and Hardy-Weinberg proportions apply with multiplicative fertilities.

For studying variation in population number (see Problem 4.3) and assigning absolute (as opposed to relative) values to the genotypic birth rates and fitnesses, it is convenient to use fertilities, $f_{ij,kl}$, relative to the number of adults, $\tilde{N}(t)$, rather than the number of matings, $M(t)$. With

$$f_{ij,kl}(t) = M(t)a_{ij,kl}(t)/\tilde{N}(t) \; ,$$

one expects $f_{ij,kl}(t)$ to be constant whenever $a_{ij,kl}(t)$ is. As (4.6) explicitly shows, the genotypic frequencies are unaffected by this scaling.

Including differential fertility in (3.6), we see that the zygotic proportions in the next generation are

$$P'_{ij} = \bar{f}^{-1} \sum_{kl} X_{ik,lj}f_{ik,lj} \; , \tag{4.6}$$

where

$$\bar{f} = \sum_{ijkl} X_{ik,lj}f_{ik,lj} \tag{4.7}$$

designates the mean fertility. The birth rate, b_{ij}, of A_iA_j adults is defined by

$$\tilde{P}_{ij}b_{ij} = \sum_{kl} X_{ij,kl}f_{ij,kl} \ . \tag{4.8}$$

It is essential to distinguish the average number of juvenile offspring of an A_iA_j adult, b_{ij}, from the number of A_iA_j zygotes born per mating,

$$\sum_{kl} X_{ik,lj}f_{ik,lj} \ .$$

The analogues of these two quantities coincide in asexual species. The fertility of an adult carrying A_i is given by

$$\tilde{p}_i b_i = \sum_{j} \tilde{P}_{ij}b_{ij} \ , \tag{4.9}$$

and using (4.7) and (4.8), we can reexpress the average fecundity as

$$\overline{f} = \overline{b} = \sum_{ij} b_{ij}\tilde{P}_{ij} \ . \tag{4.10}$$

Returning to (4.6) and employing (4.1), (4.8), (4.9), and (4.4), we find

$$p_i' = p_i(v_ib_i)/(\overline{vb}) \ . \tag{4.11}$$

Let us define the expected number of progeny of an A_iA_j zygote,

$$w_{ij} = v_{ij}b_{ij} \ , \tag{4.12}$$

as the fitness of A_iA_j. The fitness of A_i and the mean fitness are defined by

$$p_iw_i = \sum_{j} P_{ij}w_{ij} \ , \tag{4.13}$$

$$\overline{w} = \sum_{ij} P_{ij}w_{ij} \ . \tag{4.14}$$

We expect

$$w_i = v_ib_i \ , \tag{4.15a}$$
$$\overline{w} = \overline{vb} \ . \tag{4.15b}$$

To prove (4.15a) we utilize (4.13), (4.2), (4.12), (4.9), and (4.4):

$$p_iw_i = \sum_{j} \overline{v}\tilde{P}_{ij}b_{ij} = \overline{v}\tilde{p}_ib_i = p_iv_ib_i \ . \tag{4.16}$$

For (4.15b), we have from (4.14), (4.13), (4.16), (4.9), and (4.10)

$$\overline{w} = \sum_{i} p_iw_i = \overline{v}\sum_{i} \tilde{p}_ib_i = \overline{vb} \ .$$

Substituting (4.15) into (4.11), we deduce the basic difference equation

$$p'_i = p_i w_i / \overline{w} . \tag{4.17}$$

The change in the gene frequency reads

$$\Delta p_i = p_i (w_i - \overline{w}) / \overline{w} . \tag{4.18}$$

Note carefully that (4.15) follows from (4.12) only because we have averaged viabilities with respect to zygotic frequencies and fertilities with respect to adult frequencies.

Although (4.17) has the same form as the corresponding equation, (2.6), for haploids, the fitnesses in (4.17) depend on the genotypic proportions and mating frequencies. Therefore, even if the viabilities, fertilities, and mating frequencies are specified as functions of the genotypic frequencies, (4.17) does not suffice to determine the evolution of the population. For this, we require (4.6).

To describe the dynamics entirely in terms of gene frequencies, let us posit *random mating*:

$$X_{ij,kl} = \tilde{P}_{ij} \tilde{P}_{kl} . \tag{4.19}$$

This reduces (4.8) to

$$b_{ij} = \sum_{kl} \tilde{P}_{kl} f_{ij,kl} . \tag{4.20}$$

We can proceed further if we confine ourselves to *multiplicative fertilities* (Penrose, 1949; Bodmer, 1965):

$$f_{ij,kl} = \beta_{ij} \beta_{kl} , \tag{4.21}$$

for some β_{ij}. Inserting (4.21) into (4.20) yields

$$b_{ij} = \beta_{ij} \overline{\beta}, \qquad \text{where} \qquad \overline{\beta} = \sum_{kl} \beta_{kl} \tilde{P}_{kl} , \tag{4.22}$$

and substituting this into (4.10) gives $\overline{b} = \overline{\beta}^2$, so that $\beta_{ij} = b_{ij} / \sqrt{\overline{b}}$. Therefore, (4.21) becomes

$$f_{ij,kl} = b_{ij} b_{kl} / \overline{b} . \tag{4.23}$$

Employing (4.19), (4.23), and (4.9) in (4.6) produces

$$P'_{ij} = \tilde{p}_i b_i \tilde{p}_j b_j / \overline{b}^2 . $$

Substituting (4.4), (4.15), and (4.17) informs us that zygotic Hardy-Weinberg proportions,

$$P'_{ij} = p'_i p'_j , \tag{4.24}$$

are attained in a single generation.

Thus, we have derived the *standard selection model*,

$$P_{ij} = p_i p_j , \qquad p_i' = p_i w_i / \overline{w} , \tag{4.25a}$$

$$w_i = \sum_j w_{ij} p_j , \quad \overline{w} = \sum_{ij} w_{ij} p_i p_j . \tag{4.25b}$$

For multiplicative fitnesses, (4.25) reduces to the haploid case, (2.6) (Li, 1959; Nagylaki, 1975). We leave the easy proof of this assertion for Problem 4.2. A straightforward argument shows that (4.25) applies also if, instead of mating, each genotype sheds gametes and these gametes fuse at random. In this case, $2b_{ij}$ represents the number of gametes produced by an $A_i A_j$ adult. Gametic selection that acts among all gametes, rather than independently among gametes in different matings, can easily be incorporated into (4.25). If p_i designates the frequency of A_i in gametes *before* selection, then $w_{ij} = \alpha_i \alpha_j \gamma_{ij}$, the factors being the gametic and zygotic fitnesses (Wright, 1969). The simplest case of (4.25) is that of constant w_{ij}, analyzed in the following two sections. The constant-fitness assumption gains in plausibility from the observation that, since a genotype-independent factor cancels out of (4.25), if the $f_{ij,kl}$ are constants, β_{ij} will be constant, and consequently (4.22) permits us to use constant b_{ij} in (4.25).

If $w_{ij} = \frac{1}{2}(w_{ii} + w_{jj})$ for all i and j, we say there is no dominance. If $w_{ij} = w_{ii}$ for all j, then A_i is completely dominant, the other alleles being recessive to A_i.

In (4.25), we understand that the fitnesses are averaged over the genetic and environmental background. Suppose $x_{ij}(y)\Delta y$ is the probability that $A_i A_j$ has a fitness between y and $y + \Delta y$. Then (4.25) applies with

$$w_{ij} = \int_0^\infty y x_{ij}(y) dy .$$

Note that if $x_{ij}(y) = x(y)$, independent of genotype, then w_{ij} is independent of genotype and there is no selection, as expected.

If there are only two alleles, (4.18) may be rewritten in the form

$$\Delta p = pq(w_1 - w_2)/\overline{w} , \tag{4.26}$$

where, as usual, $p = p_1$ and $q = p_2$. It is often convenient to set $w_{11} = 1 + s$, $w_{12} = 1 + hs$, and $w_{22} = 1 - s$. Then s represents the intensity of selection; h specifies the degree of dominance: $h = 0$ if there is no dominance and $h = 1$ (-1) if A_1 is dominant (recessive). We designate the situations $w_{12} > w_{11}, w_{22}$ $(h > 1)$ and $w_{12} < w_{11}, w_{22}$ $(h < -1)$ by the terms *overdominance* and *underdominance*.

4.2 Dynamics with Two Alleles

The difference equation (4.26) with constant fitnesses w_{ij} is known to be solvable only in two cases. As mentioned above, if fitnesses are multiplicative, it reduces to the haploid case, and this was solved for any number of alleles in Sect. 2.1. We can also calculate $p(t)$ if a genotype has zero fitness (Problem 4.4). In homozygotes, this can happen either because the allele involved, say a $(= A_2)$, is *lethal*, which means aa is inviable $(v_{22} = 0)$, or because aa is *sterile* $(b_{22} = 0)$. Here, we shall determine the fate of the population for all possible fitnesses (Fisher, 1922, 1930). This global analysis is a special case of that in Sect. 4.4; we present it separately in order to expound the fundamental ideas below in their simplest form. We shall also calculate the rates of convergence to equilibrium for the various fitness patterns.

We rewrite (4.26) in the form

$$\Delta p = pqg(p)/\overline{w} , \tag{4.27}$$

where

$$g(p) = w_{12} - w_{22} + (w_{11} - 2w_{12} + w_{22})p . \tag{4.28}$$

We exclude the trivial case of no selection by assuming that the w_{ij} are not all equal. There are four distinct cases, depending on the signs of

$$g(0) = w_{12} - w_{22} \quad \text{and} \quad g(1) = w_{11} - w_{12} .$$

1. $w_{22} \leq w_{12} \leq w_{11}$

We have $g(0) \geq 0$ and $g(1) \geq 0$, so the linear function $g(p) > 0$ for $0 < p < 1$. Since $\operatorname{sgn} \Delta p = \operatorname{sgn} g(p)$, therefore $p(t) \to 1$ as $t \to \infty$. This is not surprising, since A is favored over a for all gene frequencies.

2. $w_{11} \leq w_{12} \leq w_{22}$

This is just Case 1 with the sign of $g(p)$ reversed. Therefore, $p(t) \to 0$ as $t \to \infty$. The deleterious allele is again eliminated.

3. $w_{12} < w_{11}, w_{22}$

In the underdominant case, we have an equilibrium

$$\hat{p} = \frac{w_{22} - w_{12}}{w_{11} - 2w_{12} + w_{22}} \tag{4.29}$$

in $(0, 1)$. From (4.27), (4.28), and (4.29) we obtain

$$\Delta p = (w_{11} - 2w_{12} + w_{22})pq(p - \hat{p})/\overline{w} , \tag{4.30}$$

whence $\operatorname{sgn} \Delta p = \operatorname{sgn} (p - \hat{p})$. Consequently, the equilibrium $\hat{p}$ is unstable: $p(t) \to 0$ if $p(0) < \hat{p}$ and $p(t) \to 1$ if $p(0) > \hat{p}$. Thus, the fate of the population depends on its initial condition. This is frequently the situation with an unstable polymorphism.

4. $w_{12} > w_{11}, w_{22}$

The equilibrium $\hat{p}$ is still in $(0,1)$ in the overdominant case, but now (4.30) informs us that $\operatorname{sgn} \Delta p = \operatorname{sgn}(\hat{p} - p)$. Owing to the *a priori* possibility of diverging oscillation about $\hat{p}$, this result does not prove that $\hat{p}$ is stable. Let us show that $p(t) \to \hat{p}$ without oscillation, provided $0 < p(0) < 1$. Since our method will be used repeatedly, we shall describe it quite generally.

Suppose the continuous mapping $p' = f(p)$ of $[a, b]$ into itself has a fixed point $\hat{p} = f(\hat{p})$ in (a, b). We define $x = p - \hat{p}$ and assume that the mapping can be rewritten in the form $x' = \lambda(p)x$, with $0 \le \lambda(p) < 1$ for $a < p < b$. Then $\hat{p}$ is the only fixed point in (a, b), and $p(t) \to \hat{p}$ as $t \to \infty$ without oscillation, provided $a < p(0) < b$.

The proof is easy. If $\bar{p} \ne \hat{p}$ is a fixed point in (a, b), then $\bar{x} = \bar{p} - \hat{p} \ne 0$, so $\lambda(\bar{p}) = 1$, contradicting $\lambda(p) < 1$. To prove convergence, note that $\{|x(t)|\}$ is a monotone decreasing sequence of positive numbers. Hence, it tends to a limit. By our uniqueness argument, this limit must be zero. Since $\lambda(p) > 0$, the convergence is nonoscillatory.

We state a more general theorem of this type in Problem 4.9.

To apply the above result to our problem, it is convenient to choose $w_{11} = 1 - r$, $w_{12} = 1$, and $w_{22} = 1 - s$, with $0 < r, s \le 1$. In terms of the selection coefficients r and s, (4.29) and (4.30) become

$$\hat{p} = \frac{s}{r + s} \, , \tag{4.31}$$

$$\Delta p = -(r + s)pqx/\bar{w} \, , \tag{4.32}$$

where $x = p - \hat{p}$ and

$$\bar{w} = 1 - rp^2 - sq^2 \, . \tag{4.33}$$

Now,

$$x' = x + \Delta x = x + \Delta p = \lambda(p)x \, , \tag{4.34}$$

where (4.32) yields

$$\lambda(p) = 1 - \frac{(r + s)pq}{\bar{w}} \, . \tag{4.35}$$

Obviously, for $0 < p < 1$, $\lambda(p) < 1$. Substituting (4.33) into (4.35) gives

$$\lambda(p) = \frac{1 - rp - sq}{1 - rp^2 - sq^2} \, . \tag{4.36}$$

Since $0 < r, s \le 1$, therefore

$$rp^2 + sq^2 < rp + sq \le 1, \qquad 0 < p < 1 \, ,$$

and (4.36) shows that $\lambda(p) \ge 0$.

We conclude that the overdominant polymorphism is globally stable. This is one of the basic mechanisms for maintaining genetic variability. Observe that it has no analogue in a haploid population.

The example of sickle-cell anemia in Negroes is discussed in detail by Cavalli-Sforza and Bodmer (1971). Briefly, the genotype AA is normal; Aa is slightly anemic, having both normal and abnormal hemoglobin, but is probably more resistant to malaria; and aa, having only abnormal hemoglobin, has severe anemia. Typical selection coefficients in African populations are $r \approx 0.15$ and $s \approx 1$, whence $\hat{q} \approx 0.13$.

We shall now calculate the *rates of approach to equilibrium*. Clearly, it will suffice to investigate $p(t) \to \hat{p}$ and $p(t) \to 0$.

(i) $p(t) \to \hat{p}$

This is Case 4 above. As $p \to \hat{p}$, (4.34) simplifies to $x' \sim \lambda(\hat{p})x$, whence we obtain as $t \to \infty$ convergence at the geometric rate (see Problem 5.12)

$$x(t) \sim (\text{const})\,[\lambda(\hat{p})]^t = (\text{const})\left(\frac{r+s-2rs}{r+s-rs}\right)^t . \tag{4.37}$$

The single wavy line means that the ratio of the two sides tends to one. For weak selection, (4.37) leads to the approximation

$$x(t) \approx (\text{const})\exp\left(-\frac{rst}{r+s}\right) , \qquad r,s \ll 1 ,$$

in which case the characteristic convergence time is $(r+s)/(rs)$.

See Problem 4.10 for the calculation of corrections to geometric rates such as (4.37).

(ii) $p(t) \to 0$

We can have either Case 2 or Case 3 above. It is necessary to separate the case of recessive A.

(a) $w_{12} < w_{22}$

As $p \to 0$, the exact equation

$$p' = pw_1/\overline{w} \tag{4.38}$$

reduces to $p' \sim pw_{12}/w_{22}$, which yields the geometric rate

$$p(t) \sim (\text{const})\,(w_{12}/w_{22})^t , \qquad t \to \infty .$$

If

$$\frac{w_{12}}{w_{22}} = 1 - k ,$$

for weak selection ($k \ll 1$) the rate is roughly $(\text{const})\,e^{-kt}$.

(b) $w_{12} = w_{22} > w_{11}$

To examine the elimination of a recessive, we must approximate (4.38) to second order in p. The failure of the linear approximation that leads to a geometric rate is not surprising: selection is only against AA, and the frequency of these individuals, p^2, is extremely small for $p \ll 1$. Therefore, we expect a very slow rate of ultimate elimination of the recessive allele A. With $w_{11} = 1 - \sigma$ and $w_{12} = w_{22} = 1$, (4.38) reads

$$p' = \frac{p(1 - \sigma p)}{1 - \sigma p^2} \ . \tag{4.39}$$

It is helpful to rewrite (4.39) in terms of a new variable, y, with the property $\Delta y \to$ const as $p \to 0$. We choose $y = 1/p$, for which

$$y' = y \left[\frac{1 - (\sigma/y^2)}{1 - (\sigma/y)} \right] \ .$$

Therefore (the order symbol O is defined more formally in the next section),

$$\Delta y = \sigma - \sigma(1 - \sigma)y^{-1} + O(y^{-2}) \tag{4.40}$$

as $y \to \infty$. Omitting the terms of $O(y^{-2})$ in (4.40) will be a good approximation for sufficiently long times : $t \geq T$ for some fixed $T \gg 1$. But

$$y(t) = y(T) + \sum_{\tau=T}^{t-1} [y(\tau + 1) - y(\tau)]$$

$$= \sigma t + \text{const} + \sum_{\tau=T}^{t-1} \left[-\frac{\sigma(1 - \sigma)}{y(\tau)} + O(y^{-2}) \right]$$

from (4.40). The dominant contribution to the sum must come from the larger y^{-1} term, and $y \approx \sigma t$ for $t \gg 1$. Since

$$\sum_{\tau=T}^{t-1} \tau^{-1} = \ln t + O(1)$$

as $t \to \infty$, we obtain

$$y(t) = \sigma t - (1 - \sigma)\ln t + O(1)$$

as $t \to \infty$. Therefore,

$$p(t) = \frac{1}{\sigma t} + \frac{(1 - \sigma)\ln t}{\sigma^2 t^2} + O(t^{-2}) \ . \tag{4.41}$$

The algebraic decay (4.41) is quite slow: it takes about $100/\sigma$ generations to reduce $p(t)$ to 0.01; the corresponding time for exponential convergence would be about $4.6/\sigma$. It is at first sight puzzling to observe that, in contrast to the leading term for geometric convergence, the first two terms in (4.41) have no arbitrary constant corresponding to $p(T)$. This happens because to this order t may be replaced by $t + \text{const}$.

4.3 Dynamics with Multiple Alleles

We shall now analyze (4.25) with constant w_{ij} and k alleles. Let us investigate first the *existence of polymorphic equilibria* (Mandel, 1959; Hughes and Seneta, 1975). If $\hat{p}_i = 0$ for some set of subscripts, we just have a system of lower dimension, so we may assume that all the alleles are present: $\hat{p}_i > 0$ for all i. Then (4.25) gives the equilibrium condition $\hat{w}_i = \bar{\hat{w}}$. Writing W for the $k \times k$ matrix of the w_{ij} and $\mathbf{u}$ for the $k \times 1$ vector of ones, we obtain $W\hat{\mathbf{p}} = \bar{\hat{w}}\mathbf{u}$. This relation becomes more illuminating with the introduction of $\operatorname{adj} W$, the adjugate matrix of W; $\det W$, the determinant of W; and I, the $k \times k$ identity matrix. These entities satisfy

$$(\operatorname{adj} W)W = (\det W)I \ . \tag{4.42}$$

Multiplying our vector equilibrium equation by $\operatorname{adj} W$ and using (4.42) yields

$$(\det W)\hat{\mathbf{p}} = \bar{\hat{w}}(\operatorname{adj} W)\mathbf{u} \ .$$

We denote the ith component of the vector $(\operatorname{adj} W)\mathbf{u}$ by W_i, i.e., $W_i = [(\operatorname{adj} W)\mathbf{u}]_i$. Hence,

$$(\det W)\hat{p}_i = \bar{\hat{w}}W_i \ . \tag{4.43}$$

If $\det W \neq 0$, $W_i \neq 0$ for any i, and all the W_i have the same sign, then there exists a unique internal equilibrium

$$\hat{p}_i = \frac{\bar{\hat{w}}W_i}{\det W} = \frac{W_i}{\sum_j W_j} \ . \tag{4.44}$$

The second equation follows by substituting for $\det W$ from

$$\det W = \bar{\hat{w}} \sum_i W_i \ ,$$

which just normalizes the allelic frequencies in (4.43). We can also see from (4.44) that if $\det W \neq 0$ and an internal equilibrium exists, then it is unique, $W_i \neq 0$ for any i, and all the W_i have the same sign.

The degenerate case $\det W = 0$ is treated by Hughes and Seneta (1975).

If the fitness matrix is completely nondegenerate, the maximum possible number of equilibria is $2^k - 1$. To see this, choose any subset S of the integers $1, 2, \ldots, k$. The principal minors of W are the matrices W_S formed from the rows and columns corresponding to the elements of S. We suppose that each of these principal minors has a nonzero determinant. Then the number of equilibria with the alleles corresponding to the elements of S present, and only those alleles present, is zero or one. Since each of the k alleles must be present or absent, but at least one must be present, the number of possible equilibria is $2^k - 1$.

If at least one of the principal minors W_S has determinant zero, then lines, surfaces, etc. of equilibria are possible. The most degenerate case is no selection, w_{ij} independent of i and j, for which every point in the *simplex* $p_i \geq 0$, $\sum_i p_i = 1$ is an equilibrium.

We shall now show that the stationary points of the mean fitness are internal equilibria of the population, and *vice versa* (Mandel, 1959). We put

$$f(\mathbf{p}) = \overline{w}(\mathbf{p}) - \lambda \sum_i p_i \ ,$$

where λ is a Lagrange multiplier. Therefore,

$$\frac{\partial f}{\partial p_i} = 2w_i - \lambda = 0$$

at the stationary points. But

$$0 = \sum_i p_i \frac{\partial f}{\partial p_i} = 2\overline{w} - \lambda \ .$$

So the stationary points are precisely the points where $w_i = \overline{w}$, i.e., the internal equilibria.

Let us demonstrate now that *the mean fitness is nondecreasing* and the change in mean fitness is zero only at equilibrium. This extremely important and useful result has been proved by Scheuer and Mandel (1959); Mulholland and Smith (1959); Atkinson, Watterson, and Moran (1960); and, most elegantly, Kingman (1961), whom we follow.

We shall require Jensen's inequality: For a convex function g defined on an open interval I and a random variable x with a probability distribution concentrated on I,

$$\mathcal{E}[g(x)] \geq g[\mathcal{E}(x)] \ . \tag{4.45}$$

For a concave function, the inequality is reversed. Equation (4.45) assumes the existence of $\mathcal{E}(x)$ and $\mathcal{E}[g(x)]$.

The proof is immediate. If $g(x)$ is convex, then at every point $P[\xi, g(\xi)]$ on its graph, there exists a line lying entirely below or on $g(x)$. Therefore,

$$g(x) \geq g(\xi) + m(x - \xi)$$

for all x in I, where m is the slope of the line. Choosing $\xi = \mathcal{E}(x)$ and taking expectations, we obtain (4.45).

If x takes values x_i with probabilities p_i, (4.45) becomes

$$\sum_i p_i g(x_i) \geq g\left(\sum_i p_i x_i\right) \ . \tag{4.46}$$

For $g(x) = x^\mu$, where $\mu \geq 1$, (4.46) reads

$$\sum_i p_i x_i^\mu \geq \left(\sum_i p_i x_i\right)^\mu \ , \tag{4.47}$$

the form we shall use in the ensuing proof.

We write a series of inequalities for the mean fitness in the next generation and explain the manipulations below:

$$\overline{w}' = \sum_{ij} p_i' p_j' w_{ij}$$

$$= \overline{w}^{-2} \sum_{ij} p_i p_j w_i w_j w_{ij} \tag{4.48}$$

$$= \overline{w}^{-2} \sum_{ijk} p_i p_j p_k w_{ij} w_{ik} (\tfrac{1}{2})(w_j + w_j) \tag{4.49}$$

$$= \overline{w}^{-2} \sum_{ijk} p_i p_j p_k w_{ij} w_{ik} (\tfrac{1}{2})(w_j + w_k)$$

$$\geq \overline{w}^{-2} \sum_{ijk} p_i p_j p_k w_{ij} w_{ik} (w_j w_k)^{1/2} \tag{4.50}$$

$$= \overline{w}^{-2} \sum_i p_i \left[\sum_j p_j w_{ij} (w_j)^{1/2} \right]^2$$

$$\geq \overline{w}^{-2} \left[\sum_i p_i \sum_j p_j w_{ij} (w_j)^{1/2} \right]^2 \tag{4.51}$$

$$= \overline{w}^{-2} \left(\sum_j p_j w_j^{\,3/2} \right)^2 \tag{4.52}$$

$$\geq \overline{w}^{-2} \left[\left(\sum_j p_j w_j \right)^{3/2} \right]^2 \tag{4.53}$$

$$= \overline{w} \ .$$

The numbered equations in the above series come, respectively, from (4.25a), (4.25b), the elementary fact that $a + b \geq 2(ab)^{1/2}$ for $a, b \geq 0$, (4.47) with $\mu = 2$, (4.25b), and (4.47) with $\mu = \tfrac{3}{2}$. Equality holds in (4.50) if and only if $w_j = w_k$ for every j and k, i.e., at equilibrium.

Notice how much easier it was to prove that $\Delta\overline{w} \geq 0$ for haploids (see Sect. 2.1). We can gain considerable evolutionary insight by following the approach in that section to evaluate $\Delta\overline{w}$. For the purpose of this computation, we shall permit the fitnesses to vary. We have

$$\Delta\overline{w} = \sum_{ij} (p_i' p_j' w_{ij}' - p_i p_j w_{ij})$$

$$= \overline{\Delta w} + \sum_{ij} (2p_j \Delta p_i + \Delta p_i \Delta p_j) w_{ij} \ , \tag{4.54}$$

where

$$\overline{\Delta w} = \sum_{ij} p_i' p_j' \Delta w_{ij} \tag{4.55}$$

is the mean of the fitness changes over the next generation. Substituting (4.18) into (4.54) leads to (cf. Li, 1969)

$$\Delta \overline{w} = \overline{w}^{-1} V_g + \overline{\Delta w} + \overline{w}^{-2} \sum_{ij} p_i p_j (w_i - \overline{w})(w_j - \overline{w})(w_{ij} - \overline{w}) \ , \qquad (4.56)$$

where

$$V_g = 2 \sum_i p_i (w_i - \overline{w})^2 \qquad (4.57)$$

represents the *genic* (or *additive genetic*) variance in fitness. It will be shown in Sect. 4.10 (4.179 to 4.189) that V_g is, in fact, the additive component of the *genotypic* (or *total genetic*) variance. The factor 2 in (4.57) is due to diploidy. Note, too, that the last term in (4.56) was absent in the haploid result, (2.16). It is easy to prove that this term is zero if there is no dominance (Li, 1969). See Problem 4.26 for a general bound.

It is very instructive to examine (4.56) for weak selection. Let us define more formally the symbol O that we have already introduced. The notation

$$\phi(x) = O[\psi(x)] \qquad \text{as } x \to a$$

means

$$\varlimsup_{x \to a} | \phi(x)/\psi(x) | < \infty \ ,$$

i.e., $|\phi(x)/\psi(x)|$ is bounded for $x \neq a$ but x sufficiently close to a. If

$$\lim_{x \to a} | \phi(x)/\psi(x) | = 0 \ ,$$

we shall write

$$\phi(x) = o[\psi(x)] \qquad \text{as } x \to a \ .$$

Of course, this means that $|\phi(x)/\psi(x)|$ can be made arbitrarily small for $x \neq a$ but x sufficiently close to a.

Since the scale of w_{ij} is irrelevant, we may take

$$w_{ij} = 1 + O(s) \qquad \text{as } s \to 0 \ , \qquad (4.58)$$

for all i and j, where s is the intensity of selection. For instance, if the genotypic fitnesses are constant, we can choose

$$\max_{i,j} w_{ij} = 1, \qquad s = 1 - \min_{i,j} w_{ij} \ .$$

Now, (4.57) shows that $V_g = O(s^2)$, and the last term in (4.56) is of $O(s^3)$. Since (4.58) implies that $\overline{w} = 1 + O(s)$, therefore (4.56) has the form

$$\Delta \overline{w} = V_g + \overline{\Delta w} + O(s^3) \ . \qquad (4.59)$$

If the w_{ij} are constant or change so slowly that $\overline{\Delta w} = o(s^2)$, then we have the approximate form of the Fundamental Theorem of Natural Selection: $\Delta \overline{w} \approx V_g$. Thus, as the mean fitness increases (proved exactly above), it does so at a rate approximately equal to the additive component of the genetic variance. From (4.25a) and (4.57) we infer that $V_g = 0$ if and only if the population is at equilibrium. By Problem 4.3, the increase of the mean absolute fitness is equivalent to maximization of the relative rate of population growth.

Owing to resource limitations, the absolute fitnesses can be constant only for rather small population sizes. Otherwise, they must be *density-dependent*, i.e., functions of the population number. The relative fitnesses may still be constant, however; e.g., we might have $w_{ij} = z_{ij}g(N)$, with z_{ij} constant and $w_{ij} < 1$ for sufficiently large N, for all i and j. Then it is the mean relative fitness, $\bar{z}$, rather than the mean absolute fitness, $\bar{w}$, that will be maximized by natural selection.

The biological significance of the increase of the mean relative fitness is the following. Let $A_i A_j$ have higher fitness than any other genotype, for some i and j. Then selection minimizes $(w_{ij} - \bar{w})/w_{ij} = (z_{ij} - \bar{z})/z_{ij}$, the relative reproductive excess of the optimal genotype. For pure viability selection, this reproductive excess, or genetic load (Crow, 1970), is just the relative amount of selective mortality.

Our proof that the mean fitness $\bar{w} = \bar{v}\bar{b}$ is nondecreasing was based on the posited constancy of the genotypic fitnesses w_{ij}. Suppose the viabilities v_{ij} are constant. Then (4.12) shows that constancy of the w_{ij} is equivalent to constancy of the *genotypic* fertilities b_{ij}. By contrast, (4.21) implies that the fertilities of *matings*, $f_{ij,kl}$, are constant if and only if the β_{ij} are constant. In this case, (4.22) enables us to replace w_{ij} in (4.25) by the scaled fitnesses $U_{ij} = v_{ij}\beta_{ij}$, and now we conclude that the modified mean fitness $\bar{U} = \bar{v}\bar{\beta} = \bar{v}\bar{b}^{1/2}$ is nondecreasing (Nagylaki, 1987). If selection is weak, more elaborate arguments establish these results in both cases for general fertilities and also demonstrate that after two generations $\Delta\bar{w} \approx V_g$ and $\Delta\bar{U} \approx V_g$ (Nagylaki, 1987), respectively. The essential point is that the fertilities of genotypes and of matings cannot both be constant, and therefore the two different assumptions of constancy lead to different results.

The *stability properties* of the basic selection model were studied by Kimura (1956) and Penrose, S. Maynard Smith, and Sprott (1956) in continuous time. The discrete model has been investigated by Mandel (1959) and Kingman (1961a). Mandel (1970) and Hughes and Seneta (1975) examine the relations among the different stability criteria. The following results are either proved in or can be deduced from Kingman's (1961a) paper. See also an der Heiden (1975), Karlin (1978), Lyubich, Maistrovskii, and Ol'khovskii (1980), and Losert and Akin (1983).

1. The stable equilibria are the local maxima of $\bar{w}$. (These correspond to the peaks, ridges, etc. of the $\bar{w}$ surface, which the system climbs because $\Delta\bar{w} \geq 0$.) The asymptotically stable equilibria are the strict (or isolated) local maxima of $\bar{w}$.

   Let S be a subset of the integers $1, 2, \ldots, k$. Suppose there exists an equilibrium $\hat{\mathbf{p}}(S)$ with $\hat{p}_i > 0$ if and only if i is in S. Then (4.44) tells us that $\hat{\mathbf{p}}(S)$ is isolated if and only if the principal minor W_S corresponding to S has nonzero determinant.

2. If the completely polymorphic equilibrium (i.e., the one with all alleles present) exists, it is stable if and only if, counting multiplicity, the fitness matrix W has exactly one positive eigenvalue. If $\det W \neq 0$, then by Result 1 stability is necessarily asymptotic. (If $W \neq 0$, then W has at least one positive eigenvalue.)

3. Let D_i, $i = 1, 2, \ldots, k$, be the determinant of the $i \times i$ submatrix in the upper left-hand corner of W. Suppose without loss of biological generality that $w_{11} = D_1 > 0$ and assume that $D_i \neq 0$ for any i. Then the stability of the completely polymorphic equilibrium is necessarily asymptotic, and a necessary and sufficient condition for this stability is

$$(-1)^i D_i < 0 \qquad \text{for all } i \; .$$

Of course, this criterion is a special case of Result 2; we state it because it is much easier to apply.

4. The fitness of A_i and the mean fitness at $\hat{\mathbf{p}}(S)$ are

$$\hat{w}_i(S) = \sum_j w_{ij} \hat{p}_j(S) \; ,$$

$$\hat{\overline{w}}(S) = \sum_{ij} w_{ij} \hat{p}_i(S) \hat{p}_j(S) \; .$$

The equilibrium $\hat{\mathbf{p}}(S)$ is stable if

$$\hat{w}_i(S) < \hat{\overline{w}}(S) \qquad \text{for all } i \text{ not in } S$$

and W_S has exactly one positive eigenvalue. Observe that these criteria require that alleles not present at equilibrium die out if introduced at low frequencies and that the alleles present be in stable equilibrium (see Result 2). The equilibrium is unstable if

$$\hat{w}_i(S) > \hat{\overline{w}}(S) \qquad \text{for some } i \text{ not in } S$$

or the number of positive eigenvalues of W_S is not exactly one.

5. If W has m positive eigenvalues, at least $m - 1$ alleles must be absent at a stable equilibrium.

6. Let S and T be distinct subsets of $\{1, 2, \ldots, k\}$ such that T is contained in S. Suppose $\theta\hat{\mathbf{p}}(S) + (1 - \theta)\hat{\mathbf{p}}(T)$ is *not* an equilibrium for some θ in $(0, 1)$. Then the equilibria $\hat{\mathbf{p}}(S)$ and $\hat{\mathbf{p}}(T)$ cannot both be stable. Hence, if $\hat{\mathbf{p}}(S)$ is stable and $\hat{\mathbf{p}}(S)$ and $\hat{\mathbf{p}}(T)$ *cannot* be connected by a line of equilibria, then $\hat{\mathbf{p}}(T)$ is unstable. This result often eliminates much calculation in testing the stability of equilibria.

In particular, if the completely polymorphic equilibrium is stable, then no other equilibrium is stable, and therefore (provided all alleles are initially present) the population converges globally to the completely polymorphic equilibrium. Note that if this equilibrium is degenerate, it will include boundary points, lines, etc., where at least one allele is absent, and these boundary equilibria are stable. The population converges to these boundary equilibria only if at least one allele is initially absent.

7. Assume that the population is in the stable completely polymorphic equilibrium $\hat{\mathbf{p}}$. The mean fitness at this equilibrium reads

$$\hat{\bar{w}} = \sum_{i,j=1}^{k} w_{ij}\hat{p}_i\hat{p}_j \ ,$$

and a new allele A_{k+1} has equilibrium fitness

$$\hat{w}_{k+1} = \sum_{i=1}^{k} w_{k+1,i}\hat{p}_i \ .$$

It follows from Results 4 and 6 that A_{k+1} persists in the population (i.e., a new stable equilibrium with A_{k+1} present is eventually established) if $\hat{w}_{k+1} > \hat{\bar{w}}$. If $\hat{w}_{k+1} < \hat{\bar{w}}$, then A_{k+1} will be lost.

Since the necessary and sufficient conditions above (4.44) and in Result 3 for the existence of an asymptotically stable, completely polymorphic equilibrium are rather complicated and unilluminating, it is desirable to have simpler conditions that are necessary but not sufficient. We present three; the second and third are due to Lewontin, Ginzburg, and Tuljapurkar (1978). See also Ginzburg (1983).

1. If the completely polymorphic equilibrium exists and is asymptotically stable, Result 6 implies that the corner equilibria ($\hat{p}_i = 1$) must be unstable. Therefore, by Result 4, for every i there exists some j such that $w_{ii} \leq w_{ij}$.

2. Existence of an asymptotically stable internal equilibrium implies concavity of the quadratic form $\bar{w}$. Enforcing this on the line $p_i + p_j = 1$ demonstrates that $w_{ij} > \frac{1}{2}(w_{ii} + w_{jj})$ for every i and j such that $i \neq j$.

3. If there are at least three alleles, then for every i and j with $i \neq j$, there exists some l, where $l \neq i$ and $l \neq j$, such that $w_{ij} < w_{il} + w_{jl}$.

We end this section with a brief discussion of the relationship between overdominance and stable polymorphism for more than two alleles. Overdominance may be pairwise ($w_{ij} > w_{ii}, w_{jj}$ for every i and j such that $i \neq j$) or total ($w_{ij} > w_{ll}$ for every i, j, and l, where $i \neq j$). On the one hand, as triallelic examples of Kimura (1956), Mandel (1959), and Lewontin, Ginzburg, and Tuljapurkar (1978) show, even pairwise overdominance is unnecessary for the existence of an asymptotically stable internal equilibrium. On the other, since the third of the above necessary conditions may be violated even if total overdominance obtains, we see that not even total overdominance suffices to guarantee the existence of an asymptotically stable internal equilibrium. Lewontin, Ginzburg, and Tuljapurkar (1978) offer a triallelic example.

4.4 Two Alleles with Inbreeding

We return to the general model, (4.13), (4.14), (4.17), and consider two alleles with inbreeding. We must specify the genotypic proportions, P_{ij}, in terms of the allelic frequencies, p_i. The required relations read [see (9.1)]

$$P_{11} = p^2 + Fpq \,, \quad P_{12} = (1-F)pq \,, \quad P_{22} = q^2 + Fpq \,. \tag{4.60}$$

For the purpose of this section, the inbreeding coefficient F, where $0 \le F \le 1$, is just a parameter that determines the excess of homozygotes and deficiency of heterozygotes relative to Hardy-Weinberg proportions. Observe that (4.60) satisfies (4.1). Assuming that F is constant obviates explicit consideration of the mating system: an example of the latter approach is treated in Sect. 5.1.

The model with constant inbreeding coefficient is due to Wright (1942). Lewontin (1958) assumed that a polymorphism exists and gave conditions for its local stability. We shall determine globally the fate of the population for all values of the fitnesses and the inbreeding coefficient. The method is that of Sect. 4.2.

Substituting (4.60) into (4.13), we calculate w_1 and w_2, and hence $\overline{w} = pw_1 + qw_2$. Then (4.26) still has the form (4.27), but now

$$g(p) = \beta - F\alpha - (1-F)\gamma p \,, \tag{4.61}$$

$$\overline{w} = w_{11}p^2 + (2w_{12} - F\gamma)pq + w_{22}q^2 \,, \tag{4.62}$$

where

$$\alpha = w_{12} - w_{11} \,, \qquad \beta = w_{12} - w_{22} \,, \tag{4.63}$$

$$\gamma = \alpha + \beta = 2w_{12} - w_{11} - w_{22} \,. \tag{4.64}$$

We remark that $\alpha = 0$, $\beta = 0$, and $\gamma = 0$ correspond to A dominant, A recessive, and no dominance, respectively. Since sgn $\Delta p = $ sgn $g(p)$ and $g(p)$ is again linear, as in Sect. 4.2, there are four cases, depending on the signs of

$$g(0) = \beta - F\alpha \qquad \text{and} \qquad g(1) = -\alpha + F\beta \,. \tag{4.65}$$

We exclude two trivial situations: no selection ($w_{11} = w_{12} = w_{22}$) and complete inbreeding ($F = 1$). The latter is easily seen from (4.27) and (4.61) to be equivalent to the diallelic haploid model (see Sect. 2.1) with $w_1 \to w_{11}$ and $w_2 \to w_{22}$. The fitness of Aa individuals becomes irrelevant because there are no heterozygotes in the population.

1. $\beta \ge F\alpha$ and $F\beta \ge \alpha$

Since $\Delta p > 0$ for $0 < p < 1$, therefore $p(t) \to 1$ as $t \to \infty$. If $w_{11} \ge w_{12} \ge w_{22}$, then $\alpha \le 0$ and $\beta \ge 0$, so this case necessarily applies for all F, as expected.

2. $\beta \le F\alpha$ and $F\beta \le \alpha$

Now $\Delta p < 0$ for $0 < p < 1$, so $p(t) \to 0$. If $w_{11} \le w_{12} \le w_{22}$, then $\alpha \ge 0$ and $\beta \le 0$, whence the requirements for $p(t) \to 0$ hold for all F.

3. $\beta < F\alpha$ and $F\beta > \alpha$

There is a polymorphism

$$\hat{p} = \frac{\beta - F\alpha}{\gamma(1 - F)} \tag{4.66}$$

in $(0, 1)$. From (4.27), (4.61), and (4.66) we find

$$\Delta p = \gamma(1 - F)pq(\hat{p} - p)/\overline{w} . \tag{4.67}$$

Since

$$0 > g(0) - g(1) = \gamma(1 - F) , \tag{4.68}$$

therefore $\operatorname{sgn} \Delta p = \operatorname{sgn}(p - \hat{p})$. Hence, $p(t) \to 1$ if $p(0) > \hat{p}$ and $p(t) \to 0$ if $p(0) < \hat{p}$.

We solve (4.65) for α and β to obtain

$$\alpha = \frac{Fg(0) - g(1)}{1 - F^2} , \qquad \beta = \frac{g(0) - Fg(1)}{1 - F^2} . \tag{4.69}$$

Thus, the conditions for Case 3, $g(0) < 0$ and $g(1) > 0$, imply $\alpha, \beta < 0$, i.e., $w_{12} < w_{11}, w_{22}$. Consequently, this unstable equilibrium requires underdominance for all F. In contrast to panmixia, however, underdominance need not lead to an unstable polymorphism.

4. $\beta > F\alpha$ and $F\beta < \alpha$

The nontrivial equilibrium $\hat{p}$ still exists, but now $\operatorname{sgn} \Delta p = \operatorname{sgn}(\hat{p} - p)$. We expect $p(t) \to \hat{p}$ globally. Before proving this, we note from the previous paragraph that our conditions now require overdominance for all F, but overdominance is not sufficient for a stable polymorphism.

We put $x = p - \hat{p}$ and obtain (4.34) with

$$\lambda(p) = 1 - \frac{\gamma(1 - F)pq}{\overline{w}} . \tag{4.70}$$

Since the inequality in (4.68) is reversed here, it follows at once that $\lambda(p) < 1$, $0 < p < 1$.

To prove that $\lambda(p) \geq 0$, we take $w_{12} = 1$. Then $0 < \alpha, \beta \leq 1$. Substituting (4.63) into (4.62) leads to

$$\overline{w} = 1 - \alpha p^2 - F\gamma pq - \beta q^2 ,$$

and inserting this into (4.70) yields

$$\overline{w}\lambda(p) = 1 - \alpha p - \beta q \geq 0 .$$

This proves global nonoscillating convergence.

4.5 Variable Environments

We shall investigate in this section the effect of deterministic time dependence of the fitnesses. We assume that the $w_{ij}(t)$ are known functions of time and restrict ourselves to two alleles. With arbitrary $w_{ij}(t)$, not only is a global analysis impossible, but we cannot even locate the possible equilibria. We can, however, still find sufficient conditions for the maintenance of genetic diversity. If, no matter what the initial conditions are, the population cannot become monomorphic, we shall say there is a *protected polymorphism* (Prout, 1968). In general, a protected polymorphism is neither necessary nor sufficient for the existence of a stable polymorphic equilibrium. For instance, on the one hand, there may be a (locally) stable polymorphic equilibrium, but selection may remove a rare allele from the population. On the other hand, a stable limit cycle to which there is global convergence exemplifies the maintenance of polymorphism without a polymorphic equilibrium.

This problem was studied first by Haldane and Jayakar (1963). In an interesting paper, however, Cornette (1981) has shown that Haldane and Jayakar (1963) failed to state their assumptions and conclusions with adequate precision. We shall adopt essentially the same assumptions as Cornette, but shall present different proofs and a new example.

Following Haldane and Jayakar (1963), we shall use relative fitnesses,

$$u_t = w_{11}(t)/w_{12}(t), \qquad v_t = w_{22}(t)/w_{12}(t) ,$$

and the gene-frequency ratio

$$x_t = p(t)/q(t) .$$

From (4.25) we obtain

$$x_{t+1} = x_t \left(\frac{u_t x_t + 1}{x_t + v_t} \right) . \tag{4.71}$$

We posit that all three fitnesses are uniformly bounded away from zero. Hence, hereafter we make

Assumption 1. There exist α and β, independent of t, such that $0 < \alpha \leq u_t, v_t \leq \beta < \infty$.

It is convenient to start with generation one. Then the geometric-mean fitness ratios read

$$u_t^* = \left(\prod_{\tau=1}^{t} u_\tau \right)^{1/t} , \qquad v_t^* = \left(\prod_{\tau=1}^{t} v_\tau \right)^{1/t} .$$

We add

Assumption 2. $\varlimsup_{t\to\infty} u_t^* < 1, \qquad \varlimsup_{t\to\infty} v_t^* < 1 .$

[For various properties of the limits superior and inferior, consult Apostol (1974, pp. 184–185, 210–211).]

Let us comment on Assumption 2 before proceeding. Since $-\ln x$ is convex, the special case (4.46) of Jensen's inequality yields

$$\ln u_t^* = \frac{1}{t}\sum_{\tau=1}^{t}\ln u_\tau \leq \ln \overline{u}_t \ ,$$

where

$$\overline{u}_t = \frac{1}{t}\sum_{\tau=1}^{t} u_\tau$$

represents the arithmetic mean. Therefore,

$$u_t^* \leq \overline{u}_t \ , \tag{4.72}$$

which shows that Assumption 2 is weaker than the corresponding condition for the arithmetic-mean fitness ratios. Assumption 2 is more or less stringent than overdominance for the arithmetic means $\overline{w}_{ij}(t)$ of the fitnesses roughly according as the heterozygote or homozygote fitnesses are more variable. With constant homozygote fitnesses, Assumption 2 becomes (see Problem 4.19)

$$\varliminf_{t\to\infty} w_{12}^*(t) = \left[\varlimsup_{t\to\infty}\frac{1}{w_{12}^*(t)}\right]^{-1} > w_{11}, w_{22} \ ,$$

which is stronger than

$$\varliminf_{t\to\infty} \overline{w}_{12}(t) > w_{11}, w_{22} \ .$$

If $w_{12}(t)$ is constant, however, Assumption 2 is equivalent to

$$\varlimsup_{t\to\infty} w_{11}^*(t) < w_{12}, \qquad \varlimsup_{t\to\infty} w_{22}^*(t) < w_{12} \ ,$$

and this can hold even if

$$\varlimsup_{t\to\infty} \overline{w}_{11}(t) < w_{12}, \qquad \varlimsup_{t\to\infty} \overline{w}_{22}(t) < w_{12}$$

does not.

For the sickle-cell anemia example of Sect. 4.2, the abnormal homozygotes are quite unfit. If heterozygotes were generally slightly less fit than normal homozygotes, but occasional malaria epidemics eliminated a large proportion of normal homozygotes, Assumption 2 would be satisfied without arithmetic-mean overdominance.

We prove now that $\{p(t)\}$ does not converge to 0 or 1, i.e., $\{x_t\}$ does not tend to 0 or ∞. By symmetry, it suffices to demonstrate that $x_t \not\to \infty$. We rearrange (4.71) to isolate the dominant term for large x_t (Haldane and Jayakar, 1963):

$$x_{t+1}/x_t = u_t F_t \ ,$$

where
$$F_t = 1 + \frac{1 - u_t v_t}{u_t(x_t + v_t)} \ .$$

This gives
$$\frac{x_{t+1}}{x_1} = \left(\prod_{\tau=1}^{t} u_\tau\right)\left(\prod_{\tau=1}^{t} F_\tau\right) \ ,$$

from which we deduce
$$(x_{t+1}/x_1)^{1/t} = u_t^* G_t \ , \tag{4.73a}$$

with
$$G_t = \left(\prod_{\tau=1}^{t} F_\tau\right)^{1/t} - \exp\left(\frac{1}{t}\sum_{\tau=1}^{t}\ln F_\tau\right) \ . \tag{4.73b}$$

We assume that $x_t \to \infty$ as $t \to \infty$ (i.e., for every X, there exists an integer T such that $x_t > X$ whenever $t > T$) and obtain a contradiction. By Assumption 1, as $x_t \to \infty$, $F_t \to 1$, so $\ln F_t \to 0$. It follows easily (see, e.g., Titchmarsh, 1939, p. 10) that the exponent in G_t tends to zero as $t \to \infty$. Therefore, $G_t \to 1$ as $t \to \infty$. Hence, for every ε in $(0,1)$, there exists T_1 such that $t > T_1$ implies $G_t < (1-\varepsilon)^{-1}$. But Assumption 2 informs us that for ε positive and sufficiently small, there exists T_2 such that $u_t^* < 1 - \varepsilon$ whenever $t > T_2$. Consequently, for $t > \max(T_1, T_2)$

$$(x_{t+1}/x_1)^{1/t} < (1-\varepsilon)(1-\varepsilon)^{-1} = 1 \ ,$$

or $x_{t+1} < x_1$. The contradiction establishes that $x_t \not\to \infty$ as $t \to \infty$. Cornette (1981) has proved that this conclusion holds even if Assumption 1 is weakened to $0 < u_t, v_t < \infty$.

Cornette (1981) has pointed out that the above result is too weak to guarantee biologically the maintenance of genetic variability. For if a subsequence, $\{p(\tau_t), t = 1, 2, \ldots\}$, of gene frequencies converged to 0 or 1, random drift would cause eventual loss of one of the alleles even in a very large population. We present below an example, due to J.L. Bona, that shows that Assumptions 1 and 2 do not rule out the existence of such a subsequence. (The analysis in Sect. 4.2 demonstrates that this pathology cannot occur with constant fitnesses. See also Problem 4.20.)

Suppose $v_t = \alpha < 1$ for all t. Then from (4.71) and Assumption 1 we find

$$x_{t+1} = (u_t x_t + 1)/(1 + \alpha x_t^{-1}) \tag{4.74a}$$
$$\geq (\alpha x_t + 1)/(1 + \alpha x_t^{-1}) \ . \tag{4.74b}$$

Now
$$\xi_{t+1} = (\alpha \xi_t + 1)/(1 + \alpha \xi_t^{-1}) \tag{4.74c}$$

describes selection with the constant overdominant fitness pattern $\alpha, 1, \alpha$. Section 4.2 informs us that $\xi_t \geq \delta \equiv \min(1, \xi_1)$ for all t. Choosing $\xi_1 = x_1$ and comparing (4.74b) with (4.74c) inductively, we conclude that $x_t \geq \xi_t \geq \delta$. Then (4.74a) yields

$$x_{t+1} \geq (u_t x_t + 1)/(1 + \alpha \delta^{-1}) \ .$$

With the definitions

$$\gamma = 1 + \alpha \delta^{-1} , \qquad \mu_t = u_t/\gamma , \qquad y_t = \gamma x_t \ ,$$

we have

$$y_{t+1} \geq \mu_t y_t + 1 \ .$$

Comparing this inductively with

$$z_{t+1} = \mu_t z_t + 1 , \qquad z_1 = y_1 \ , \qquad (4.75a)$$

we see that $y_t \geq z_t$ for all t.

Our task is reduced to finding a sequence $\{\mu_t\}$, subject to

Assumption 1*. $0 < \alpha/\gamma \leq \mu_t \leq \beta/\gamma < \infty$,

Assumption 2*. $\overline{\lim_{t \to \infty}} \mu_t^* < 1/\gamma$,

such that a subsequence of $\{z_t\}$ tends to infinity. Iterating (4.75a) leads to ($t \geq 1$; the sum is absent for $t = 1$)

$$z_{t+1} = 1 + z_1 U_{t,t} + \sum_{\tau=1}^{t-1} U_{t,\tau} \ , \qquad (4.75b)$$

where $(1 \leq \tau \leq t)$

$$U_{t,\tau} = \prod_{n=t-\tau+1}^{t} \mu_n \ . \qquad (4.75c)$$

Let

$$\{\mu_1, \mu_2, \ldots\} = \{a, b; \ a, a, b, b; \ a, a, a, b, b, b; \ \ldots\} \ ,$$

use $i = 1, 2, \ldots$ to label the groups separated by semicolons, take $a < 1 < b$, and define

$$j = \tfrac{1}{2} i (i+1) , \qquad k = i (i+1) , \qquad l = i (i-1) \ .$$

Thus, the ith a-run and ith b-run both have length i, for all $i = 1, 2, \ldots$

$$\mu_t = \begin{cases} a , & l < t \leq l + i \ , \\ b , & l + i < t \leq k \ . \end{cases}$$

Clearly, for $l < t \leq k$, $\mu_t^* \leq \mu_k^* = \sqrt{ab}$. Then it suffices to require

Assumption 1**. $0 < \alpha/\gamma \leq a < 1 < b \leq \beta/\gamma < \infty$,

Assumption 2**. $\sqrt{ab} < 1/\gamma$.

These constraints on the constants a and b are easily satisfied. For instance, we can choose $a = \alpha/\gamma$ and $b = (1-\varepsilon)/(\gamma^2 a) = (1-\varepsilon)/(\alpha\gamma)$, where $0 < \varepsilon < 1$. If α

is sufficiently small and β is sufficiently large, both double-starred assumptions are fulfilled.

Suppose $1 \geq \eta > 0$ and $[\eta i] \leq m \leq i$ (the brackets denote the greatest integer in ηi) and consider for $i = 1, 2, \ldots$ the values of t satisfying $t = l+i+m$. Evidently,

$$z_{t+1} \geq U_{l+i+m,m} = b^m \geq b^{[\eta i]} \to \infty$$

as $i \to \infty$. Not only have we found a divergent subsequence, but it is not even sparse: asymptotically, it includes a fraction $\frac{1}{2}(1-\eta)$, or arbitrarily close to $\frac{1}{2}$, of the entire sequence. In practice, this implies eventual fixation of the allele A_1.

How can we strengthen Assumption 2 sufficiently to ensure a protected polymorphism? From (4.71) we infer

$$x_{t+1} \leq u_t x_t + 1 \;,$$

which shows that $x_t \leq \zeta_t$, where ζ_t is the solution of

$$\zeta_{t+1} = u_t \zeta_t + 1, \qquad \zeta_1 = x_1 \;.$$

The solution (4.75b), (4.75c) informs us that

$$\zeta_{t+1} = 1 + \zeta_1 (u^*_{t,t})^t + \sum_{\tau=1}^{t-1} (u^*_{t,\tau})^\tau \;, \tag{4.76a}$$

where the geometric means $u^*_{t,\tau}$ $(1 \leq \tau \leq t)$ are given by

$$u^*_{t,\tau} = \left(\prod_{n=t-\tau+1}^{t} u_n \right)^{1/\tau} . \tag{4.76b}$$

Instead of Assumption 2, we make

Assumption 3.   There exist $c_0 < 1$ and positive integers ρ and σ such that

$$u^*_{t,\rho} \leq c_0 , \qquad v^*_{t,\sigma} \leq c_0$$

for all $t \geq \rho$ and $t \geq \sigma$, respectively.

This means that all subsequences of $\{u_t\}$ and $\{v_t\}$ of respective lengths ρ and σ have geometric means not exceeding c_0. By Assumption 1, for all t such that $1 \leq t \leq T_0 < \infty$, the gene frequency is bounded strictly away from zero and one. Hence, if Assumption 3 held only for $t \geq T_0$ for some $T_0 < \infty$, we would simply start the analysis at $t = T_0$.

To demonstrate that Assumptions 1 and 3 together guarantee that $\{x_t\}$ is bounded above and has a positive lower bound, by symmetry it will suffice to find an upper bound. We show first that there exist $c < 1$ and T such that $u^*_{t,\tau} \leq c$ for all $\tau \geq T$ and $t \geq \tau$. Let $\tau = j\rho + k$, where the integers j and k satisfy $j \geq 1$ and $0 \leq k < \rho$. From (4.76b) we find

$$(u_{t,\tau}^*)^\tau = \left(\prod_{l=1}^{j} u_{t-\tau+l\rho,\rho}^*\right)^\rho \left(\prod_{m=t-k+1}^{t} u_m\right) \le c_0^{j\rho}\beta^k ,$$

whence

$$u_{t,\tau}^* \le c_0(\beta/c_0)^{k/\tau} .$$

It follows that for any c in $(c_0, 1)$, there exists T sufficiently large to ensure that $u_{t,\tau}^* \le c$ for $\tau \ge T$.

Returning to (4.76a), we have for $t > T$

$$\zeta_{t+1} = 1 + \zeta_1(u_{t,t}^*)^t + \sum_{\tau=1}^{T-1}(u_{t,\tau}^*)^\tau + \sum_{\tau=T}^{t-1}(u_{t,\tau}^*)^\tau$$

$$\le 1 + \zeta_1 c^t + \sum_{\tau=1}^{T-1}\beta^\tau + \sum_{\tau=T}^{\infty}c^\tau$$

$$\le 1 + \zeta_1 + (\beta^T - \beta)(\beta - 1)^{-1} + c^T(1-c)^{-1} \equiv Z_1 .$$

Put $Z_2 = \max\{\zeta_t, t = 1, 2, \ldots, T+1\}$. Then $\zeta_t \le \max(Z_1, Z_2)$, as required.

Periodic environments, applicable to seasonal variation in many organisms, can be analyzed much more easily than those that vary arbitrarily. Consult Karlin and Liberman (1974), Hoekstra (1975), and Nagylaki (1975).

If A_2 *is dominant*, then $v_t = 1$, so Assumption 2 fails. We replace it by

Assumption 4. $\overline{\lim\limits_{t\to\infty}} u_t^* < 1, \qquad \underline{\lim\limits_{t\to\infty}} \overline{u}_t > 1 .$

We establish that Assumptions 1 and 4 preclude both $x_t \to 0$ and $x_t \to \infty$. Our previous proof applies to $x_t \to \infty$. To show that $x_t \not\to 0$, let $r_t = 1/x_t$. Then (4.71) becomes

$$r_{t+1} = r_t(r_t + 1)/(r_t + u_t) , \qquad (4.77a)$$

or, isolating the leading term as $r_t \to \infty$, we obtain (Haldane and Jayakar, 1963)

$$r_{t+1} - r_t = 1 - u_t + H_t ,$$

where

$$H_t = u_t(u_t - 1)/(r_t + u_t) .$$

Therefore,

$$(r_{t+1} - r_1)/t = 1 - \overline{u}_t + K_t , \qquad (4.77b)$$

where

$$K_t = \frac{1}{t}\sum_{\tau=1}^{t} H_\tau . \qquad (4.77c)$$

Suppose that $r_t \to \infty$ as $t \to \infty$. By Assumption 1, as $t \to \infty$, $H_t \to 0$, and therefore $K_t \to 0$ (see, e.g., Titchmarsh, 1939, p. 10). Hence, for every $\varepsilon > 0$, there exists T_1 such that $K_t < \varepsilon$ whenever $t > T_1$. But Assumption 4 implies that for sufficiently small ε, there exists T_2 such that $\bar{u}_t > 1 + \varepsilon$ for all $t > T_2$. We conclude that for $t > \max(T_1, T_2)$,

$$(r_{t+1} - r_1)/t < 1 - (1 + \varepsilon) + \varepsilon = 0 \ ,$$

or $r_{t+1} < r_1$, which demonstrates that $r_t \nrightarrow \infty$, i.e., $x_t \nrightarrow 0$, by contradiction.

Assumptions 1 and 4 do not guarantee a protected polymorphism, because our previous counterexample still applies. For if $\{x_t\}$ has a subsequence that converges to zero, we lose genetic variability. If it does not, then there exists $\delta > 0$ such that $x_t \geq \delta$, in which case (4.71) yields

$$x_{t+1} \geq \gamma^{-1}(u_t x_t + 1) \ ,$$

where now $\gamma = 1 + \delta^{-1}$. Furthermore, for $l < t \leq k$,

$$\bar{\mu}_t \geq \bar{\mu}_{l+i} = \tfrac{1}{2}[a + b - i^{-1}(b - a)] \to \tfrac{1}{2}(a + b)$$

as $i \to \infty$. It follows that Assumptions 1 and 4 hold if in addition to Assumption 1**, we require

Assumption 4.** $\sqrt{ab} < \gamma^{-1} , \qquad \tfrac{1}{2}(a + b) > \gamma^{-1} .$

With our previous choice of a and b, since $\tfrac{1}{2}(a+b) > \gamma^{-1}$ for sufficiently small α, therefore $\{x_t\}$ has a subsequence that tends to infinity.

Cornette (1981) has established the following sufficient condition for a protected polymorphism. Define the arithmetic means ($1 \leq \tau \leq t$)

$$\bar{u}_{t,\tau} = \frac{1}{\tau} \sum_{n=t-\tau+1}^{t} u_\tau$$

and require

Assumption 5.   There exist $c_0 < 1$, $d_0 > 1$, and positive integers ρ and σ such that

$$u_{t,\rho}^* \leq c_0 , \qquad \bar{u}_{t,\sigma} \geq d_0$$

for all $t \geq \rho$ and $t \geq \sigma$, respectively.

Assumptions 1 and 5 together ensure that $\{x_t\}$ is bounded above and has a positive lower bound.

Refer to Dempster (1955), Karlin and Liberman (1974), and Hoekstra (1975) for cyclic environments.

4.6 Intra-Family Selection

Let us suppose that the number of adults contributed by a family to the next generation is independent of the parental genotypes. Within families, however,

$A_i A_j$ individuals have a relative viability x_{ij}. Therefore, an $A_i A_k \times A_l A_j$ mating contributes $A_i A_j$ adults in proportion to $x_{ij} c_{ik,lj}$, where

$$c_{ik,lj} = 4(x_{ij} + x_{il} + x_{kl} + x_{kj})^{-1} . \tag{4.78}$$

For random mating, our basic equation reads

$$P'_{ij} = \sum_{kl} P_{ik} P_{lj} x_{ij} c_{ik,lj} , \tag{4.79}$$

where P_{ij} is the frequency of $A_i A_j$ adults.

To verify that the normalization in (4.79) is correct, note that

$$c_{ik,lj} = c_{ki,lj} = c_{lj,ik} , \tag{4.80}$$

whence

$$\sum_{ij} P'_{ij} = \tfrac{1}{4} \sum_{ijkl} P_{ik} P_{lj} (x_{ij} + x_{il} + x_{kl} + x_{kj}) c_{ik,lj}$$

$$= \sum_{ijkl} P_{ik} P_{lj}$$

$$= 1 .$$

To study weak selection, let us put $x_{ij} = 1 + s_{ij}$, the intra-family selection coefficients s_{ij} being small. From (4.79) we obtain at once

$$P'_{ij} = \sum_{kl} P_{ik} P_{lj} + O(s) = p_i p_j + O(s) , \tag{4.81}$$

i.e., the deviation of the adults from Hardy-Weinberg proportions is reduced to the order of the selection intensity in one generation of panmixia. The zygotes, of course, are exactly in Hardy-Weinberg proportions, but, since the selection is mating dependent, this fact is not useful in the formulation or analysis. The gene-frequency change may be computed by using (4.78), (4.79), and (4.80):

$$\Delta p_i = \sum_{jkl} P_{ik} P_{lj} (x_{ij} c_{ik,lj} - 1)$$

$$= \tfrac{1}{4} \sum_{jkl} P_{ik} P_{lj} c_{ik,lj} (3x_{ij} - x_{il} - x_{kl} - x_{kj})$$

$$= \tfrac{1}{2} \sum_{jkl} P_{ik} P_{lj} c_{ik,lj} (x_{ij} - x_{kj})$$

$$= \tfrac{1}{2} \sum_{jkl} P_{ik} P_{lj} (s_{ij} - s_{kj}) + O(s^2) . \tag{4.82}$$

If we assume that there has been at least one generation of random mating, (4.81) reduces (4.82) to

$$\Delta p_i = \tfrac{1}{2} p_i (s_i - \bar{s}) + O(s^2) , \tag{4.83}$$

where

$$s_i = \sum_j s_{ij} p_j , \qquad \bar{s} = \sum_{ij} s_{ij} p_i p_j .$$

For the same set of gene frequencies and selection coefficients, the lowest-order term in (4.83) is just one-half of the corresponding result for the standard model, (4.25). Wright (1955) derived this result for two alleles by assuming that the population was in Hardy-Weinberg proportions, a procedure justified by (4.81). Since a population will evolve quite differently under selection within the entire population and selection only within families, therefore, even with the same selection coefficients, the gene frequencies in the two cases will differ. Consequently, the rate of gene-frequency change in the second case will not be one-half that in the first.

For two alleles, with the notation $P = P_{11}$, $Q = P_{12}$, and $R = P_{22}$, the exact equations (4.79) become (Haldane and Jayakar, 1965; King, 1965)

$$P' = P^2 + 4x_{11}Q \left(\frac{P}{x_{11} + x_{12}} + \frac{Q}{x_{11} + 2x_{12} + x_{22}} \right) , \qquad (4.84a)$$

$$Q' = PR + 2x_{12}Q \left(\frac{P}{x_{11} + x_{12}} + \frac{2Q}{x_{11} + 2x_{12} + x_{22}} + \frac{R}{x_{12} + x_{22}} \right) , \quad (4.84b)$$

$$R' = R^2 + 4x_{22}Q \left(\frac{Q}{x_{11} + 2x_{12} + x_{22}} + \frac{R}{x_{12} + x_{22}} \right) . \qquad (4.84c)$$

From (4.84) we find the gene-frequency change

$$\Delta p = Q \left[P \left(\frac{x_{11} - x_{12}}{x_{11} + x_{12}} \right) + 2Q \left(\frac{x_{11} - x_{22}}{x_{11} + 2x_{12} + x_{22}} \right) + R \left(\frac{x_{12} - x_{22}}{x_{12} + x_{22}} \right) \right] .$$
$$(4.85)$$

For constant x_{ij}, we conclude from (4.85) at a glance that (as long as there is selection) $x_{11} \geq x_{12} \geq x_{22}$ implies $\Delta p > 0$ for $0 < p < 1$, so $p(t) \to 1$, as in the usual model. Similarly, if $x_{11} \leq x_{12} \leq x_{22}$, then $p(t) \to 0$. It is easy to check, however, that the Hardy-Weinberg overdominant or underdominant equilibrium (4.29) never satisfies (4.84). We shall show, nevertheless, that overdominance is sufficient for a protected polymorphism.

Consider $p \to 0$. Since $p = P + Q \geq P, Q$, we infer from (4.84a) that $P = O(p^2)$ after one generation. Therefore, $p = Q + O(p^2)$ and (4.85) becomes

$$\Delta p = p \left(\frac{x_{12} - x_{22}}{x_{12} + x_{22}} \right) + O(p^2) ,$$

whence we require $x_{12} > x_{22}$. Obviously, a will be protected if $x_{12} > x_{11}$, which proves our assertion.

See Haldane and Jayakar (1965) for other results.

4.7 Maternal Inheritance

Characters are frequently affected by the maternal genotype. We shall discuss the extreme case of traits that depend entirely on the mother's genotype. Consult Srb, Owen, and Edgar (1965, Ch. 11) for examples. Let y_{ij} be the viability of an individual whose mother is $A_i A_j$. We postulate random mating. In terms of the frequency, P_{ij}, of $A_i A_j$ adults, our difference equations read

$$P'_{ij} = (2\overline{y})^{-1} \sum_{kl} P_{ik} P_{lj}(y_{ik} + y_{lj}) \tag{4.86}$$

$$= (2\overline{y})^{-1} p_i p_j (y_i + y_j) \ , \tag{4.87}$$

where

$$p_i y_i = \sum_j P_{ij} y_{ij} \ , \qquad \overline{y} = \sum_{ij} P_{ij} y_{ij} \ . \tag{4.88}$$

We set $y_{ij} = 1 + s_{ij}$ and infer immediately from (4.87)

$$P'_{ij} = p_i p_j + O(s) \ . \tag{4.89}$$

Thus, as for intra-family selection, the deviation of the adults from Hardy-Weinberg proportions is reduced to the order of the selection intensity in one generation of random mating. Owing to the parental effect, the fact that zygotes are in Hardy-Weinberg proportions is again useless. From (4.87) we obtain

$$\Delta p_i = (2\overline{y})^{-1} p_i (y_i - \overline{y}) \ , \tag{4.90}$$

which, for the same fitnesses and genotypic frequencies, is one-half the gene-frequency change in the standard model. For two alleles, Wright (1969) made this observation in the Hardy-Weinberg approximation. The comments below (4.83) apply here.

With two alleles, (4.86) reduces to (Wright, 1969)

$$P' = \overline{y}^{-1} p (P y_{11} + Q y_{12}) \ , \tag{4.91a}$$

$$Q' = (2\overline{y})^{-1} (q P y_{11} + Q y_{12} + p R y_{22}) \ , \tag{4.91b}$$

$$R' = \overline{y}^{-1} q (Q y_{12} + R y_{22}) \ , \tag{4.91c}$$

where $P = P_{11}$, $Q = P_{12}$, and $R = P_{22}$. From (4.90) we find

$$\Delta p = (2\overline{y})^{-1} [q P (y_{11} - y_{12}) + p R (y_{12} - y_{22})] \ . \tag{4.92}$$

For constant y_{ij}, it follows readily that $y_{11} \geq y_{12} \geq y_{22}$ implies $p(t) \to 1$, and $y_{11} \leq y_{12} \leq y_{22}$ implies $p(t) \to 0$. It is easily verified that the overdominant or underdominant equilibrium $\hat{P} = \hat{p}^2$, $\hat{Q} = \hat{p}\hat{q}$, $\hat{R} = \hat{q}^2$,

$$\hat{p} = \frac{y_{22} - y_{12}}{y_{11} - 2y_{12} + y_{22}} \ ,$$

satisfies (4.91). As in Sect. 4.6, we see that $P' = O(p^2)$ as $p \to 0$. Therefore, (4.92) informs us that overdominance, $y_{12} > y_{11}, y_{22}$, is sufficient for a protected polymorphism.

4.8 Meiotic Drive

Meiotic drive refers to non-Mendelian segregation at meiosis. Its effect is manifestly the same as that of selection acting independently among the gametes produced by each individual. The most thoroughly investigated examples are in *Drosophila* and the house mouse. Non-Mendelian segregation can maintain a deleterious allele in the population. Indeed, the "driven" allele is often lethal in the homozygous state. We shall show only how to formulate the single-locus equations that describe meiotic drive. Even the analysis of fairly idealized models requires considerable algebra (Hartl, 1970, 1970a; Charlesworth and Hartl, 1978).

Let α_{ij} be the probability that a successful gamete from an $A_i A_j$ male carries A_i (Hartl, 1970b). Thus, $\alpha_{ji} = 1 - \alpha_{ij}$, whence $\alpha_{ii} = \frac{1}{2}$. For Mendelian segregation, $\alpha_{ij} = \frac{1}{2}$ for all i and j. In most cases, the amount of segregation distortion is different in the two sexes. We denote the female segregation parameter corresponding to α_{ij} by β_{ij}. Let p_i and q_i, and x_{ij} and y_{ij} represent the frequencies of A_i in sperm and eggs, and the viabilities of $A_i A_j$ males and females, respectively. With random mating, the zygotic frequencies in both sexes are $\frac{1}{2}(p_i q_j + p_j q_i)$. Therefore,

$$p_i' = \bar{x}^{-1} \sum_j (p_i q_j + p_j q_i) x_{ij} \alpha_{ij} \ ,$$

$$q_i' = \bar{y}^{-1} \sum_j (p_i q_j + p_j q_i) y_{ij} \beta_{ij} \ ,$$

where

$$\bar{x} = \sum_{ij} (p_i q_j + p_j q_i) x_{ij} \alpha_{ij}$$

$$= \frac{1}{2} \sum_{ij} (p_i q_j + p_j q_i) x_{ij} (\alpha_{ij} + \alpha_{ji})$$

$$= \frac{1}{2} \sum_{ij} (p_i q_j + p_j q_i) x_{ij} \ ,$$

$$\bar{y} = \frac{1}{2} \sum_{ij} (p_i q_j + p_j q_i) y_{ij} \ .$$

Note that we require equal viabilities, $x_{ij} = y_{ij}$, and equal segregation parameters, $\alpha_{ij} = \beta_{ij}$, in order to have the same allelic frequencies in the two sexes, $p_i' = q_i'$.

4.9 Mutation and Selection

To introduce mutation, we follow the procedure of Sect. 2.2. Let P_{ij} be the frequency of $A_i A_j$ zygotes. According to (4.6), the zygotic proportions in the next generation in the absence of mutation would read

$$P^{\star}_{ij} = \overline{f}^{-1} \sum_{kl} X_{ik,lj} f_{ik,lj} \; , \tag{4.93}$$

where $\overline{f}$ is given by (4.7). While selection acts on the phenotype, which develops from the zygotic genotype, the germ cells mutate with no phenotypic effect. Hence, if R_{ij} designates the probability that an A_i allele in a zygote appears as A_j in a gamete, the zygotic frequencies in the next generation are

$$P'_{ij} = \sum_{kl} P^{\star}_{kl} R_{ki} R_{lj} \; . \tag{4.94}$$

Equations (4.93) and (4.94) constitute our model.

The allelic frequencies satisfy much simpler equations. From (4.17) we see that

$$p^{\star}_i = p_i w_i / \overline{w} \; , \tag{4.95}$$

where the fitnesses are specified by (4.13) and (4.14). Since

$$\sum_j R_{ij} = 1 \; ,$$

(4.94) yields

$$p'_i = \sum_k p^{\star}_k R_{ki} \; . \tag{4.96}$$

Denoting the mutation rates by u_{ij} ($u_{ii} = 0$) and substituting (2.26) into (4.96), we find

$$p'_i = p^{\star}_i \left(1 - \sum_k u_{ik} \right) + \sum_k p^{\star}_k u_{ki} \; . \tag{4.97}$$

If mating is random and fertilities are multiplicative, (4.24) informs us that

$$P^{\star}_{ij} = p^{\star}_i p^{\star}_j \; . \tag{4.98}$$

Inserting (4.98) into (4.94) and employing (4.96) gives

$$P'_{ij} = p'_i p'_j \; . \tag{4.99}$$

Thus, (4.25b), (4.95), and (4.96) determine the evolution of the population in this case.

With weak selection and mutation, the argument in Sect. 2.2 again shows that the changes due to selection and mutation are approximately additive.

Assuming only that mutation is much weaker than selection, we shall relate the total frequency of mutants at equilibrium to the total mutation rate and the mean selection coefficient. This was done in continuous time by Crow and Kimura (1970). Let A_1 be the normal allele; A_i is deleterious for $i > 1$. Since mutation is much weaker than selection, $p_i \ll p_1$ for $i > 1$. Then we may neglect the last term in (4.97) for $i = 1$. Without reverse mutation, (4.97) yields

$$p'_1 = p^{\star}_1 (1 - u) \; , \tag{4.100}$$

where

$$u = \sum_k u_{1k}$$ (4.101)

is the total forward mutation rate. Substituting (4.95) into (4.100), at equilibrium we find

$$\overline{w} = w_1(1 - u) .$$ (4.102)

We must approximate the fitnesses in (4.102). There are two cases.

1. A_1 Completely Dominant

Choosing $w_{11} = 1$, we have $w_{1i} = 1$ for all i. We write $w_{ij} = 1 - s_{ij}$, where $s_{1i} = 0$ and $s_{ij} \geq 0$, for all i and j. It suffices to suppose that $u_{1i} \ll s_{ii}$ for all i. The total frequency of mutant phenotypes is

$$Q = \sum_{i,j>1} P_{ij} .$$ (4.103)

From (4.13) we find $w_1 = 1$, and from (4.14) we obtain

$$\overline{w} = \sum_{ij} P_{ij}(1 - s_{ij}) = 1 - Q\overline{s} ,$$ (4.104)

where

$$\overline{s} = Q^{-1} \sum_{ij} P_{ij}s_{ij}$$ (4.105)

is the mean selection coefficient of the mutant phenotypes. Substituting our results into (4.102), we conclude

$$Q = u/\overline{s} ,$$ (4.106)

i.e., the total frequency of mutant phenotypes is the ratio of the total forward mutation rate to the average selection coefficient.

This formula is approximate only because we ignored reverse mutation. For two alleles, recalling (4.101), (4.103), and (4.105), we obtain from (4.106)

$$P_{22} = u_{12}/s_{22} ;$$ (4.107)

in Hardy-Weinberg proportions, this reduces to

$$p_2 = (u_{12}/s_{22})^{1/2} .$$ (4.108)

2. A_1 Incompletely Dominant

We take again $w_{ij} = 1 - s_{ij}$, where $s_{11} = 0$ and $s_{1i} > 0$, for all i and j. We assume that $u_{1i} \ll s_{1i}$ for all i. From (4.13) we derive

$$p_1w_1 = \sum_i P_{1i}(1 - s_{1i}) = p_1 - T\tilde{s} ,$$ (4.109)

where

$$T = \sum_{i>1} P_{1i} = p_1 - P_{11} \qquad (4.110)$$

and

$$\tilde{s} = T^{-1} \sum_i P_{1i} s_{1i} \qquad (4.111)$$

are one-half the total frequency and the average selection coefficient of normal-mutant heterozygotes. The mean fitness reads

$$\begin{aligned}
\overline{w} &= \sum_{ij} P_{ij}(1 - s_{ij}) \\
&= 1 - 2 \sum_i P_{1i} s_{1i} - \sum_{i,j>1} P_{ij} s_{ij} \\
&\approx 1 - 2T\tilde{s} \ . \qquad (4.112)
\end{aligned}$$

Since $p_i \ll 1$, we shall normally have $P_{ij} \ll P_{1i}$ for $i, j > 1$. As we can see from (4.60), however, this will fail with substantial inbreeding. Equation (4.112) is inaccurate also if A_1 is almost completely dominant, $s_{1i} \ll s_{ii}$. Substituting (4.109) and (4.112) into (4.102) leads to

$$u \approx T\tilde{s}[2 - p_1^{-1}(1 - u)] \approx T\tilde{s} \ ,$$

whence

$$T \approx u/\tilde{s} \ . \qquad (4.113)$$

But

$$p_i = P_{i1} + \sum_{j>1} P_{ij} \approx P_{1i} \ , \qquad (4.114)$$

so the sum in (4.110) yields

$$T \approx \sum_{i>1} p_i = 1 - p_1 \ . \qquad (4.115)$$

Combining this with (4.113), we obtain the desired result:

$$1 - p_1 \approx u/\tilde{s} \ , \qquad (4.116)$$

where (4.111), (4.114), and (4.115) give

$$\tilde{s} \approx (1 - p_1)^{-1} \sum_i p_i s_{1i} \ . \qquad (4.117)$$

Thus, the total frequency of mutant alleles is approximately equal to the ratio of the total forward mutation rate to the mean normal-mutant heterozygote selection coefficient. For two alleles, (4.116) reduces to

$$p_2 \approx u_{12}/s_{12} \ , \qquad (4.118)$$

which, for comparable parameters, is a much lower frequency than the corresponding one, (4.108), for a recessive mutant. This is hardly surprising: selection against very rare recessive homozygotes is much less effective than selection against heterozygotes.

For dynamical results with two alleles, refer to Problems 4.6 to 4.8.

4.10 Continuous Model with Overlapping Generations

In this section, we shall construct a model of a continuously evolving population with age structure. It will be possible to carry over some of the results of Sect. 2.4 with only obvious modifications. With some simplifying assumptions, we shall reduce the general model to one that involves only the total numbers of the various genotypes. We shall investigate the dynamics of the simplified model, with particular attention to weak selection and the Fundamental Theorem of Natural Selection.

Our formulation of the age-structured model will be based on that of Cornette (1975). See also Norton (1928) and Charlesworth (1970, 1974, 1980). As far as possible, the notation will be the diploid version of that of Sect. 2.4. Let $\nu_{ij}(t,x)\Delta x$ be the number of A_iA_j individuals between the ages of x and $x + \Delta x$ at time t. We measure x and t in the same units. Then the total number of A_iA_j individuals and the total population size at time t are

$$n_{ij}(t) = \int_0^\infty \nu_{ij}(t,x)dx \tag{4.119}$$

and

$$N(t) = \sum_{ij} n_{ij}(t) \; . \tag{4.120}$$

As in Sect. 2.4, we start observing the population at $t = 0$. For $t \geq x$, the life table $l_{ij}(t,x)$ represents the probability that an A_iA_j individual born at time $t - x$ survives to age x. If $t < x$, $l_{ij}(t,x)$ designates the probability that an A_iA_j individual aged $x - t$ at time 0 survives to age x. Therefore,

$$l_{ij}(t,0) = 1 \quad \text{and} \quad l_{ij}(0,x) = 1 \; .$$

Since the analysis and discussion between (2.46) and (2.54) do not depend on the mode of reproduction, they apply equally well to a diploid population. Thus, as in (2.47), the age distribution $\nu_{ij}(t,x)$ is determined by the initial age distribution $\nu_{ij}(0,x)$, the life table $l_{ij}(t,x)$, and the rate of birth of A_iA_j individuals per unit time, $\nu_{ij}(t,0)$:

$$\nu_{ij}(t,x) = \begin{cases} \nu_{ij}(t - x, 0)l_{ij}(t,x), & t \geq x \; , \\ \nu_{ij}(0, x - t)l_{ij}(t,x), & t < x \; . \end{cases}$$

If $d_{ij}(t,x)\Delta x$ is the probability that an A_iA_j individual aged x at time t dies between ages x and $x + \Delta x$, from (2.49) we have

$$\frac{\partial l_{ij}}{\partial t} + \frac{\partial l_{ij}}{\partial x} = -d_{ij}(t,x)l_{ij}(t,x) \ ,$$

with the solution [see (2.52)]

$$l_{ij}(t,x) = \begin{cases} \exp\left[-\int_0^x d_{ij}(\xi+t-x,\xi)d\xi\right], & t \geq x \ , \\[2mm] \exp\left[-\int_{x-t}^x d_{ij}(\xi+t-x,\xi)d\xi\right] \ , & t < x \ . \end{cases}$$

We shall not write the obvious analogues of (2.58) and (2.59). Finally, (2.53) and (2.54) yield

$$\frac{\partial \nu_{ij}}{\partial t} + \frac{\partial \nu_{ij}}{\partial x} = -d_{ij}(t,x)\nu_{ij}(t,x)$$

and

$$\dot{n}_{ij}(t) = \nu_{ij}(t,0) - \int_0^\infty d_{ij}(t,x)\nu_{ij}(t,x)dx \ . \tag{4.121}$$

To reduce the birth term, $\nu_{ij}(t,0)$, in (4.121), we must consider the mating structure of the population. We designate by $Y_{ij,kl}(t,x,y)\Delta t \Delta x \Delta y$ the number of matings from time t to $t+\Delta t$ between A_iA_j individuals aged x to $x+\Delta x$ and A_kA_l individuals aged y to $y+\Delta y$. Let $a_{ij,kl}(t,x,y)$ be the average number of progeny from one such union. Then

$$\nu_{ij}(t,0) = \sum_{kl} \int_0^\infty dx \int_0^\infty dy Y_{ik,lj}(t,x,y)a_{ik,lj}(t,x,y) \ . \tag{4.122}$$

We wish to impose on the mating frequencies, fertilities, and mortalities restrictions that will lead to a closed system of differential equations for $n_{ij}(t)$. As in Sect. 2.4, we shall do this in two ways. First, we assume *age independence*. For the fertilities and mortalities, this means

$$a_{ij,kl}(t,x,y) = a_{ij,kl}(t) \ , \qquad d_{ij}(t,x) = d_{ij}(t) \ . \tag{4.123}$$

We cannot suppose that $Y_{ij,kl}(t,x,y) = Y_{ij,kl}(t)$, because $Y_{ij,kl}(t,x,y) = 0$ if $\nu_{ij}(t,x) = 0$ or $\nu_{kl}(t,y) = 0$. Therefore, we hypothesize

$$Y_{ij,kl}(t,x,y) = \overline{Y}_{ij,kl}(t)\nu_{ij}(t,x)\nu_{kl}(t,y) \ . \tag{4.124}$$

Let $M(t)$ be the total number of matings per unit time, and $X_{ij,kl}(t)$ the fraction of these matings between A_iA_j and A_kA_l. Thus, from (4.119) and (4.124) we obtain

$$M(t)X_{ij,kl}(t) = \int_0^\infty dx \int_0^\infty dy \overline{Y}_{ij,kl}(t)\nu_{ij}(t,x)\nu_{kl}(t,y)$$

$$= \overline{Y}_{ij,kl}(t)n_{ij}(t)n_{kl}(t) \ , \tag{4.125}$$

which shows that with the *Ansatz* (4.124), $X_{ij,kl}(t)$ is independent of age structure. Substituting (4.119), (4.123), (4.124), and (4.125) into (4.121) and (4.122) yields

$$\dot{n}_{ij}(t) = M(t) \sum_{kl} X_{ik,lj}(t)a_{ik,lj}(t) - d_{ij}(t)n_{ij}(t) \ . \qquad (4.126)$$

The dependence on time in our basic equation (4.126) includes possible dependence on the genotypic numbers $\mathbf{n}(t)$.

Now let us derive (4.126) by assuming that the population has reached a *stable age distribution*. We posit that the proportion of A_iA_j individuals in the age range x to $x + \Delta x$ is $\lambda_{ij}(x)\Delta x$, i.e., $\nu_{ij}(t,x) = n_{ij}(t)\lambda_{ij}(x)$. The hypothesis is that $\lambda_{ij}(x)$ is a known function that depends only on age. Since

$$\int_0^\infty \lambda_{ij}(x)dx = 1 \ ,$$

the average death rate of A_iA_j individuals reads

$$d_{ij}(t) = \int_0^\infty d_{ij}(t,x)\lambda_{ij}(x)dx \ . \qquad (4.127)$$

We suppose that the age and time dependence of the mating frequencies are uncoupled:

$$Y_{ij,kl}(t,x,y) = M(t)X_{ij,kl}(t)R_{ij,kl}(x,y) \ , \qquad (4.128)$$

where

$$\int_0^\infty dx \int_0^\infty dy R_{ij,kl}(x,y) = 1 \ .$$

Integrating (4.128) over x and y justifies interpreting $X_{ij,kl}(t)$ again as the proportion of matings, regardless of age, between A_iA_j and A_kA_l. Substituting the results in this paragraph into (4.121) and (4.122) leads to (4.126), with

$$a_{ij,kl}(t) = \int_0^\infty dx \int_0^\infty dy R_{ij,kl}(x,y)a_{ij,kl}(t,x,y) \ , \qquad (4.129)$$

which is analogous to (4.127). Clearly, $a_{ij,kl}(t)$ is the average fertility of an $A_iA_j \times A_kA_l$ union.

The discrete model formulated in Sect. 4.1 applies exactly to some organisms, such as annual plants and some insects. Our discussion of the conditions required for the validity of the continuous model (4.126) suggests that this model may describe some populations of higher organisms more accurately than the discrete one.

We proceed now to derive from (4.126) equations for the genotypic and gene frequencies, P_{ij} and p_i. Our treatment follows Nagylaki and Crow (1974), who present also analyses of some special cases and generalizations to two loci and dioecious populations. It is convenient to use fertilities, $f_{ij,kl}$, relative to the number of individuals, rather than to the number of matings. With

$$f_{ij,kl}(t) = M(t)a_{ij,kl}(t)/N(t) \ , \qquad (4.130)$$

one expects $f_{ij,kl}(t)$ to be constant whenever $a_{ij,kl}(t)$ is. Inserting (4.130) into (4.126), we get

$$\dot{n}_{ij} = N \sum_{kl} X_{ik,lj} f_{ik,lj} - d_{ij} n_{ij} \qquad (4.131)$$

The mating frequencies are normalized,

$$\sum_{ijkl} X_{ij,kl} = 1 \ ,$$

and satisfy the obvious symmetries

$$X_{ij,kl} = X_{kl,ij} = X_{ji,kl} \ .$$

The fertilities possess the same symmetries. Therefore, (4.131) informs us that $n_{ij}(0) = n_{ji}(0)$ implies $\dot{n}_{ij}(0) = \dot{n}_{ji}(0)$, whence $n_{ij}(t) = n_{ji}(t)$ for $t \geq 0$, as is necessary.

Recalling (4.120), we deduce from (4.131)

$$\dot{N} = \overline{m} N \ , \qquad (4.132)$$

where the genotypic frequencies are

$$P_{ij} = n_{ij}/N \ , \qquad (4.133)$$

and the mean mortality, fertility, and fitness read

$$\overline{d} = \sum_{ij} P_{ij} d_{ij} \ , \qquad (4.134)$$

$$\overline{f} = \sum_{ijkl} X_{ij,kl} f_{ij,kl} \ , \qquad (4.135)$$

$$\overline{m} = \overline{f} - \overline{d} \ . \qquad (4.136)$$

Differentiating (4.133) and substituting (4.131) and (4.132) gives

$$\dot{P}_{ij} = \sum_{kl} X_{ik,lj} f_{ik,lj} - (d_{ij} + \overline{m}) P_{ij} \ . \qquad (4.137)$$

This is analogous to the basic equation (4.6) of the discrete model; like (4.6), it requires specification of the mating frequencies and the birth and death rates.

The number of offspring of a single $A_i A_j$ individual per unit time, b_{ij}, is given by

$$P_{ij} b_{ij} = \sum_{kl} X_{ij,kl} f_{ij,kl} \ . \qquad (4.138)$$

The mean fecundity of the population is

$$\overline{b} = \sum_{ij} P_{ij} b_{ij} = \overline{f} \ . \qquad (4.139)$$

To derive the differential equation satisfied by the allelic frequencies p_i, we introduce allelic fertilities, mortalities, and fitnesses by the definitions

$$p_i b_i = \sum_j P_{ij} b_{ij} \ , \tag{4.140}$$

$$p_i d_i = \sum_j P_{ij} d_{ij} \ , \tag{4.141}$$

$$m_i = b_i - d_i \ . \tag{4.142}$$

Summing (4.137) over j and substituting (4.138), (4.140), (4.141), and (4.142), we obtain the fundamental equation for gene-frequency change:

$$\dot{p}_i = p_i(m_i - \overline{m}) \ . \tag{4.143}$$

This has the same form as (2.63), but the definition of the fitnesses is more complicated. Defining the Malthusian parameter of $A_i A_j$ as

$$m_{ij} = b_{ij} - d_{ij} \ , \tag{4.144}$$

we infer from (4.140), (4.141), and (4.142) that

$$p_i m_i = \sum_j P_{ij} m_{ij} \ , \tag{4.145}$$

and from (4.136), (4.139), and (4.144) that

$$\overline{m} = \sum_{ij} m_{ij} P_{ij} \ . \tag{4.146}$$

Thus, (4.143) involves the fertilities and mortalities explicitly only in the Malthusian-parameter combination (4.144). This observation, however, is deceptive: the evolution of the genotypic proportions, controlled by (4.137), depends separately on the birth and death rates, and P_{ij} must be specified as a function of allelic frequencies in order to complete (4.143).

Henceforth, we shall suppose that *mating is random*:

$$X_{ij,kl} = P_{ij} P_{kl} \ . \tag{4.147}$$

This simplifies (4.138) to

$$b_{ij} = \sum_{kl} P_{kl} f_{ij,kl} \ , \tag{4.148}$$

but does not lead to Hardy-Weinberg proportions. Just as in the discrete case, to obtain Hardy-Weinberg proportions, we must restrict the fertilities. The correct condition in continuous time is additivity of the birth rates (Nagylaki and Crow, 1974). As expected intuitively, however, even if the population is originally in Hardy-Weinberg proportions, dominance in the mortalities produces deviations from Hardy-Weinberg proportions (Nagylaki and Crow, 1974). The variables

$$Q_{ij} = P_{ij} - p_i p_j \tag{4.149}$$

are suitable measures of these departures.

We shall now analyze (4.137) for *weak selection* (Nagylaki, 1976):

$$f_{ij,kl} = b + O(s), \qquad d_{ij} = d + O(s) , \qquad (4.150)$$

for all i, j, k, and l. Genotype-independent mortality clearly has no effect on genotypic proportions; indeed, d cancels out of (4.137). We assume that the fertility b is constant. At this stage, the selection pattern is completely arbitrary and may depend on time and the genotypic frequencies. The positive constant s gives the order of magnitude of the intensity of selection. If the selective differences are constant, we can always take s to be the largest selection coefficient. We are interested in the most common biological situtation, $s \ll b$. Since the population size will usually be approximately stabilized, if time is measured in generations, we shall have $b \approx 1$, whence $s \ll 1$.

Substituting (4.147) and (4.150) into (4.137) yields

$$\dot{P}_{ij} = -bQ_{ij} + O(s) , \qquad (4.151)$$

and (4.143) and (4.150) tell us

$$\dot{p}_i = O(s) . \qquad (4.152)$$

From (4.149), (4.151), and (4.152) we conclude

$$\dot{Q}_{ij} = -bQ_{ij} + sg_{ij}(\mathbf{P}, t) , \qquad (4.153)$$

where $g_{ij}(\mathbf{P}, t)$ is a complicated function of order unity of birth and death rates, genotypic frequencies, and possibly time. More precisely, we may assume that $g_{ij}(\mathbf{P}, t)$ is uniformly bounded for $t \geq 0$ as $s \to 0$. Our analysis will be based on (4.152) and (4.153). Observe that in the absence of selection ($s = 0$), $Q_{ij}(t)$ is reduced to zero at the rapid exponential rate e^{-bt}. We shall exploit this behavior for weak selection.

By the variation-of-parameters formula (see, e.g., Brauer and Nohel, 1969, p. 72), (4.153) has the unique solution

$$Q_{ij}(t) = Q_{ij}(0)e^{-bt} + se^{-bt} \int_0^t e^{b\tau} g_{ij}[\mathbf{P}(\tau), \tau]d\tau . \qquad (4.154)$$

We choose t_1 as the shortest time such that

$$|Q_{ij}(0)|e^{-bt_1} \leq s \qquad (4.155)$$

for all i and j. Of course, if the system starts sufficiently close to Hardy-Weinberg proportions, t_1 will be zero. Otherwise, we infer from (4.155) that roughly $t_1 \approx -b^{-1} \ln s$, which will usually be about 5 or 10 generations. Since (4.154) and (4.155) yield $Q_{ij}(t) = O(s)$ for $t \geq t_1$, we write $Q_{ij}(t) = sQ_{ij}^0(t)$. Thus, the population approaches the Hardy-Weinberg surface $Q_{ij} = 0$, very rapidly, but at $t = t_1$ it is still very far from gene-frequency equilibrium, because the gene-frequency change from $t = 0$ to $t = t_1$ is quite small (crudely, $st_1 \approx -b^{-1}s \ln s$). For $t \geq t_1$, (4.153) reads

$$\dot{Q}_{ij}^0 = -bQ_{ij}^0 + g_{ij}(\mathbf{P}, t) \ . \tag{4.156}$$

From (4.149), (4.152), and (4.156) we conclude that $\dot{P}_{ij} = O(s)$ for $t \geq t_1$. Assuming

$$\frac{\partial g_{ij}}{\partial t} = O(s) \ , \tag{4.157}$$

from (4.156) we deduce

$$\frac{d}{dt}[Q_{ij}^0 - b^{-1}g_{ij}(\mathbf{P}, t)] = -b[Q_{ij}^0 - b^{-1}g_{ij}(\mathbf{P}, t)] + O(s) \ . \tag{4.158}$$

Recalling (4.153), we note that (4.157) requires that the explicit dependence of the selection coefficients on time, if any, be of $O(s^2)$. Arbitrary dependence on the genotypic frequencies is permitted. Writing the analogue of (4.154) for the bracket in (4.158), we obtain

$$Q_{ij}^0(t) - b^{-1}g_{ij}[\mathbf{P}(t), t] = \{Q_{ij}^0(t_1) - b^{-1}g_{ij}[\mathbf{P}(t_1), t_1]\}e^{-b(t-t_1)} + O(s) \ . \tag{4.159}$$

Therefore, we define t_2 $(\geq t_1)$ as the shortest time such that

$$|Q_{ij}^0(t_1) - b^{-1}g_{ij}[\mathbf{P}(t_1), t_1]|e^{-b(t_2-t_1)} \leq s \tag{4.160}$$

for all i and j. Since the coefficient of the exponential in (4.160) is of order unity, roughly $t_2 - t_1 \approx t_1$, or $t_2 \approx 2t_1$. From (4.159) we find for $t \geq t_2$

$$Q_{ij}^0(t) = b^{-1}g_{ij}[\mathbf{P}(t), t] + O(s) \ . \tag{4.161}$$

But $\dot{P}_{ij}(t) = O(s)$ for $t \geq t_1$. Hence, with the aid of (4.157), we deduce from (4.161) that $\dot{Q}_{ij}^0(t) = O(s)$, or

$$\dot{Q}_{ij}(t) = O(s^2), \qquad t \geq t_2 \ , \tag{4.162}$$

which is the desired result.

Notice that the system is not close to equilibrium until a time $t_3 \approx s^{-1} \gg t_2$ has elapsed, so (4.162) cannot be obtained by local analysis. During the period $t_1 \leq t < t_2$, $\dot{p}_i = O(s)$ and $\dot{Q}_{ij} = O(s)$. Consequently, these quantities change by only about $s(t_2 - t_1) \approx -b^{-1}s\ln s$ during this epoch. For $t_2 \leq t < t_3$, by (4.162), Q_{ij} changes only by roughly $s^2(t_3 - t_2) \approx s$, whereas p_i approaches equilibrium closely.

Our results are easy to display diagrammatically in the two-allele case. We put $p = p_1$, $q = p_2$, and $Q = Q_{12}$ and use Q and p as coordinates (see Problem 3.2). It is trivial to see that $Q_{11} = Q_{22} = -Q$. The constraints $P_{11} \geq 0$, $P_{22} \geq 0$, and $P_{11} + P_{22} \leq 1$ imply that the population evolves in the region bounded by the three parabolas in Fig. 4.1. Hardy-Weinberg proportions correspond to the dashed line $Q = 0$, $0 \leq p \leq 1$. The times t_1, t_2, and t_3 are indicated on a typical trajectory, which is nearly horizontal for $0 \leq t < t_1$ and close to vertical for $t \geq t_2$. The system moves much faster on the "horizontal" than on the "vertical" part of the trajectory, though the rate of gene-frequency change is always $\dot{p} = O(s)$. The equilibrium satisfies $Q(\infty) = O(s)$.

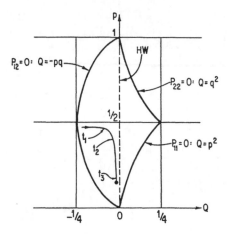

Fig. 4.1. The evolution of a typical diallelic single-locus system.

Before applying (4.162) to the evolution of a character controlled by a single locus, we shall demonstrate that the dynamics of the genotypic frequencies can be approximated with an error of $O(s)$ by the behavior of the much simpler system on the Hardy-Weinberg surface. The exact gene frequencies, $p_i(t)$, evolve according to the complicated law derived from (4.137) and (4.147); at time t_1 they have the values $p_i(t_1)$. The gene frequencies, $\pi_i(t)$, on the Hardy-Weinberg surface evolve according to the much simpler law obtained by imposing Hardy-Weinberg proportions on (4.143); we choose $\pi_i(t_1) = p_i(t_1)$ for all i. For example, in Fig. 4.1, $\pi(t)$ evolves on the p-axis and $\pi(t_1) = p(t_1)$ is found by horizontal projection from the trajectory. We shall show that $P_{ij}(t) = \pi_i(t)\pi_j(t) + O(s)$ for $t \geq t_1$.

We display the possible dependence of the genotypic fitnesses on the genotypic frequencies and time and set

$$m_{ij}(\mathbf{P}, t) = m(\mathbf{P}, t) + sr_{ij}(\mathbf{P}, t) \ , \tag{4.163}$$

where $r_{ij} = O(1)$. From (4.145) and (4.146) we obtain

$$m_i = m + sr_i, \qquad \overline{m} = m + s\overline{r} \ , \tag{4.164}$$

in which

$$p_i r_i(\mathbf{P}, t) = \sum_j r_{ij}(\mathbf{P}, t)P_{ij}, \qquad \overline{r}(\mathbf{P}, t) = \sum_{ij} r_{ij}(\mathbf{P}, t)P_{ij} \ . \tag{4.165}$$

Substituting (4.164) into (4.143), we find

$$\dot{p}_i = sp_i[r_i(\mathbf{P}, t) - \overline{r}(\mathbf{P}, t)] \ . \tag{4.166}$$

Recalling (4.149), we see that

$$r_{ij}(\mathbf{P}, t) = r_{ij}(\mathbf{p} * \mathbf{p}, t) + O(\mathbf{Q}) , \qquad (4.167)$$

where the asterisk indicates evaluation at $P_{ij} = p_i p_j$. Inserting (4.149) and (4.167) into (4.165) produces

$$r_i(\mathbf{P}, t) = c_i(\mathbf{p}, t) + O(\mathbf{Q}), \qquad \bar{r}(\mathbf{P}, t) = \bar{c}(\mathbf{p}, t) + O(\mathbf{Q}) , \qquad (4.168)$$

where

$$c_i(\mathbf{p}, t) = \sum_j r_{ij}(\mathbf{p} * \mathbf{p}, t) p_j , \qquad \bar{c}(\mathbf{p}, t) = \sum_i c_i(\mathbf{p}, t) p_i . \qquad (4.169)$$

Substituting (4.168) into (4.166) gives

$$\dot{p}_i = s p_i [c_i(\mathbf{p}, t) - \bar{c}(\mathbf{p}, t)] + O(s\mathbf{Q}) . \qquad (4.170)$$

Equation (4.170) shows that the allelic frequencies, $\pi_i(t)$, on the Hardy-Weinberg surface satisfy

$$\dot{\pi}_i = s \pi_i [c_i(\boldsymbol{\pi}, t) - \bar{c}(\boldsymbol{\pi}, t)] . \qquad (4.171)$$

For $t \geq t_1$, we have proved that $Q_{ij} = O(s)$, which implies that the error in (4.170) is of $O(s^2)$. If the Hardy-Weinberg system (4.171) tends to isolated equilibrium points, it follows by comparing (4.170) with (4.171) that

$$p_i(t) = \pi_i(t) + O(s), \qquad t \geq t_1 . \qquad (4.172)$$

Combining (4.149) and (4.172), we deduce

$$P_{ij}(t) = \pi_i(t)\pi_j(t) + O(s), \qquad t \geq t_1 , \qquad (4.173)$$

as claimed. If $\boldsymbol{\pi}(t)$ does not necessarily converge to some equilibrium point or if $\boldsymbol{\pi}(t_1)$ happens to be on the stable manifold of an unstable equilibrium, then small perturbations may cause large deviations in its ultimate state. In this case, (4.173) still holds if $t_1 \leq t \leq k/s$ for some constant k.

Finally, let us examine the *evolution of the mean* value, $\bar{Z}$, of a single-locus trait. For the same calculation for two loci, the reader may consult Nagylaki (1989). We shall slightly generalize Kimura's (1958) analysis of fitness, which we shall discuss as a special case. If Z_{ij} designates the character value of $A_i A_j$, then

$$\bar{Z} = \sum_{ij} Z_{ij} P_{ij} . \qquad (4.174)$$

Differentiating (4.174) yields

$$\dot{\bar{Z}} = \dot{\bar{Z}} + \sum_{ij} Z_{ij} \dot{P}_{ij} , \qquad (4.175)$$

where

$$\bar{\dot{Z}} = \sum_{ij} \dot{Z}_{ij} P_{ij} \tag{4.176}$$

is the mean rate of change of the genotypic values. Introducing the deviations from the mean, $z_{ij} = Z_{ij} - \bar{Z}$, and noting that

$$\sum_{ij} \dot{P}_{ij} = \frac{d}{dt} \sum_{ij} P_{ij} = 0 \; ,$$

we obtain from (4.175)

$$\dot{\bar{Z}} = \bar{\dot{Z}} + \sum_{ij} z_{ij} \dot{P}_{ij} \; . \tag{4.177}$$

It is convenient to measure the departure from Hardy-Weinberg proportions by $\theta_{ij} = P_{ij}/(p_i p_j)$. Substituting

$$\dot{P}_{ij} = \dot{p}_i p_j \theta_{ij} + p_i \dot{p}_j \theta_{ij} + p_i p_j \dot{\theta}_{ij}$$

into (4.177) leads to

$$\dot{\bar{Z}} = \bar{\dot{Z}} + \sum_{ij} (2\dot{p}_i p_j \theta_{ij} + p_i p_j \dot{\theta}_{ij}) z_{ij} \; . \tag{4.178}$$

We can reduce (4.178) by performing a least-squares decomposition of the *total genetic variance*

$$V = \sum_{ij} P_{ij} z_{ij}^2 \; . \tag{4.179}$$

We write

$$z_{ij} = \alpha_i + \alpha_j + \Delta_{ij} \; , \tag{4.180}$$

where α_i is the *average effect* of A_i on the character, and Δ_{ij}, which is zero in the absence of dominance, is the *dominance deviation*. The procedure requires minimizing the *dominance variance*

$$V_d = \sum_{ij} P_{ij} \Delta_{ij}^2 = \sum_{ij} P_{ij} (z_{ij} - \alpha_i - \alpha_j)^2 \tag{4.181}$$

with respect to α_i. We have

$$-\frac{1}{4} \frac{\partial V_d}{\partial \alpha_i} = \sum_j P_{ij} (z_{ij} - \alpha_i - \alpha_j) = \sum_j P_{ij} \Delta_{ij} = 0 \; . \tag{4.182}$$

Although this fact is not needed here, we shall show that the variances are additive. The *genic* (or *additive genetic*) variance is

$$V_g = \sum_{ij} P_{ij} (\alpha_i + \alpha_j)^2 \; . \tag{4.183}$$

Inserting (4.180) into (4.179) and using (4.181) and (4.183), we find

$$V = V_g + V_d + 2 \sum_{ij} P_{ij}(\alpha_i + \alpha_j)\Delta_{ij} = V_g + V_d \ , \tag{4.184}$$

in which the sum vanishes on account of (4.182). Using (4.180) and (4.182), we can express the *average excess*, z_i, of A_i as

$$p_i z_i = \sum_j P_{ij} z_{ij} = p_i \alpha_i + \sum_j P_{ij} \alpha_j \ . \tag{4.185}$$

Observe that

$$\sum_i p_i z_i = \sum_{ij} P_{ij} z_{ij} = 0 \ , \tag{4.186}$$

and

$$2 \sum_i p_i \alpha_i = \sum_{ij} P_{ij} z_{ij} = 0 \ , \tag{4.187}$$

i.e., the means of the average effects and excesses are zero. Furthermore, from (4.183) and (4.185) we deduce

$$V_g = \sum_{ij} P_{ij}(\alpha_i + \alpha_j)(\alpha_i + \alpha_j)$$

$$= 2 \sum_{ij} P_{ij}(\alpha_i + \alpha_j)\alpha_i$$

$$= 2 \sum_i p_i z_i \alpha_i \ . \tag{4.188}$$

In Hardy-Weinberg proportions, (4.185) and (4.187) inform us that $z_i = \alpha_i$, so

$$V_g = 2 \sum_i p_i z_i^2 \ . \tag{4.189}$$

Let us return to $\dot{\overline{Z}}$. We substitute $P_{ij} = p_i p_j \theta_{ij}$ into (4.185), and the result,

$$\sum_j p_j \theta_{ij} z_{ij} = \alpha_i + \sum_j p_j \theta_{ij} \alpha_j \ , \tag{4.190}$$

and (4.180) into (4.178):

$$\dot{\overline{Z}} = \overline{\overline{Z}} + 2 \sum_i \dot{p}_i \left(\alpha_i + \sum_j p_j \theta_{ij} \alpha_j \right) + \sum_{ij} p_i p_j \dot{\theta}_{ij}(\alpha_i + \alpha_j + \Delta_{ij})$$

$$= \overline{\overline{Z}} + 2 \sum_i \dot{p}_i \alpha_i + 2 \sum_{ij} p_j \alpha_j(\dot{p}_i \theta_{ij} + p_i \dot{\theta}_{ij}) + \sum_{ij} p_i p_j \theta_{ij} \overset{\circ}{\theta}_{ij} \Delta_{ij} \ . \tag{4.191}$$

In (4.191),

$$\overset{\circ}{\theta} = \frac{d}{dt} \ln \theta$$

is the logarithmic derivative. But

$$\sum_i p_i \theta_{ij} = \sum_i p_j^{-1} P_{ij} = 1 \ .$$

Therefore,

$$0 = \frac{d}{dt} \sum_i p_i \theta_{ij} = \sum_i (\dot{p}_i \theta_{ij} + p_i \dot{\theta}_{ij}) \ ,$$

which tells us that the third term in (4.191) is zero. Recalling (4.143), we arrive at the desired formula:

$$\dot{\overline{Z}} = \text{Cov}\,(\mu, \alpha) + \overline{\dot{Z}} + \overline{\dot{\theta}\Delta} \ , \tag{4.192}$$

where

$$\text{Cov}\,(\mu, \alpha) = 2 \sum_i p_i \mu_i \alpha_i \ , \tag{4.193}$$

$$\overline{\dot{\theta}\Delta} = \sum_{ij} P_{ij} \dot{\theta}_{ij} \Delta_{ij} \ , \tag{4.194}$$

and $\mu_i = m_i - \overline{m}$ is the average excess for fitness of A_i.

The covariance in (4.192) is the *genic covariance* of the average excess of fitness and the average effect on the character. The fitnesses appear explicitly only in this term; selection influences the other terms through the genotypic frequencies. With constant genotypic values, the second term in (4.192) is zero. The last term vanishes if either there is no dominance for the trait or the population is in Hardy-Weinberg proportions.

Since $\mu_i = O(s)$, the covariance in (4.192) is of $O(s)$. Now,

$$\theta_{ij} = 1 + \frac{Q_{ij}}{p_i p_j} \ ,$$

and for $t \geq t_2$, $\dot{Q}_{ij} = O(s^2)$, $\dot{p}_i = O(s)$, and $Q_{ij} = O(s)$. Hence, $\dot{\theta}_{ij} = O(s^2)$ (except possibly if p_i or p_j becomes arbitrarily small), and (4.192) has the form

$$\dot{\overline{Z}} = \text{Cov}\,(\mu, \alpha) + \overline{\dot{Z}} + O(s^2), \qquad t \geq t_2 \ . \tag{4.195}$$

If the genotypic values change very slowly, so that $\overline{\dot{Z}} = o(s)$, we obtain the approximation $\dot{\overline{Z}} \approx \text{Cov}\,(\mu, \alpha)$. This may fail if the covariance is particularly small. At equilibrium, $\dot{p}_i = p_i \mu_i = 0$, so $\text{Cov}\,(\mu, \alpha) = 0$; therefore, the failure is most likely to occur near equilibrium.

If the trait is fitness, (4.188) and (4.193) give $\text{Cov}\,(\mu, \alpha) = V_g$, the genic variance in fitness, and (4.192) becomes

$$\dot{\overline{m}} = V_g + \overline{\dot{m}} + \overline{\dot{\theta}\Delta} \ , \tag{4.196}$$

the form of the Fundamental Theorem of Natural Selection due to Kimura (1958). Of course, $V_g = O(s^2)$ and $\Delta_{ij} = O(s)$, whence

$$\dot{\overline{m}} = V_g + \dot{\overline{m}} + O(s^3) , \qquad t \geq t_2 . \tag{4.197}$$

With nearly constant fitnesses, i.e., $\dot{\overline{m}} = o(s^2)$, we get the approximation $\dot{\overline{m}} \approx V_g$. Again, this approximation may occasionally fail, particularly in the neighborhood of an equilibrium. Observe that $\dot{m}_{ij} = o(s^2)$ is a much stronger condition than the requirement [see (4.157)]

$$\frac{\partial m_{ij}}{\partial t} = O(s^2) , \tag{*}$$

which we imposed in the analysis, for $(*)$ involves only the explicit time dependence. We conclude from the result $\dot{\overline{m}} \approx V_g$ that the mean fitness will generally increase for $t \geq t_2$. The remarks after (2.70) and (4.59) are pertinent here.

For further results, the reader may refer to Nagylaki (1976, 1977a, 1987).

4.11 Density and Frequency Dependence

In this section, we shall study the dynamics of a multiallelic locus with absolute fitnesses that depend on the population number and the gene frequencies. We assume that generations are discrete and nonoverlapping and selection is weak. The model with stationary age distribution in continuous time is analyzed in Nagylaki (1979).

For pure density dependence, Fisher (1930, 1958) asserted without proof that instead of the mean fitness, it is the population size itself that is increasing. Kimura (1978) noted the approach of the mean fitness to rough constancy in various situations. By intuitive arguments, he obtained for two alleles some special cases of the conclusions in this section. Léon and Charlesworth (1976, 1978) examined the evolution of two interacting species and derived intuitively special cases of some of the single-species results proved below. We follow Nagylaki (1979) in seeking conclusions corresponding to some of those in Sects. 4.3 and 4.10.

We posit multiplicative fertilities, in which case Sect. 4.1 informs us that the zygotes are in Hardy-Weinberg proportions in each generation. We denote the selection intensity by s ($s > 0$) and assume $s \ll 1$. Let N and p_i represent the number of zygotes and the frequency of the allele A_i in zygotes in generation t. The population number and gene frequencies evolve according to the difference equations

$$\Delta N = f(N) + sg(N, \mathbf{p}) , \tag{4.198a}$$
$$\Delta p_i = sh_i(N, \mathbf{p}) . \tag{4.198b}$$

The function $f(N)$ describes population regulation in the absence of selection. Without selection, all the genotypic fitnesses must be the same, but, in principle, they may still be functions of the gene frequencies. Since biologically we expect this to happen extremely rarely, we suppose that f depends only on N. Selection, described by g and h_i, may be density and frequency dependent; g

and h_i may be functions of s, but they are bounded as $s \to 0$. We suppose that functions such as f, g, and h_i are smooth.

We assume that without selection the population size converges exponentially to a unique, globally stable equilibrium $\hat{N}$, so $f(\hat{N}) = 0$. Our first task is to demonstrate that $N(t)$ rapidly approaches and remains close to $\hat{N}$. Define $x(t) = N(t) - \hat{N}$ and $F(x) = 1 + x^{-1}f(\hat{N} + x)$. From (4.198a) we deduce

$$x(t+1) = xF(x) + sg(\hat{N} + x, \mathbf{p}) , \tag{4.199}$$

in which we suppressed the argument t of x and $\mathbf{p}$ on the right-hand side. For any fixed initial values $N(0)$ and $\mathbf{p}(0)$, by global convergence for sufficiently small s, $N(t)$ is bounded uniformly in s for all t. Therefore, we may suppose that g is uniformly bounded on the trajectory, say $|g(N, \mathbf{p})| \le B$. If we posit that the convergence of $|x(t)|$ to zero is monotone, then there exists c, where $0 < c < 1$, such that $|F(x)| \le c$. Consequently, (4.199) yields

$$|x(t+1)| \le c|x(t)| + sB . \tag{4.200}$$

Now consider a sequence $\{y(t)\}$ that satisfies

$$y(t+1) = cy(t) + sB , \tag{4.201}$$

with $y(0) = |x(0)|$. Inductive comparison of (4.200) with (4.201) shows that $|x(t)| \le y(t)$. But (4.201) informs us that

$$y(t) = [y(0) - sB(1-c)^{-1}]c^t + sB(1-c)^{-1} ,$$

whence

$$|x(t)| \le |x(0)|c^t + sB(1-c)^{-1} . \tag{4.202}$$

From (4.202) we infer that

$$N(t) = \hat{N} + O(s) \tag{4.203}$$

uniformly for $t \ge t_1 = \ln[s/|x(0)|]/\ln c$. Thus, in a short time, typically 5 to 10 generations, $N(t)$ closely approaches $\hat{N}$. The convergence time can be much longer if the reproductive rate is very close to unity even if N is far from $\hat{N}$, for then c is only slightly less than one. This is, however, biologically extremely unlikely. Observe that the definition of $F(x)$ gives $c \ge F(0) = 1 + f'(\hat{N})$, the prime here denoting the derivative.

We remark in passing that (4.198b) and (4.203) permit us to approximate the gene frequencies $\mathbf{p}$ by the simpler set $\boldsymbol{\pi}$ obtained by setting $N = \hat{N}$. Inserting (4.203) into (4.198b), we find for $t \ge t_1$

$$\Delta p_i = sh_i(\hat{N}, \mathbf{p}) + O(s^2) . \tag{4.204}$$

Choose $\pi_i(t_1) = p_i(t_1)$ and

$$\Delta \pi_i = sh_i(\hat{N}, \boldsymbol{\pi}) . \tag{4.205}$$

Then (4.204) and (4.205) show that

$$p_i(t) = \pi_i(t) + O(s) ,\qquad(4.206)$$

provided $t_1 \leq t \leq k/s$ for some constant k. If $\pi(t)$ does not necessarily converge to some equilibrium point or if $\pi(t_1)$ happens to be on the stable manifold of an unstable equilibrium, small perturbations may cause large deviations in its ultimate state. Excluding these biologically unlikely situations, (4.206) holds uniformly for $t \geq t_1$. For pure density dependence, (4.206) supports the ubiquitous constant-fitness approximation in population genetics.

Let us establish next that $N(t)$ is nearly constant, and hence the mean absolute fitness $\overline{w}$, which is even closer to constancy, deviates only slightly from unity. Writing $\Delta^2 N = \Delta(\Delta N)$ and expanding $f(N)$ in (4.198a) in a Taylor series leads to

$$\Delta^2 N = f'(\hat{N})\Delta N + s\Delta g(N, \mathbf{p}) + O[(N - \hat{N})\Delta N] .\qquad(4.207)$$

In view of (4.198b) and (4.203), this reduces for $t \geq t_1$ to

$$\Delta^2 N = f'(\hat{N})\Delta N + O(s^2) .\qquad(4.208)$$

Arguing either as in the derivation of (4.203) or appealing to the discrete version of the variation-of-parameters formula as in (8.20), we deduce readily for $t \geq t_2 \approx 2t_1$ that

$$\Delta N = O(s^2) .\qquad(4.209)$$

Since (see Problem 4.3)

$$\Delta N = (\overline{w} - 1)N ,\qquad(4.210)$$

(4.209) is equivalent to

$$\overline{w} = 1 + O(s^2) .\qquad(4.211)$$

Perhaps the result (4.203) is intuitively obvious; untutored intuition would suggest, however, only the weaker conclusion $O(s)$ in (4.209) and (4.211). The approximate constancy of the population number is reminiscent of the corresponding behavior of the deviations from random combination of alleles with nearly constant fitnesses, discussed in Sects. 4.10, 8.2, and 8.4. Note that (4.211) allows us to set $\overline{w} = 1$ wherever the mean fitness appears as a factor, thereby simplifying (4.205) without altering (4.206).

By dint of (4.203) and (4.209), the last term in (4.207) is of $O(s^3)$, so (4.207) yields for $t \geq t_2$

$$\Delta^3 N = f'(\hat{N})\Delta^2 N + O(s^3) .\qquad(4.212)$$

From (4.212) we obtain

$$\Delta^2 N = O(s^3) ,\qquad(4.213)$$

$t \geq t_3 \approx 3t_1$. Combining (4.209), (4.210), and (4.211) yields

$$\Delta^2 N = (\overline{w} - 1)\Delta N + (N + \Delta N)\Delta \overline{w} = N\Delta \overline{w} + O(s^4) .\qquad(4.214)$$

Hence, (4.213) may be rewritten as

$$\Delta\overline{w} = O(s^3) \ . \tag{4.215}$$

Substituting (4.209), (4.210), and (4.213) into (4.207), we obtain

$$\overline{w} = 1 - s[Nf'(\hat{N})]^{-1}\Delta g(N, \mathbf{p}) + O(s^3) \ . \tag{4.216}$$

Invoking (4.209) again, we rearrange (4.216) for $t \geq t_3$ as

$$\overline{w} = 1 - s[f'(\hat{N})]^{-1}\Delta[N^{-1}g(N, \mathbf{p})] + O(s^3) \ . \tag{4.217}$$

We can reduce (4.217) to an analogue of the Fundamental Theorem of Natural Selection, (4.59), by positing that frequency dependence is weaker than the selection intensity. If the absolute fitness of $A_i A_j$ individuals reads

$$. \ w_{ij}(N, \mathbf{p}) = 1 + N^{-1}f(N) + sG_{ij}(N, \mathbf{p}) \ , \tag{4.218}$$

we assume

$$\frac{\partial G_{ij}}{\partial p_l} = o(1) \tag{4.219}$$

as $s \to 0$ for all l, i.e., the partial derivatives in (4.219) tend to 0 as $s \to 0$. From (4.198a), (4.210), and (4.218) we have

$$N^{-1}g(N, \mathbf{p}) = \overline{G} = \sum_{ij} G_{ij}p_i p_j \ , \tag{4.220}$$

whence (4.198b), (4.209), and (4.219) tell us that for $t \geq t_2$

$$\Delta[N^{-1}g(N, \mathbf{p})] = 2\sum_{ij} G_{ij}p_j \Delta p_i + o(s) \ . \tag{4.221}$$

With the fitnesses (4.218), the gene-frequency changes are given by

$$\overline{w}\Delta p_i = p_i(G_i - \overline{G}) \ , \tag{4.222a}$$

where

$$G_i = \sum_j G_{ij}p_j \ . \tag{4.222b}$$

An easy calculation that uses (4.222) converts (4.221) to

$$\Delta[N^{-1}g(N, \mathbf{p})] = 2s\overline{w}^{-1}\sum_j p_i(G_i - \overline{G})^2 + o(s) \ . \tag{4.223}$$

Identifying the genic variance

$$V_g = 2\sum_i p_i(w_i - \overline{w})^2 \ , \tag{4.224a}$$

$$w_i = \sum_j w_{ij}p_j \ , \tag{4.224b}$$

in (4.223), recalling (4.211), and substituting into (4.217) produces

$$\overline{w} = 1 - [f'(\hat{N})]^{-1}V_g + o(s^2) \ . \tag{4.225}$$

Thus, after a time t_3 (about 15 to 30 generations), the mean fitness will generally be greater than one, the excess being proportional to the additive component of the genetic variance. In the absence of selection, $N(t)$ ultimately converges to $\hat{N}$ at the rate $[1 + f'(\hat{N})]^t$; with selection, the rate is still very close to this. The decay in the genic variance to zero leads not to an increase in the mean absolute fitness, as for constant absolute fitnesses, but to an increase in population size at the rate

$$\Delta N = -N[f'(\hat{N})]^{-1}V_g + o(s^2) \ . \tag{4.226}$$

The mean relative fitness, $\overline{z}$, may still increase, as would happen, e.g., if $w_{ij} = z_{ij}H(N)$, with z_{ij} constant for all i and j. Consult the discussion below (4.59).

If frequency dependence is no stronger than the square of the selection intensity, we may replace $o(1)$ by $O(s)$ in (4.219); this leads to an error of $O(s^2)$ in (4.221) and (4.223), and therefore to one of $O(s^3)$ in (4.225) and (4.226). These are, of course, also the errors for pure density dependence.

It is interesting to note that the ultimate increase, (4.226), of the population size implies that if its initial value exceeds its equilibrium value, then the population size will overshoot this equilibrium value at least once.

In spite of the validity of (4.225) and (4.226), the mean fitness may be less than one and the population size may decrease in some special cases with extremely small genic variance. Such a failure will usually occur if there is no gene-frequency change, for then the genic variance is identically zero. A simple example is the case of two alleles in overdominant equilibrium with constant G_{ij}. For suitable $f(N)$, $N(t)$ may decrease monotonically to its equilibrium value. The exceptions are quite analogous to those to the Fundamental Theorem of Natural Selection (Nagylaki, 1976, 1977a, 1987).

The first term of (4.226) was derived heuristically for pure density dependence by Léon and Charlesworth (1976, 1978). With two alleles and pure density dependence, Kimura (1978) obtained formulae equivalent of (4.225) for haploids and for diploids with complete dominance and logarithmic population control.

4.12 Problems

4.1 For the model of Sect. 4.1, assume that the zygotes are in Hardy-Weinberg proportions. Show that the adults will be in Hardy-Weinberg proportions if and only if the viabilities are multiplicative. Express v_{ij} in terms of v_i.

4.2 Show that for multiplicative fitnesses, (4.25) reduces to the haploid model, (2.6).

4.3   Referring to Sect. 4.1, show that the number of $A_i A_j$ zygotes in the next generation is

$$n'_{ij} = \tilde{N} \sum_{kl} X_{ik,lj} f_{ik,lj} \ .$$

If N is the number of zygotes in the previous generation, prove that $\tilde{N} = \bar{v}N$ and $N' = \bar{w}N$.

4.4   Assume that $w_{11} = 1$, $w_{12} = w$, and $w_{22} = 0$ in (4.26). Calculate $q(t)$ for constant w. Compare $q(\infty)$ and the rate of convergence with the results in Sect. 4.2. Repeat for $w_{11} = a$, $w_{12} = 0$, and $w_{22} = b$.

4.5   The monotonicity property of the haploid model described in Problem 2.2 holds also for the diallelic diploid model, (4.26), with constant fitnesses. Prove this by deducing that

$$\frac{dp'}{dp} = \frac{q w_{22} w_1 + p w_{11} w_2}{\bar{w}^2} > 0 \ .$$

4.6   Assume that the continuous mapping $p' = f(p)$ of $[a, b]$ into itself is monotone increasing and start iterating with $p(0) = p_0$.

(a) Prove that if $f(p_0) < p_0$, there exists at least one fixed point in $[a, p_0)$ and $p(t)$ converges without oscillation to the largest of these fixed points.

(b) Similarly, if $f(p_0) > p_0$, there exists at least one fixed point in $(p_0, b]$ and $p(t)$ converges monotonically to the smallest of these fixed points.

Thus, iteration of a monotone increasing mapping always produces non-oscillatory global convergence, though the limiting value may depend on the starting point. Note that, by virtue of Problem 4.5, this conclusion applies to diallelic mutation and selection for $u_{12} + u_{21} < 1$.

4.7   Choose $w_{11} = 1 + s_1$, $w_{12} = 1 + s_2$, $w_{22} = 1$, $u_{12} = u$, and $u_{21} = v$ in the diallelic, Hardy-Weinberg mutation-selection model. Show that either conditions (a) or (b) below suffice for the existence of a unique equilibrium, which is globally asymptotically stable for $u + v < 1$ by Problem 4.6.

(a) $u, v > 0$ and $-1 < s_1 \leq 2s_2$.
  This condition includes the following important special cases:
  (i) no dominance: $s_1 = 2s_2$;
  (ii) favored allele completely dominant: $s_1 = s_2 > 0$, or $s_2 = 0$ and $s_1 < 0$;
  (iii) overdominance: $s_1 < s_2$ and $s_2 > 0$.
(b) $0 < v < 1 + u$, $s_1 = -1$, and $s_2 > -1$.
  Here mutation maintains the lethal allele A_1 in the population. The

upper limit on v is always satisfied if $u > 0$; it is biologically trivial in any case.

For weak selection and mutation, the criterion (a) is due to Norman (1974).

4.8 In the diallelic, Hardy-Weinberg mutation-selection model, take $w_{11} = w_{12} = 1$, $w_{22} = 1 - s$, $u_{12} = u$, and $u_{21} = 0$, where $0 < u < s \leq 1$. Prove that there is global nonoscillatory convergence to a unique equilibrium. What is the asymptotic rate of convergence? Show that for $u \ll 1$ this is approximately $\exp[-2(\sqrt{us} - u)t]$. Of course, this reduces to $\exp(-2\sqrt{ust})$ in the usual case, $u \ll s$.

4.9 Suppose that the continuous mapping $p' = f(p)$ of $[a, b]$ into itself has a fixed point $\hat{p} = f(\hat{p})$ in (a, b). Define $x = p - \hat{p}$ and assume that the mapping can be rewritten in the form $x' = \lambda(p)x$, with $-1 < \lambda(p) < 1$ for $a < p < b$. Suppose also that there is no p in (a, b) such that $f(p) = a$ or $f(p) = b$. Prove that $\hat{p}$ is unique and $p(t) \to \hat{p}$ as $t \to \infty$, provided $a < p(0) < b$.

4.10 To see how to approximate geometric rates of convergence to the second order, consider the difference equation

$$x' = \alpha x(1 + \beta x), \qquad |\alpha| < 1 .$$

Suppose $|\alpha|[1 + |\beta x(0)|] < 1$, which is certainly true for sufficiently small $|x(0)|$. Show that

$$x(t) \approx c\alpha^t - \frac{\beta c^2 \alpha^{2t}}{1 - \alpha}$$

as $t \to \infty$, where c is a constant.

4.11 For the fitness patterns displayed below, locate all the equilibria; state whether they are stable, asymptotically stable, or unstable; and, as far as possible with the results in Sect. 4.3, sketch the convergence pattern.

(a)
$$W = \begin{pmatrix} 1 & 2 & 3 \\ 2 & 2 & 1 \\ 3 & 1 & 1 \end{pmatrix} ,$$

(b)
$$W = \begin{pmatrix} 1 & 3 & 2 \\ 3 & 1 & 2 \\ 2 & 2 & 2 \end{pmatrix} ,$$

(c)
$$W = \begin{pmatrix} 1 & 2 & 2 \\ 2 & 1 & 2 \\ 2 & 2 & 1 \end{pmatrix} ,$$

(d)

$$W = \begin{pmatrix} 1 & 2 & 2 \\ 2 & 1 & 1 \\ 2 & 1 & 1 \end{pmatrix} ,$$

(e)

$$W = \begin{pmatrix} 2 & 2 & 3 \\ 2 & 2 & 0 \\ 3 & 0 & 1 \end{pmatrix} .$$

4.12 The alleles A_1 and A_2 are in stable polymorphic equilibrium. A small proportion of A_3 is introduced into the population. If the fitness pattern is

$$W = \begin{pmatrix} 1 & 3 & 1 \\ 3 & 2 & 4 \\ 1 & 4 & . \end{pmatrix} ,$$

will A_3 be established?

4.13 Assume that $w_{11} \geq w_{1i}$ and $w_{1i} + w_{1j} > 2w_{ij}$ for $i, j = 2, \ldots, k$ in (4.25). Show that A_1 is fixed. Biologically interesting special cases are
  (i) no dominance: $w_{ij} = c_i + c_j$, with $c_1 > c_i$ for $i > 1$;
  (ii) the dominance series with A_i dominant to A_j for all $j > i$:
  $w_{ij} = w_{ii} > w_{jj}$.

4.14 Suppose that $w_{ii} = 1 - s_i$, where $s_i \leq 1$ and $s_i \neq 0$ for some i, and $w_{ij} = 1$ for $i \neq j$ and $1 \leq i, j \leq k$. Demonstrate that the completely polymorphic equilibrium exists if and only if $s_i s_j > 0$ for every i and j, in which case it is given by (Wright, 1949a)

$$\hat{p}_i = \frac{1}{s_i} \Big/ \sum_{j=1}^{k} \frac{1}{s_j} .$$

Prove that this equilibrium is globally asymptotically stable if $0 < s_i \leq 1$ for every i and is unstable otherwise.

4.15 Consider the model (4.26) with frequency-dependent fitnesses $w_{ij}(p)$. Assuming that the heterozygote fitness, $w_{12}(p)$, cannot vanish simultaneously with either homozygote fitness, obtain sufficient conditions for a protected polymorphism if the allele A is
  (i) neither completely dominant nor recessive,
  (ii) completely dominant.

4.16 Consider (4.26) with complete dominance and frequency-dependent fitnesses $w_{11}(p) = w_{12}(p)$ and $w_{22}(p)$. Show that at a nontrivial equilibrium ($0 < p < 1$), the two phenotypes have the same fitness. For much more general results of this type, consult Lloyd (1977) and Slatkin (1979, 1979a).

4.17 Analyze completely (4.26) with $w_{11} = w_{12} = 1$ and $w_{22} = 1 + s - aq^2$, where $s > -1$ and $0 < a < 1 + s$. Obtain as explicitly as you can the asymptotic behavior of the gene frequency.

4.18 For the model (4.25), show that (even with frequency and density dependence) at equilibrium the average fitness of the progeny of *each* genotype is the mean fitness of the population.

4.19 Let $\{a_n\}$ be a sequence of real numbers. Show that if $f(x)$ is continuous and all the limits below exist, then the following results hold.

(a) If $f(x)$ is monotone nondecreasing,

$$\overline{\lim_{n \to \infty}} f(a_n) = f\left(\overline{\lim_{n \to \infty}} a_n\right) .$$

(b) If $f(x)$ is monotone nonincreasing,

$$\varliminf_{n \to \infty} f(a_n) = f\left(\varliminf_{n \to \infty} a_n\right) .$$

4.20 Suppose that the mapping $x_{t+1} = f_t(x_t)$, where $t = 1, 2, \ldots$, of $[0, a]$ into itself has the property that there exists $\xi > 0$, independent of t, such that $f_t(x) \geq \min(x, \xi)$. Prove that $x_t \geq \min(x_1, \xi)$ for all t.

4.21 The genotypic frequencies for inbreeding with multiple alleles are [see (9.2)]

$$P_{ij} = p_i F \delta_{ij} + p_i p_j (1 - F) .$$

Use (4.194) to show that if F is constant,

$$\overset{\circ}{\theta}\Delta = -F \sum_i p_i \mu_i \Delta_{ii} .$$

4.22 For weak selection, the discrete model (4.25) may be approximated by the continuous Hardy-Weinberg model (4.171). Assume that $w_{ij} = 1 + s r_{ij}$, where r_{ij} may depend on the gene frequencies, but not on time or the selection intensity s ($s > 0$). Rescale time according to $\tau = st$ and set $\pi_i(\tau) = p_i(t)$ for $t = 0, 1, \ldots$. Deduce from (4.25) that in the limit $s \to 0$

$$\frac{d\pi_i}{d\tau} = \pi_i(r_i - \bar{r}), \qquad r_i = \sum_j r_{ij} \pi_j, \qquad \bar{r} = \sum_i r_i \pi_i .$$

4.23 Extend Sect. 4.5 to the island model of migration by assuming that a variable proportion, m_t, of randomly chosen adults is replaced by $A_2 A_2$ immigrants in each generation. Suppose that there exist α and β such that $0 < \alpha \leq u_t, v_t \leq \beta < \infty$ and $m_t \leq 1 - \alpha$.

(a) Prove that if

$$\overline{\lim_{t \to \infty}} \, \sigma_t^* < 1 \,, \qquad (4.227a)$$

where

$$\sigma_t = v_t/(1 - m_t) \,, \qquad \sigma_t^* = \left(\prod_{\tau=1}^{t} \sigma_\tau \right)^{1/t} \,, \qquad (4.227b)$$

then $p_t \nrightarrow 0$.

(b) By considering a fixed, sufficiently small, positive migration rate, conclude from the example in Sect. 4.5 that with the above hypotheses, $\{p_t\}$ may have a subsequence that converges to zero.

(c) Define for $1 \le \tau \le t$

$$\sigma_{t,\tau}^* = \left(\prod_{n=t-\tau+1}^{t} \sigma_n \right)^{1/\tau} \,; \qquad (4.228a)$$

instead of (4.227), assume that there exist $c < 1$ and T such that

$$\sigma_{t,T}^* \le c \,. \qquad (4.228b)$$

Demonstrate that $\{p_t\}$ has a positive lower bound.

4.24 If we are not interested in deriving the exact result (4.192) but are satisfied with the approximation (4.195), then we can greatly simplify the analysis in Sect. 4.10.

(a) From (4.177), the result that

$$Q_{ij}(t) = O(s) \qquad (4.229)$$

for $t \ge t_1$ [see below (4.155)], and (4.162), show that

$$\dot{\bar{Z}} = \mathrm{Cov}\,(\mu, z) + \bar{\dot{Z}} + O(s^2)\,, \qquad t \ge t_2 \,, \qquad (4.230)$$

where

$$\mathrm{Cov}\,(\mu, z) = 2 \sum_i p_i \mu_i z_i \,. \qquad (4.231)$$

(b) From (4.185) and (4.229) deduce that

$$\mathrm{Cov}\,(\mu, z) = \mathrm{Cov}\,(\mu, \alpha) + O(s^2)\,, \qquad t \ge t_1 \,. \qquad (4.232)$$

Substituting (4.232) into (4.230) proves (4.195).

4.25 The investigation of the evolution of the approximate mean

$$\tilde{Z} = \sum_{ij} Z_{ij} p_i p_j \qquad (4.233)$$

requires only the simple fact (4.299); the more refined result (4.162) is not needed.

(a) Establish that

$$\dot{\tilde{Z}} = \text{Cov}\,(\mu, z) + \dot{\tilde{Z}} + O(s^2), \qquad t \geq t_1 \ . \tag{4.234}$$

(b) Demonstrate that

$$\text{Cov}\,(\mu, z) = \text{Cov}\,(\tilde{\mu}, \tilde{z}) + O(s^2), \qquad t \geq t_1 \ , \tag{4.235}$$

where

$$\tilde{z}_i = \sum_j z_{ij} p_j, \qquad \tilde{z} = \sum_i \tilde{z}_i p_i \ , \tag{4.236}$$

$\tilde{m}_i$ and $\tilde{m}$ are defined similarly, and $\tilde{\mu}_i = \tilde{m}_i - \tilde{m}$.
Inserting (4.235) into (4.234) yields

$$\dot{\tilde{Z}} = \text{Cov}\,(\tilde{\mu}, \tilde{z}) + \dot{\tilde{Z}} + O(s^2), \qquad t \geq t_1 \ , \tag{4.237}$$

which depends only on gene frequencies.

4.26 Suppose that the fitnesses w_{ij} are all constant and positive. Denote the largest w_{ij} by W and the smallest by w, define the selection intensity as $s = (W - w)/w$, and write $D/\overline{w}$ for the last term in (4.56). Prove that $|D| \leq \frac{1}{2}sV_g$. Therefore, the last term in (4.56) is much smaller than the first if $s \ll 1$, i.e., if selection is weak (Nagylaki, 1991).

4.27 Assume that the continuous mapping $p' = f(p)$ of $[a, b]$ into itself is monotone decreasing. Prove the following.

(a) The mapping f has a unique fixed point $\hat{p}$.
(b) The mapping $p' = g(p) \equiv f[f(p)]$ is monotone increasing.
(c) The fixed points of g are $\hat{p}$ and possibly pairs (p^*, p^{**}), $p^* \neq p^{**}$, such that $p^{**} = f(p^*)$ and $p^* = f(p^{**})$.
(d) The fixed points of g are ordered as $\ldots r^* < q^* < \hat{p} < q^{**} < r^{**} \ldots$, where $q^{**} = f(q^*)$, etc.
(e) Either $p_t \to \hat{p}$, or $p_{2t} \to p^*$ and $p_{2t+1} \to p^{**}$ ($t = 0, 1, 2, \ldots$), where $\hat{p}$, p^*, and p^{**} are as in Part c. Thus, $\{p_t\}$ converges either to the fixed point $\hat{p}$ or to a two-point cycle (p^*, p^{**}).

5. Nonrandom Mating

As discussed in Chap. 1, systems of nonrandom mating may depend on relatedness (inbreeding) or phenotype (assortative of disassortative mating). In the former case, the effect is the same on all loci; in the absence of selection, gene frequencies are conserved. Assortation or disassortation may be directly for the trait under investigation or for a character correlated with it owing to common genetic or environmental factors. The effect on each locus, if any, depends on the genetic determination of the phenotype. In linkage equilibrium, loci that do not influence the trait are unaffected. Since some genotypes may be more likely to mate than others, gene frequencies may change even if all matings have the same fertility and all genotypes are equally viable. Inbreeding increases homozygosity and genetic variability; to the extent that genotype and phenotype are correlated, assortative mating will have similar effects.

The variety of models and the diversity of their dynamical behavior are limitless (Karlin, 1968; O'Donald, 1980). We have already considered selection with a fixed inbreeding coefficient in Sect. 4.4. We shall analyze selfing with selection in Sect. 5.1. Our first assortative mating model is equivalent to partial self-fertilization without selection (Sect. 5.2). Assortative mating with complete dominance will be treated in Sect. 5.3. The model of Sect. 5.4 is best interpreted as random mating with differential (nonmultiplicative) fertility. Sections 5.5 and 5.6 are motivated by existing incompatibility systems in plants.

Before analyzing particular models, let us return to the important question of the possible gene-frequency change due to differential probability of mating. In this chapter, we confine ourselves to a single autosomal locus and assume that the sexes need not be distinguished. As usual, we write p_i, P_{ij}, and $X_{ij,kl}$ for the allelic, genotypic, and mating frequencies. If all matings have fertility one and there are no viability differences, then (4.8) informs us that the fertility of A_iA_j is given by

$$P_{ij}b_{ij} = \sum_{kl} X_{ij,kl} \ . \tag{5.1}$$

The right-hand side of (5.1) is the total proportion of matings in which A_iA_j participates as the "first" partner. If this proportion is the same as the frequency of A_iA_j in the population, i.e., if

$$P_{ij} = \sum_{kl} X_{ij,kl} \; , \tag{5.2}$$

then we infer from (5.1) that $b_{ij} = 1$. If all genotypes have the same fertility, (4.17) tells us that the gene frequencies are constant: $p'_i = p_i$. Thus, (5.2) is sufficient for gene-frequency conservation. The necessary condition

$$p_i = \sum_{jkl} X_{ik,lj} \tag{5.3}$$

follows directly from (4.6).

5.1 Selfing with Selection

The offspring from the self-fertilization of $A_i A_j$ are the same as from the cross $A_i A_j \times A_i A_j$. We confine ourselves to two alleles and use the respective numbers x_1, x_2, and x_3 of AA, Aa, and aa zygotes as our basic variables. Since the numbers satisfy linear equations, they are more convenient than the frequencies

$$y_i = x_i \Big/ \sum_j x_j \; . \tag{5.4}$$

With viabilities v_i, the adult numbers are $v_i x_i$. Therefore, with fertilities f_i and fitnesses $w_i = f_i v_i$, the recursion relations read

$$x'_1 = w_1 x_1 + \tfrac{1}{4} w_2 x_2 \; , \tag{5.5a}$$
$$x'_2 = \tfrac{1}{2} w_2 x_2 \; , \tag{5.5b}$$
$$x'_3 = \tfrac{1}{4} w_2 x_2 + w_3 x_3 \; . \tag{5.5c}$$

We rewrite (5.5) in matrix form and suppose henceforth that the fitnesses are constant:

$$\mathbf{x}' = B\mathbf{x} \; , \tag{5.6}$$

where

$$B = \begin{pmatrix} w_1 & \tfrac{1}{4} w_2 & 0 \\ 0 & \tfrac{1}{2} w_2 & 0 \\ 0 & \tfrac{1}{4} w_2 & w_3 \end{pmatrix} \tag{5.7}$$

Since matrix difference equations of the form (5.6) are very common in population genetics, we shall explain how to analyze them in general. Regardless of the number of dimensions (n) and the form of B, (5.6) is immediately iterated to

$$\mathbf{x}(t) = B^t \mathbf{x}(0) \; . \tag{5.8}$$

If B has a complete set of eigenvectors $\mathbf{V}_i$, which it necessarily has if all the eigenvalues are distinct, we have

$$\mathbf{x}(0) = \sum_{j=1}^{n} c_j \mathbf{V}_j \ , \tag{5.9}$$

where the constants c_j are determined from the initial numbers $\mathbf{x}(0)$. Substituting (5.9) into (5.8), we obtain the desired result:

$$\mathbf{x}(t) = \sum_{j=1}^{n} c_j \lambda_j^t \mathbf{V}_j \ , \tag{5.10}$$

where the eigenvalues λ_j of B satisfy

$$B\mathbf{V}_j = \lambda_j \mathbf{V}_j \ . \tag{5.11}$$

Let λ_i, where $i = 1, 2, \ldots, m \le n$, be the eigenvalue(s) of greatest modulus, i.e., $\lambda_i = \mu \omega_i$ for $i \le m$, where μ is real and positive and $|\omega_i| = 1$, and $|\lambda_i| < \mu$ for $i > m$. As $t \to \infty$, (5.10) yields

$$\mathbf{x}(t) \sim \mu^t \sum_{j=1}^{m} c_j \omega_j^t \mathbf{V}_j \ . \tag{5.12}$$

Usually, we shall have $m = 1$, and $x_i(t)$ will decay or grow like μ^t or $(-\mu)^t$. A single wavy line always means that the ratio of the two sides tends to one.

If B does not have a complete set of eigenvectors, let the distinct eigenvalues $\lambda_1, \ldots, \lambda_k$ have multiplicities $r_1, \ldots, r_k$. We can always decompose $\mathbf{x}(0)$ as

$$\mathbf{x}(0) = \sum_{i=1}^{k} \mathbf{U}_i \ , \tag{5.13}$$

where

$$(B - \lambda_i I)^{r_i} \mathbf{U}_i = \mathbf{0}, \qquad i = 1, 2, \ldots k \ , \tag{5.14}$$

and I is the $n \times n$ identity matrix (see, e.g., Brauer and Nohel, 1969, p. 64). Substituting (5.13) into (5.8) and employing (5.14) gives

$$\begin{aligned}
\mathbf{x}(t) &= \sum_{i=1}^{k} B^t \mathbf{U}_i \\
&= \sum_{i=1}^{k} [\lambda_i I + (B - \lambda_i I)]^t \mathbf{U}_i \\
&= \sum_{i=1}^{k} \sum_{j=0}^{t} \binom{t}{j} \lambda_i^{t-j} (B - \lambda_i I)^j \mathbf{U}_i \\
&= \sum_{i=1}^{k} \lambda_i^t \left[\sum_{j=0}^{r_i-1} \binom{t}{j} \lambda_i^{-j} (B - \lambda_i I)^j \right] \mathbf{U}_i
\end{aligned} \tag{5.15}$$

for $t \ge \max_i r_i - 1$. The elements of the matrix in the brackets in (5.15) are polynomials of degree $r_i - 1$. Thus, in general, polynomials may multiply the powers λ_i^t of degenerate eigenvalues. If $r_i = 1$ for all i, then (5.15) reduces to (5.10). As $t \to \infty$, we may restrict the sum over i in (5.15) to the eigenvalue(s) of greatest absolute value.

Having developed the above machinery, we return to (5.6), with B given by (5.7). The eigenvalues are $\lambda_1 = w_1$, $\lambda_2 = \frac{1}{2}w_2$, $\lambda_3 = w_3$; the corresponding eigenvectors read

$$\mathbf{V}_1 = \begin{pmatrix} 1 \\ 0 \\ 0 \end{pmatrix} , \qquad \mathbf{V}_2 = \begin{pmatrix} \lambda_2/[2(\lambda_2 - \lambda_1)] \\ 1 \\ \lambda_2/[2(\lambda_2 - \lambda_3)] \end{pmatrix} , \qquad \mathbf{V}_3 = \begin{pmatrix} 0 \\ 0 \\ 1 \end{pmatrix} ,$$

where $\mathbf{V}_2$ does not exist if $\lambda_2 = \lambda_1$ or $\lambda_2 = \lambda_3$. If $\lambda_2 \neq \lambda_1, \lambda_3$, then (5.10) applies and (5.9) yields

$$c_1 = (\, 1 \quad -\lambda_2/[2(\lambda_2 - \lambda_1)] \quad 0\,)\mathbf{x}(0) \ ,$$
$$c_2 = (\, 0 \quad 1 \quad 0\,)\mathbf{x}(0) \ ,$$
$$c_3 = (\, 0 \quad -\lambda_2/[2(\lambda_2 - \lambda_3)] \quad 1\,)\mathbf{x}(0) \ .$$

Assuming, without loss of generality, that $\lambda_1 \geq \lambda_3$, we have the following seven cases.

1. $\lambda_1 > \lambda_2 > \lambda_3$: From (5.4), (5.10), and the eigenvectors we infer at once that $y_1(t) \to 1$, i.e., AA is fixed at the ultimate rate $(\lambda_2/\lambda_1)^t$.

2. $\lambda_1 > \lambda_3 > \lambda_2$: This is the same as Case 1, except that the asymptotic rate of convergence is $(\lambda_3/\lambda_1)^t$.

3. $\lambda_2 > \lambda_1 \geq \lambda_3$: The genotypic frequencies converge to $b\mathbf{V}_2$, where the normalization constant

$$b = \frac{2(\lambda_2 - \lambda_1)(\lambda_2 - \lambda_3)}{4\lambda_2^2 - 3\lambda_2(\lambda_1 + \lambda_3) + 2\lambda_1\lambda_3} \ .$$

Thus, all three genotypes survive, and the equilibrium frequencies are $y_i(\infty) = bV_{2i}$. The ultimate rate of convergence is $(\lambda_1/\lambda_2)^t$.

4. $\lambda_1 = \lambda_3 > \lambda_2$: The genotype Aa is eliminated at the rate $(\lambda_2/\lambda_1)^t$, which simplifies to $(\frac{1}{2})^t$ in the absence of selection. The ultimate ratio of the frequencies of AA and aa zygotes is c_1/c_3. With no selection, this reduces correctly to $p(0)/q(0)$.

Since the remaining situations have either $\lambda_2 = \lambda_1$ or $\lambda_2 = \lambda_3$ or both, we must use (5.15) or return to the original system (5.5). In view of the simplicity of (5.5), we shall do the latter. From (5.5b) we obtain

$$x_2(t) = x_2(0)\lambda_2^t \ , \tag{5.16}$$

whence (5.5c) gives

$$x_3' - \lambda_3 x_3 = \tfrac{1}{2}x_2(0)\lambda_2^{t+1} \ . \tag{5.17}$$

This is an inhomogenous linear difference equation. It will be useful to examine linear equations of arbitrary order and dimension.

Suppose L is a linear operator and the vector $\mathbf{z}(t)$ satisfies

$$L\mathbf{z}(t) = \mathbf{g}(t) , \qquad (5.18)$$

where $\mathbf{g}(t)$ is known. Examples in one dimension in discrete and continuous time are

$$Lz(t) = z(t+3) - t^3 z(t+2) + 7z(t) , \qquad (5.19a)$$

$$Lz(t) = 2\frac{d^2 z}{dt^2} + e^t \frac{dz}{dt} - z . \qquad (5.19b)$$

In more than one dimension, the left-hand side of (5.18) would be a vector with components like (5.19). Assume we know a particular solution, $\mathbf{z}_p(t)$, of (5.18) and the general solution, $\mathbf{z}_h(t)$, of the homogeneous equation

$$L\mathbf{z}_h(t) = \mathbf{0} . \qquad (5.20)$$

If $\mathbf{z}(t)$ is any solution (5.18), we have

$$L(\mathbf{z} - \mathbf{z}_p) = L\mathbf{z} - L\mathbf{z}_p = \mathbf{g} - \mathbf{g} = \mathbf{0} . \qquad (5.21)$$

Comparing (5.21) with (5.20), we conclude

$$\mathbf{z}(t) = \mathbf{z}_h(t) + \mathbf{z}_p(t) . \qquad (5.22)$$

Returning to (5.17), we try the particular solution $\alpha\lambda_2^t$ and obtain

$$\alpha = \frac{\lambda_2 x_2(0)}{2(\lambda_2 - \lambda_3)} , \qquad \lambda_2 \neq \lambda_3 .$$

For $\lambda_2 = \lambda_3$, motivated by (5.15), we substitute $\beta t \lambda_2^t$ and find

$$\beta = \tfrac{1}{2} x_2(0) , \qquad \lambda_2 = \lambda_3 .$$

The general solution of the homogeneous equation is obviously $\gamma\lambda_3^t$. Adding this to the particular solution and determining γ from the initial condition $x_3(0)$, we arrive at

$$x_3(t) = \begin{cases} \alpha\lambda_2^t + [x_3(0) - \alpha]\lambda_3^t , & \lambda_2 \neq \lambda_3 , \\ [\beta t + x_3(0)]\lambda_2^t , & \lambda_2 = \lambda_3 . \end{cases} \qquad (5.23)$$

From (5.5a) and (5.23) we deduce by symmetry

$$x_1(t) = \begin{cases} \overline{\alpha}\lambda_2^t + [x_1(0) - \overline{\alpha}]\lambda_1^t , & \lambda_2 \neq \lambda_1 , \\ [\beta t + x_1(0)]\lambda_2^t , & \lambda_2 = \lambda_1 , \end{cases} \qquad (5.24)$$

where

$$\overline{\alpha} = \frac{\lambda_2 x_2(0)}{2(\lambda_2 - \lambda_1)}, \qquad \lambda_2 \neq \lambda_1 .$$

The remaining results follow from (5.16), (5.23), and (5.24).

5. $\lambda_1 > \lambda_2 = \lambda_3$: As in Cases 1 and 2, AA is ultimately fixed, i.e., $y_1(\infty) = 1$, but the asymptotic rate of convergence is now $t(\lambda_2/\lambda_1)^t$. Heterozygotes, however, disappear faster, at the geometric rate $(\lambda_2/\lambda_1)^t$.

6. $\lambda_1 = \lambda_2 > \lambda_3$: Again $y_1(t) \to 1$, but now Aa and aa disappear at the slow algebraic rate $1/t$.

7. $\lambda_1 = \lambda_2 = \lambda_3$: Heterozygotes are lost at the rate $1/t$, and $y_1(\infty) = y_3(\infty) = \frac{1}{2}$.

The approach presented here is closest to that of Karlin (1968), who reviews some of the literature.

5.2 Assortative Mating with Multiple Alleles and Distinguishable Genotypes

We suppose that all unordered genotypes have a different phenotype. A fraction r $(0 \leq r \leq 1)$ of each unordered genotype chooses a mate of its own genotype, whereas the remainder mates at random. Clearly, this model applies equally to a population with a proportion r of selfing and $1 - r$ of random mating. We assume that all genotypes have the same fitness.

In our usual notation for ordered genotypic and mating frequencies,

$$P'_{ij} = \sum_{kl} X_{ik,lj} . \tag{5.25}$$

The contribution to $X_{ik,lj}$ from the panmictic subpopulation is $(1-r)P_{ik}P_{lj}$. In the assortatively mating subpopulation, $A_iA_i \times A_iA_i$ homozygote matings contribute rP_{ii}. The frequency of assortative unordered $A_iA_k \times A_iA_k$ $(i \neq k)$ heterozygote matings is $2P_{ik}$. Inserting a factor of $\frac{1}{2}$ to order each genotype, we obtain the contribution $r(\frac{1}{2})^2(2P_{ik}) = \frac{1}{2}rP_{ik}$. Appropriate Kronecker deltas enable us to collect these frequencies in the single formula

$$X_{ik,lj} = (1-r)P_{ik}P_{lj} + r[\delta_{ik}\delta_{lj}\delta_{il}P_{ii} + (1 - \delta_{ik})(\delta_{il}\delta_{kj} + \delta_{ij}\delta_{kl})\tfrac{1}{2}P_{ik}]$$
$$= (1-r)P_{ik}P_{lj} + \tfrac{1}{2}r(\delta_{il}\delta_{kj} + \delta_{ij}\delta_{kl})P_{ik} . \tag{5.26}$$

It is easy to see that (5.26) satisfies (5.2), so we know that the gene frequencies, p_i, will be conserved. Substituting (5.26) into (5.25) yields the recursion relations

$$P'_{ij} = (1-r)p_ip_j + \tfrac{1}{2}r(P_{ij} + \delta_{ij}p_i) , \tag{5.27}$$

which lead at once to $p'_i = p_i$. With $p_i = p_i(0)$, the solution of (5.27) is trivial. At equilibrium, the genotypic proportions are

$$\hat{P}_{ij} = \frac{2(1-r)p_i p_j + r p_i \delta_{ij}}{2-r} \ , \tag{5.28}$$

and the time dependence is given by

$$P_{ij}(t) = \hat{P}_{ij} + [P_{ij}(0) - \hat{P}_{ij}](\tfrac{1}{2}r)^t \ . \tag{5.29}$$

Using $Q_{ij}(t) = P_{ij}(t) - p_i p_j$ as indices of deviation from Hardy-Weinberg proportions, we find from (5.28) the excess of homozygotes

$$\hat{Q}_{ii} = \frac{r p_i (1 - p_i)}{2-r} > 0$$

and the deficiency of heterozygotes

$$\hat{Q}_{ij} = -\frac{r p_i p_j}{2-r} < 0, \qquad i \neq j \ .$$

From (5.28) and (5.29) we derive the heterozygosity

$$h(t) = \sum_{ij,\ i \neq j} P_{ij} = \hat{h} + [h(0) - \hat{h}](\tfrac{1}{2}r)^t \ ,$$

where the equilibrium heterozygosity reads (Kimura, 1963)

$$\hat{h} = \left[\frac{2(1-r)}{2-r} \right] \sum_{ij,\ i \neq j} p_i p_j \ . \tag{5.30}$$

Haldane (1924a) obtained the special case of our results for two alleles. If the population is initially in Hardy-Weinberg proportions, then the sum in (5.30) is just $h(0)$ and (5.30) reduces to a result of Wright (1921). If mating is random ($r = 0$), then (5.28) and (5.29) yield $P_{ij}(t) = \hat{P}_{ij} = p_i p_j$ for $t \geq 1$, as expected. For complete assortative mating ($r = 1$), $\hat{P}_{ij} = p_i \delta_{ij}$, i.e., heterozygotes are eliminated; the rate of convergence is $(\tfrac{1}{2})^t$. For two alleles, these results were derived in Sect. 5.1 (Case 4).

5.3 Assortative Mating with Two Alleles and Complete Dominance

Here A, with frequency p, is completely dominant to a, which has frequency q. A proportion α of dominant phenotypes and β of recessive ones mate assortatively, whereas the remainder mate at random. The assortative and random matings have fertilities 1 and $b = 1 - s$, respectively, where $s < 1$. The model with $\alpha = \beta$ and $s = 0$ was formulated and analyzed by O'Donald (1960). Scudo and Karlin (1969) generalized it to $\alpha \neq \beta$ and Crow and Kimura (1970, pp. 162–164) to $\alpha \neq \beta$ and $s \neq 0$. We put $P = P_{11}$, $Q = P_{12}$, and $R = P_{22}$, so that the frequency of dominants is

$$D = P + 2Q = 1 - R \ ,$$

and
$$p = P + Q , \qquad q = Q + R .$$

The proportions of the three subpopulations and the frequency of A within each will appear as follows:

1. Assortative recessive, with proportion βR, has $p_1 = 0$.

2. Assortative dominant, with proportion αD, has gene frequency

$$p_2 = (P + Q)/D = p/D .$$

3. Random mating, has frequencies

$$T = 1 - \beta R - \alpha D ,$$
$$p_3 = (1 - \alpha)(P + Q)/T = (1 - \alpha)p/T .$$

The genotypic proportions in the next generation are sums of contributions from each of the subpopulations, which mate at random within themselves. Hence,

$$XP' = \alpha D p_2^2 + bT p_3^2 , \qquad (5.31a)$$
$$XQ' = \alpha D p_2 q_2 + bT p_3 q_3 , \qquad (5.31b)$$
$$XR' = \beta R + \alpha D q_2^2 + bT q_3^2 . \qquad (5.31c)$$

We introduced X in order to normalize the genotypic frequencies in the next generation. From (5.31) and the requirement $P' + 2Q' + R' = 1$ we derive

$$X = 1 - sT . \qquad (5.32)$$

As anticipated, X is the mean fertility of the population. Of course,

$$q_2 = 1 - p_2 = Q/D ,$$
$$q_3 = 1 - p_3 = [(1 - \beta)R + (1 - \alpha)Q]/T .$$

Substituting for p_i and q_i, we reduce (5.31) to

$$XP' = p^2 \left\{ \frac{\alpha}{D} + \frac{b(1 - \alpha)^2}{T} \right\} , \qquad (5.33a)$$
$$XQ' = p \left\{ \frac{\alpha Q}{D} + \frac{b(1 - \alpha)[(1 - \beta)R + (1 - \alpha)Q]}{T} \right\} , \qquad (5.33b)$$
$$XR' = \beta R + \frac{\alpha Q^2}{D} + \frac{b[(1 - \beta)R + (1 - \alpha)Q]^2}{T} . \qquad (5.33c)$$

Calculating
$$Xp' = X(P' + Q') = p[1 - s(1 - \alpha)]$$

produces the gene-frequency change

$$\Delta p = s(\alpha - \beta)pR/X . \qquad (5.34)$$

It suffices to consider $s \geq 0$; the conclusions with $s < 0$ will clearly be the reverse of those with $s > 0$. We have the following situations.

1. $s > 0$

(a) $\alpha > \beta$

Evidently, $\Delta p \geq 0$. For $p > 0$, $\Delta p = 0$ if and only if $R = 0$. But (5.33c) informs us that $R = 0$ implies $R' = 0$ if and only if $Q = 0$, in which case $p = 1$. Therefore, $\Delta p > 0$ for $0 < p < 1$, so that $p(t) \to 1$ as $t \to \infty$. Obviously, this means $P(t) \to 1$. The reason A is ultimately fixed is that assortative matings are more fertile than random ones and dominants mate more assortatively than recessives. Hence, dominants are favored over recessives.

(b) $\alpha < \beta$

Manifestly, by the argument just presented, $p(t) \to 0$.

(c) $\alpha = \beta$

In view of the biological argument above, it is hardly surprising that with equal assortative mating, p is constant. We focus our attention on (5.33a); noting that $D = 2p - P$, $T = 1 - \alpha$, and $X = 1 - s(1 - \alpha)$, we simplify it to the linear fractional transformation

$$P' = \frac{p^2}{1 - s(1 - \alpha)} \left[\frac{\alpha}{2p - P} + (1 - s)(1 - \alpha) \right] \equiv f(P) . \tag{5.35}$$

Since $Q = p - P \geq 0$ implies $P \leq p$, and $R = q - Q = 1 - 2p + P \geq 0$ implies $P \geq 2p - 1$, therefore (5.35) is a mapping of the interval $[P^*, p]$ into itself, where $P^* = \max(0, 2p - 1)$. It is easy to verify for $0 < p < 1$ that $f(P^*) > P^*$, and $f(p) < p$ unless $\alpha = 1$. From the theory of the linear fractional transformation developed in Sect. 2.2, it follows immediately that for $0 \leq \alpha < 1$, $P(t)$ converges globally to the unique root in (P^*, p) of the quadratic $f(P) = P$. (In fact, we have Case 2a of Sect. 2.2.)

If $\alpha = 1$, (5.35) reduces to

$$P' = p^2/(2p - P) . \tag{5.36}$$

With the substitution $x = 1/(p - P)$, we easily derive

$$P(t) = p - \frac{p[p - P(0)]}{p + [p - P(0)]t} ,$$

which shows that $P(t) \to p$, $Q(t) \to 0$, and $R(t) \to q$ at the algebraic rate $1/t$. Heterozygotes are eliminated because assortative mating is complete.

2. $s = 0$

With no fertility differences, (5.34) tells us that p is constant and (5.32) yields $X = 1$. Therefore, (5.33a) becomes

$$P' = p^2 \left[\frac{\alpha}{2p - P} + \frac{(1-\alpha)^2}{1 - \beta + (\beta - \alpha)(2p - P)} \right] \equiv g(P) \ . \qquad (5.37)$$

Again, $g(P)$ is a mapping of $[P^*, p]$ into itself. For $0 < p < 1$, we have $g(P^*) > P^*$, and $g(p) < p$ unless $\alpha = 1$ or $\beta = 1$. If $\alpha = 1$ or $\beta = 1$, (5.37) reduces to (5.36). Therefore, we assume $\alpha, \beta < 1$ and exclude also the trivial case of pure panmixia, $\alpha = \beta = 0$. From (5.37) we compute

$$\frac{dg}{dP} = p^2 \left[\frac{\alpha}{(2p - P)^2} + \frac{(\beta - \alpha)(1-\alpha)^2}{[1 - \beta + (\beta - \alpha)(2p - P)]^2} \right] \ , \qquad (5.38)$$

$$\frac{d^2 g}{dP^2} = 2p^2 \left[\frac{\alpha}{(2p - P)^3} + \frac{(\beta - \alpha)^2 (1-\alpha)^2}{[1 - \beta + (\beta - \alpha)(2p - P)]^3} \right] > 0 \ . \qquad (5.39)$$

The convexity displayed by (5.39) implies

$$\frac{dg}{dP}(P) \geq \frac{dg}{dP}(P^*) \ ,$$

with equality only at $P = P^*$.

We shall show that $g(P)$ is increasing by proving that

$$\frac{dg}{dP}(P^*) \geq 0 \ .$$

(a) $p \leq \frac{1}{2}$

Here $P^* = 0$, and (5.38) yields

$$\frac{dg}{dP}(0) = \frac{\alpha}{4} + \frac{p^2(\beta - \alpha)(1-\alpha)^2}{[1 - \beta + 2p(\beta - \alpha)]^2} \ . \qquad (5.40)$$

If $\beta \geq \alpha$, obviously

$$\frac{dg}{dP}(0) > 0 \ .$$

If $\beta < \alpha$, since $p \leq \frac{1}{2}$, from (5.40) we obtain

$$\frac{dg}{dP}(0) \geq \frac{\alpha}{4} + \frac{(\frac{1}{2})^2(\beta - \alpha)(1-\alpha)^2}{[1 - \beta + 2(\frac{1}{2})(\beta - \alpha)^2]} = \frac{\beta}{4} \geq 0 \ .$$

(b) $p > \frac{1}{2}$

Now $P^* = 2p - 1$, and (5.38) gives

$$\frac{dg}{dP}(2p - 1) = \beta p^2 \geq 0 \ .$$

Collecting our results enables us to sketch $g(P)$ in Fig. 5.1, which shows immediately that $P(t)$ converges globally, without oscillation, to $\hat{P}$, the unique root in (P^*, p) of the cubic $g(P) = P$. Since mating is assortative, we expect an excess of homozygotes relative to Hardy-Weinberg proportions. According to Problem 5.3, $\hat{P} > p^2$. If $\alpha = \beta$, (5.37) reduces to the special case of (5.35) with $s = 0$.

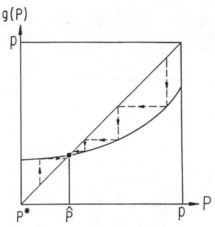

Fig. 5.1. The iteration of (5.37).

Finally, we shall show that the departure from Hardy-Weinberg proportions is approximately linear in α and β for $\alpha, \beta \ll 1$. Expanding (5.37) leads to

$$\frac{g(P)}{p^2} = 1 + \frac{(1 - 2p + P)[\alpha(1 - 2p + P) + \beta(2p - P)]}{2p - P} + O(\alpha^2, \alpha\beta, \beta^2) \ .$$

(5.41)

Therefore,

$$\hat{P} = g(\hat{P}) = p^2 + O(\alpha, \beta) \ .$$

(5.42)

Substituting (5.42) into (5.41) gives

$$\frac{\hat{P}}{p^2} = 1 + \frac{q^2[\alpha q^2 + \beta p(1 + q)]}{p(1 + q)} + O(\alpha^2, \alpha\beta, \beta^2) \ .$$

(5.43)

For sufficiently small α and β, (5.43) proves that $\hat{P} > p^2$.

5.4 Random Mating with Differential Fertility

Again, A is dominant to a; p, q, P, Q, R, and D are as in the previous section. Mating is random, but dominant $\times$ recessive matings have fertility $b = 1 - s$, where $0 < s < 1$, instead of unity. The difference equations are most easily written by collecting progeny in Table 5.1.

Table 5.1. Random mating with differential fertility

Mating	Contribution	Progeny		
		AA	Aa	aa
$AA \times AA$	P^2	1	0	0
$AA \times Aa$	$2P(2Q)$	$\frac{1}{2}$	$\frac{1}{2}$	0
$AA \times aa$	$2PRb$	0	1	0
$Aa \times Aa$	$(2Q)^2$	$\frac{1}{4}$	$\frac{1}{2}$	$\frac{1}{4}$
$Aa \times aa$	$2(2Q)Rb$	0	$\frac{1}{2}$	$\frac{1}{2}$
$aa \times aa$	R^2	0	0	1

The total contribution is just the mean fecundity,

$$X = 1 - 2sRD \ . \tag{5.44}$$

The recursion relations

$$XP' = P^2 + 2PQ + Q^2 \ ,$$
$$XQ' = PQ + PRb + Q^2 + QRb \ ,$$
$$XR' = Q^2 + 2QRb + R^2$$

simplify directly to

$$XP' = p^2 \ , \tag{5.45a}$$
$$XQ' = p(q - sR) \ , \tag{5.45b}$$
$$XR' = q^2 - 2sQR \ . \tag{5.45c}$$

From (5.45a) and (5.45b) we deduce

$$Xp' = p(1 - sR) \ . \tag{5.46}$$

We employ p and R as our two independent variables and begin by locating the *equilibria*. If $p' = p$, (5.46) yields $p = 0$, which implies $R = 1$, or

$$X = 1 - sR \ . \tag{5.47}$$

Equating (5.44) and (5.47) immediately gives $R = 0$ or $\hat{D} = \frac{1}{2}$. At equilibrium, setting $R = 0$ in (5.45c) informs us that $q = 0$. So, this is just the other trivial equilibrium. To find the gene frequency at the polymorphic equilibrium, we substitute $\hat{D} = \frac{1}{2}$ into (5.45a):

$$(1 - \tfrac{1}{2}s)(2p - \tfrac{1}{2}) = p^2 \ .$$

This leads to

$$\hat{p} = \tfrac{1}{2}\left[2 - s - \sqrt{(2 - s)(1 - s)}\right] \ . \tag{5.48a}$$

The root with a positive sign in front of the radical is trivially seen to exceed $\frac{1}{2}$, which is impossible because $\hat{p} \le \hat{D} = \frac{1}{2}$. Note that

$$\hat{q} = \frac{1}{2}\left[s + \sqrt{(2 - s)(1 - s)}\right] \ . \tag{5.48b}$$

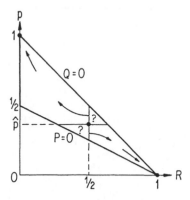

Fig. 5.2. The iteration of (5.45).

We now turn to the *evolution* of the population. Since

$$P = p - Q = 2p - 1 + R \ge 0 \tag{5.49a}$$

and

$$Q = q - R = 1 - p - R \ge 0 \ , \tag{5.49b}$$

the system is confined to the interior of the triangle with the heavily drawn boundaries in Fig. 5.2. Dominant and recessive individuals have respective probabilities R and D of mating disassortatively. Hence, an individual with the more (less) frequent phenotype has a lower (higher) probability of mating disassortatively. Since disassortative matings are less fertile, the more frequent phenotype is favored. Therefore, we expect the two trivial equilibria to be stable and the polymorphic one to be unstable. Establishing the convergence pattern displayed in Fig. 5.2 proves considerably more than this. If the population starts in either of the two regions marked with a question mark, the analysis below does not determine its dynamics. In fact, the regions of attraction of the two stable equilibria are separated by a curve that passes through the unstable equilibrium and the two small triangles, and populations that start on this separatrix converge to the unstable equilibrium (Boucher, 1991).

We wish to prove that if $p(0) > \hat{p}$ and $R(0) \le \frac{1}{2}$, then $p(t) \to 1$ as $t \to \infty$. If $R = 0$ and $p < 1$, then (5.45c) implies that $R' > 0$, so we may assume that $R(0) > 0$. From (5.44) and (5.46) we have

$$p' = \frac{p(1 - sR)}{1 - 2sR(1 - R)} \ .$$

This shows that $p' \gtreqless p$ for $R \lesseqgtr \frac{1}{2}$. Hence, it suffices to demonstrate that $p > \hat{p}$ and $R \leq \frac{1}{2}$ imply $R' < \frac{1}{2}$, for then the population cannot enter the upper triangle with the question mark. We rewrite (5.45c) in the form

$$f(q, R) \equiv X(R' - \tfrac{1}{2}) = q^2 - \tfrac{1}{2} + sR(1 + R - 2q) \ . \tag{5.50}$$

Now,

$$\frac{\partial f}{\partial R} = s(1 + R - 2q) + sR = s(P + R) > 0$$

from (5.49a). Therefore, $f(q, R)$ is monotone increasing in R for fixed q. In the region under consideration, $R \leq \min(q, \frac{1}{2})$.

If $q \leq \frac{1}{2}$,

$$f(q, R) \leq f(q, q) = (1 - s)q^2 + sq - \tfrac{1}{2} < 0$$

for $0 \leq q \leq \frac{1}{2}$. If $q > \frac{1}{2}$,

$$f(q, R) \leq f(q, \tfrac{1}{2}) = (q - \hat{q})(q + \hat{q} - s) \tag{5.51}$$

from (5.48b). But

$$q + \hat{q} - s > \tfrac{1}{2} + \hat{q} - s > 0 \ ,$$

in which the positivity follows from the trivial calculation that rules out a positive sign in front of the radical in (5.48a). Consequently, for $\frac{1}{2} < q < \hat{q}$ we have established that $f(q, R) < 0$. Together with the same result for $q \leq \frac{1}{2}$, this proves that $R' < \frac{1}{2}$ in the region of interest, as required.

To study the dynamics in the quadrilateral on the lower right in Fig. 5.2, we suppose that $p(0) < \hat{p}$ and $R(0) \geq \frac{1}{2}$. From the argument for the upper pentagon; we need to show only that $p < \hat{p}$ and $R \geq \frac{1}{2}$ imply $R' > \frac{1}{2}$. For $R \geq \frac{1}{2}$, we have (5.51) with the inequality reversed. Since $q > \hat{q}$, (5.51) becomes

$$f(q, R) > (q - \hat{q})(2\hat{q} - s) > 0 \ ,$$

where the positivity is ensured by (5.48b). Therefore, $R' > \frac{1}{2}$ and the proof is complete.

Since with $R(0) = \frac{1}{2}$, $p(t)$ tends to zero or one according as $p(0) < \hat{p}$ or $p(0) > \hat{p}$, it is clear that $R = \frac{1}{2}$ does not separate the domains of attraction of the two stable equilibria. Therefore, one cannot take completely literally the intuitive argument with which we commenced our dynamical investigation.

We owe the model discussed in this section to Scudo and Karlin (1969); our interpretation is that of Crow and Kimura (1970, pp. 164–166). Down to (5.50), our analysis followed Karlin (1968).

Let us end this section by calculating the ultimate *rates of convergence* to the pure states.

1. $p(\infty) = 0$

Since $D(\infty) = 0$, we use D instead of R. On account of (5.44), to first order in p and D (5.46) becomes

$$p' \sim (1 - s)p \ ,$$

whence

$$p(t) \sim k(1-s)^t \tag{5.52}$$

as $t \to \infty$, for some constant k. To first order, we deduce from (5.45c)

$$D' \sim 2(1-s)p .$$

Substituting (5.52) yields

$$D(t) \sim 2(1-s)p(t-1) \sim 2p(t) \tag{5.53}$$

as $t \to \infty$. Thus, the dominant allele, A, is lost at a geometric rate.

2. $p(\infty) = 1$

The convenient variables now are q and R. From (5.46) and (5.45c) we obtain

$$q' = q - sR + O(qR, R^2) , \tag{5.54a}$$
$$R' = q^2 - 2sR(q - R) + O(q^2 R, qR^2, R^3) . \tag{5.54b}$$

On the basis of (5.54b), we expect $R \sim q^2$ as $t \to \infty$. Hence, (5.54b) yields

$$R = q^2 + O(q^3)$$

as $t \to \infty$, so that (5.54a) implies

$$q' = q(1 - sq) + O(q^3) .$$

With the method applied to (4.39), we obtain, (see Problem 5.4)

$$q(t) = \frac{1}{st} + O\left(\frac{\ln t}{t^2}\right) \tag{5.55a}$$

as $t \to \infty$. Therefore, $R \sim q^2$ yields

$$R(t) = \frac{1}{s^2 t^2} + O\left(\frac{\ln t}{t^3}\right) . \tag{5.55b}$$

It is easy to check that (5.55) satisfies (5.54) with the errors indicated. As in Sect. 4.2, the recessive allele is lost at an algebraic rate.

5.5 Self-Incompatibility Alleles

Self-incompatibility systems in plants usually increase genetic variability by preventing self-fertilization. Many flowering plants possess a self-incompatibility locus such that pollen will not function on a style that carries the allele in the pollen grain. This mechanism evidently favors rare alleles. The extensive literature on the prediction of the expected number of alleles present in terms of the population size and the mutation rate is briefly reviewed in Nagylaki (1975a). In this section, we shall concentrate on one component of this interesting and difficult problem: the deterministic behavior of the system (Nagylaki, 1975a).

We denote the frequencies of the n self-incompatibility alleles S_i by p_i and those of the ordered genotypes S_iS_j by P_{ij}. Since S_i and S_j pollen will not fertilize an S_iS_j plant, there are no homozygotes: $P_{ii} = 0$. Clearly, $n \geq 3$. Let us suppose that the proportion of S_iS_j genotypes among fertilized plants is P_{ij}. This means that no failures of fertilization are to be ascribed to the different allelic frequencies in pollen, a reasonable assumption if there is excess pollen. The frequency of S_k, where $k \neq i,j$, among pollen that can fertilize S_iS_j is $p_k/(1-p_i-p_j)$. Positing random pollination, we obtain Fisher's (1958) equation:

$$2P'_{ij} = \sum_{k \neq i,j} \frac{P_{ik}p_j}{1 - p_i - p_k} + \sum_{k \neq i,j} \frac{P_{kj}p_i}{1 - p_k - p_j} \, , \qquad (5.56)$$

for $i \neq j$. Since there are no homozygotes, obviously $p_i \leq \frac{1}{2}$ for all i. Therefore, $n \geq 3$ implies that $p_i + p_j < 1$ for all i and j. Thus, the denominators in (5.56) are positive. From (5.56) we deduce the gene-frequency equation

$$2p'_i = p_i + \sum_{j \neq i} \sum_{k \neq i,j} \frac{P_{kj}p_i}{1 - p_k - p_j} \, . \qquad (5.57)$$

Another summation verifies that the normalization is correct.

It is easy to check that the symmetry point

$$\hat{P}_{ij} = 1/[n(n - 1)], \qquad \hat{p}_i = 1/n \qquad (5.58)$$

is an equilibrium of (5.56). Indeed, W. Boucher (1991a) has recently proved that (5.58) is the only completely polymorphic equilibrium of (5.56) and that it is globally asymptotically stable, as suggested by the following results.

For three alleles, and only for three alleles, (5.56) is linear and a trivial calculation (left for Problem 5.5) gives

$$P_{ij}(t) = \tfrac{1}{6} + [P_{ij}(0) - \tfrac{1}{6}](-\tfrac{1}{2})^t \, . \qquad (5.59)$$

The allelic frequencies can be computed from

$$p_k(t) = \tfrac{1}{2} - P_{ij}(t), \qquad k \neq i,j \, . \qquad (5.60)$$

Equations (5.59) and (5.60) show that the genotypic and allelic frequencies converge globally to the symmetry point and oscillate at the rate $(-\frac{1}{2})^t$ while they do so.

We shall now demonstrate that the symmetric equilibrium (5.58) is asymptotically stable. Consider the transformation

$$\mathbf{x}' = \mathbf{f}(\mathbf{x}) \qquad (5.61)$$

of a set of variables x_i. Suppose $\hat{\mathbf{x}}$ is an equilibrium:

$$\hat{\mathbf{x}} = \mathbf{f}(\hat{\mathbf{x}}) \, . \qquad (5.62)$$

Near $\mathbf{x} = \hat{\mathbf{x}}$, (5.61) has the Taylor expansion

$$x_i' = f_i(\hat{\mathbf{x}}) + \sum_j \frac{\partial f_i}{\partial x_j}(\hat{\mathbf{x}})(x_j - \hat{x}_j) + O(|\mathbf{x} - \hat{\mathbf{x}}|^2) \ . \tag{5.63}$$

Defining $\boldsymbol{\varepsilon} = \mathbf{x} - \hat{\mathbf{x}}$ and

$$b_{ij} = \frac{\partial f_i}{\partial x_j}(\hat{\mathbf{x}}) \tag{5.64a}$$

and substituting (5.62) into (5.63) produces the linearized equation

$$\boldsymbol{\varepsilon}' = B\boldsymbol{\varepsilon} \ . \tag{5.64b}$$

Glancing at (5.15), we expect $\hat{\mathbf{x}}$ to be asymptotically stable if the absolute value of every eigenvalue of B is less than unity. This assertion can be justified rigorously (Bellman, 1949; LaSalle, 1977).

To apply the above procedure to our problem, put $(i \neq j)$

$$P_{ij} = \hat{P}_{ij} + \varepsilon_{ij} , \qquad p_i = \hat{p}_i + \eta_i \ , \tag{5.65}$$

with

$$\eta_i = \sum_{j \neq i} \varepsilon_{ij} , \qquad \sum_i \eta_i = 0 \ . \tag{5.66}$$

Substituting (5.65) and (5.66) into (5.56), we find the linarized equation

$$\varepsilon_{ij}' = -\mu_n \varepsilon_{ij} + \left(\frac{\eta_i + \eta_j}{n-2} \right) \left[1 - \frac{1}{(n-2)(n-1)} \right] , \tag{5.67}$$

where

$$\mu_n = 1/(n-2) \ . \tag{5.68a}$$

From (5.66) and (5.67) we derive $\eta_i' = \lambda_n \eta_i$, with

$$\lambda_n = 1 - \frac{n}{(n-2)(n-1)} \ . \tag{5.68b}$$

Therefore,

$$\eta_i(t) = \eta_i(0)\lambda_n^t \ . \tag{5.69}$$

Inserting (5.69) into (5.67) yields a difference equation of the same form as (5.17); its solution reads

$$\varepsilon_{ij}(t) = \mu_n[\eta_i(0) + \eta_j(0)]\lambda_n^t + \{\varepsilon_{ij}(0) - \mu_n[\eta_i(0) + \eta_j(0)]\}(-\mu_n)^t \ . \tag{5.70}$$

It must be kept in mind that (5.70) is a good approximation if and only if $|\varepsilon_{ij}(0)| \ll \hat{P}_{ij}$ for all i and j. The case of three alleles is an exception: since (5.56) is linear, therefore (5.70) is exact, and it is easily seen to reduce to (5.59). Hence, (5.68) and (5.70) show that $\varepsilon_{ij}(t) \to 0$ as $t \to \infty$.

The nature of the convergence to equilibrium is rather interesting. For $n = 3$, both the genotypic and the gene frequencies oscillate at the rate $(-\frac{1}{2})^t$. If there are four alleles, $\mu_4 = \frac{1}{2} > \frac{1}{3} = \lambda_4$, so (5.69) and (5.70) inform us that ultimately the genotypic frequencies oscillate, but the gene frequencies do not. With $n \geq 5$, we have $\lambda_n > \mu_n$, which gives nonoscillatory asymptotic convergence at the rate λ_n^t. In the biologically important case of many alleles ($n \gg 1$), $\lambda_n \approx 1 - n^{-1}$ and the rate of convergence is very close to $e^{-t/n}$, which has a characteristic time equal to the number of alleles.

We shall briefly consider the fate of rare alleles. From (5.57) it can be proved that (Nagylaki, 1975a)

$$\Delta p_i > p_i[-p_i(2 - p_i) + (1 - p_i)^3(n - 2 + p_i)^{-1}] .$$

The bracket is positive at $p_i = 0$, and its smallest positive root is

$$q = 1 + a - (1 + a^2)^{1/2} ,$$

where $a = [2(n - 1)]^{-1}$. Consequently, $\Delta p_i > 0$ for $0 < p_i < q$, which shows that rare alleles are established. Note that $q > 1/(2n)$, but $q \approx 1/(2n)$ if $n \gg 1$. The condition $p_i < q$ is sufficient but not necessary for $\Delta p_i > 0$. Numerical counterexamples for four alleles dispose of the naive conjecture that $\Delta p_i > 0$ for $p_i < 1/n$ (Nagylaki, 1975a).

5.6 Pollen and Zygote Elimination

As our final example, we shall examine some disassortative mating schemes proposed for plants by Finney (1952). He assumed that only crosses of unlike phenotypes could produce offspring and formulated two models for random pollination. These differ in the specification of the phenotypic mating frequencies.

Let r_i and R_{ij} ($R_{ii} = 0$) represent the frequencies of the phenotype π_i and the ordered union $\pi_i \times \pi_j$. The proportion of pollen from π_i plants among pollen that can fertilize π_j is $r_i/(1 - r_j)$. If there are no failures of fertilization due to the phenotypic origin of the pollen, the frequency of π_j phenotypes among fertilized plants is r_j. This should be a good approximation with excess pollen. Thus, the *pollen elimination* model is based on the ordered mating frequencies

$$R_{ij} = \tfrac{1}{2}r_j[r_i/(1-r_j)] + \tfrac{1}{2}r_i[r_j/(1-r_i)] = \tfrac{1}{2}r_ir_j[(1-r_i)^{-1} + (1-r_j)^{-1}] \quad (5.71)$$

for $i \neq j$. It is easy to verify that (5.71) is correctly normalized. Note that the reasoning used to deduce (5.56) for incompatibility determined by the plant genotype and the allele in the pollen corresponds to pollen elimination.

If pollen is relatively scarce, we suppose that seeds from incompatible crosses do not germinate. Then the ordered mating frequencies read

$$R_{ij} = r_ir_j/X , \qquad i \neq j , \tag{5.72a}$$

where

$$X = \sum_{ij} R_{ij} = 1 - \sum_i r_i^2 . \tag{5.72b}$$

Following Finney (1952), we shall call (5.72) *zygote elimination*.

With only two phenotypes, since there is but one fertile mating, pollen and zygote elimination are equivalent.

We proceed to analyze (5.71) and (5.72) for a single diallelic locus. It will be convenient to use P, H, and R for the respective frequencies of AA, Aa, and aa; the heterozygote frequency, $H = 2P_{12}$, is unordered here.

1. Complete Dominance

Suppose A is dominant to a. As explained above, pollen and zygote elimination are the same. The permitted crosses, $AA \times aa$ and $Aa \times aa$, have frequencies proportional to P and H, respectively. Therefore, our recursion relations are

$$TP' = 0 \ , \tag{5.73a}$$

$$TH' = P + \tfrac{1}{2}H \ , \tag{5.73b}$$

$$TR' = \tfrac{1}{2}H \ , \tag{5.73c}$$

where normalization requires

$$T = P + H \ . \tag{5.74}$$

From (5.73a), we have $P(t) = 0$ for $t \geq 1$. Of course, the elimination of dominant homozygotes in one generation is quite obvious. Hence, (5.74) gives $T' = H'$. Substituting this into (5.73b) informs us that $H'H'' = \tfrac{1}{2}H'$. But (5.73b) shows also that $H' = 0$ only if $R = 1$, which leads to extinction of the population in one generation, so $H'' = \tfrac{1}{2}$. Thus, the equilibrium $R = H = \tfrac{1}{2}$ is attained in two generations (Finney, 1952).

2. Identical Homozygotes

Again, there is only one scheme. Now the fertile unions, $AA \times Aa$ and $aa \times Aa$, have frequencies proportional to P and R, respectively; this yields the difference equations

$$TP' = \tfrac{1}{2}P \ , \tag{5.75a}$$

$$TH' = \tfrac{1}{2}P + \tfrac{1}{2}R \ , \tag{5.75b}$$

$$TR' = \tfrac{1}{2}R \ , \tag{5.75c}$$

where

$$T = P + R \ . \tag{5.76}$$

Adding (5.75a) and (5.75c) and excluding $H = 1$ to prevent extinction in a single generation, we obtain $P' + R' = \tfrac{1}{2}$, i.e., $H(t) = \tfrac{1}{2}$ for $t \geq 1$. Dividing (5.75a) by (5.75c) shows that $P(t)/R(t)$ is constant, whence we derive for $t \geq 1$

$$P(t) = \tfrac{1}{2}P(0)/[P(0) + R(0)], \qquad R(t) = \tfrac{1}{2}R(0)/[P(0) + R(0)] \ . \tag{5.77}$$

This equilibrium is reached in just one generation, but depends on the initial conditions (Finney, 1952).

3. Pollen Elimination with Three Distinct Phenotypes

If the three genotypes have distinguishable phenotypes, pollen and zygote elimination lead to different models with quite dissimilar dynamical behavior. The permitted matings are $AA \times Aa$, $AA \times aa$, and $Aa \times aa$. The equations for pollen elimination follow directly from (5.71) and Mendelian segregation:

$$P' = \tfrac{1}{2}PH[(1-P)^{-1} + (1-H)^{-1}] \ ,$$

$$H' = \tfrac{1}{2}PH[(1-P)^{-1} + (1-H)^{-1}] + PR[(1-P)^{-1} + (1-R)^{-1}]$$
$$+ \tfrac{1}{2}HR[(1-H)^{-1} + (1-R)^{-1}] \ ,$$

$$R' = \tfrac{1}{2}HR[(1-H)^{-1} + (1-R)^{-1}] \ .$$

This system reduces to

$$P' = \frac{PH(1+R)}{2(1-P)(1-H)} \ , \tag{5.78a}$$

$$H' = \tfrac{1}{2} + \frac{PR(1+H)}{2(1-P)(1-R)} \ , \tag{5.78b}$$

$$R' = \frac{HR(1+P)}{2(1-H)(1-R)} \ . \tag{5.78c}$$

We cannot allow any genotypic frequency to be unity because the population would be extinguished in the next generation. We conclude at once from (5.78b) that the frequency of heterozygotes will be at least $\tfrac{1}{2}$ after one generation. Hence, for $R \neq 0$ the ratio of (5.78a) to (5.78c) gives

$$\frac{P'}{R'} = \frac{P(1-R^2)}{R(1-P^2)} \ . \tag{5.79}$$

Let us locate the equilibria of (5.78). If $R \neq 0$, (5.79) implies $\hat{P} = \hat{R}$ or $P = 0$. Assuming that $P \neq 0$ and substituting $\hat{P} = \hat{R} = \tfrac{1}{2}(1-\hat{H})$ into (5.78b) at equilibrium, we deduce

$$\hat{H}^2 + 3\hat{H} - 2 = 0 \ ,$$

whence

$$\hat{H} = (\sqrt{17} - 3)/2 \approx 0.562 \ . \tag{5.80}$$

From (5.78a) and (5.78b) we infer that $P = 0$, $H = \tfrac{1}{2}$ (from which $R = \tfrac{1}{2}$) is an equilibrium. Since (5.78) is symmetric under the interchange $P \leftrightarrow R$, it is not surprising that with $R = 0$ we find the corresponding equilibrium: $H = \tfrac{1}{2}$, $P = \tfrac{1}{2}$. The three equilibria are shown in Fig. 5.3.

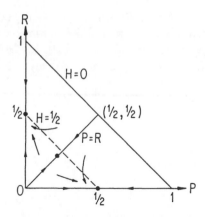

Fig. 5.3. The iteration of (5.78).

We continue with the demonstration of the convergence pattern displayed in Fig. 5.3. If $P = 0$, then $P' = 0$ and $H' = \frac{1}{2}$; if $R = 0$, then $R' = 0$ and $H' = \frac{1}{2}$. Henceforth, we may suppose that $P(0) > 0$ and $R(0) > 0$. We may also assume that $H(0) > 0$, for otherwise the population will be extinguished in the second generation. Therefore, we may analyze (5.78) for $P, H, R > 0$.

If $P < R$, we infer from (5.79) that $P'/R' < P/R$. Suppose $P(0) < R(0)$. Since the monotone decreasing sequence $\{P(t)/R(t)\}$ is bounded below by 0, it must have a limit, say α, where $0 \leq \alpha < 1$. If $\alpha > 0$, (5.79) yields

$$1 = \lim_{t \to \infty} \frac{1 - [R(t)]^2}{1 - [P(t)]^2} = \lim_{t \to \infty} \frac{1 - [R(t)]^2}{1 - \alpha^2 [R(t)]^2} \ . \tag{5.81}$$

Hence, the sequence $\{R(t)\}$ has a limit, $R(\infty)$. In view of the equilibrium structure, $R(\infty) = 0$ would imply $P(\infty) = \frac{1}{2} > R(\infty)$. Thus, $R(\infty) > 0$, and (5.81) is impossible. Therefore, $\alpha = 0$ and $P(\infty) = 0$. Since $H(t) \geq \frac{1}{2}$ for $t \geq 1$, $R(t)$ cannot tend to 1. Then (5.78b) informs us that $H(t) \to \frac{1}{2}$. We conclude from (5.79) that P/R converges to 0 at the asymptotic rate $(\frac{3}{4})^t$.

If $P(0) > R(0)$, the same argument applied to R/P establishes that $P(\infty) = H(\infty) = \frac{1}{2}$ and R/P converges to 0 at the ultimate rate $(\frac{3}{4})^t$.

If $P = R$, then (5.79) gives $P' = R'$. Therefore, if $P(0) = R(0)$, we have $P(t) = R(t)$ and (5.78b) simplifies to

$$H' = \frac{1}{2} + \frac{(1 - H)^2}{2(1 + H)} \equiv f(H), \qquad 0 \leq H \leq 1 \ . \tag{5.82}$$

We exclude $H(0) = 0, 1$ to prevent extinction. Observe that there is no H in $(0, 1)$ such that $f(H) = 0$ or $f(H) = 1$. We note that $f(0) = 1$ and $f(1) = \frac{1}{2}$, and easily verify that $f(H)$ is monotone decreasing for $0 \leq H < 1$ and convex for $0 \leq H \leq 1$. Therefore, the mapping (5.82) has the form depicted in Fig. 5.4, which has the crucial feature

$$H < f(H) < 2\hat{H} - H, \qquad 0 < H < \hat{H}, \tag{5.83a}$$
$$2\hat{H} - H < f(H) < H, \qquad \hat{H} < H < 1. \tag{5.83b}$$

Subtracting $\hat{H}$ from (5.83), we obtain

$$|f(H) - \hat{H}| < |H - \hat{H}|, \qquad 0 < H < 1.$$

It follows from Problem 4.9 that $H(t) \to \hat{H}$ globally. Clearly, the convergence is oscillatory. From (5.64), (5.80), and (5.82) we deduce the ultimate rate of convergence λ^t, where

$$\lambda = \frac{df}{dH}(\hat{H}) = -\frac{1}{\sqrt{17} - 1} \approx -0.320.$$

Of course, in practice, small perturbations would drive the population into the region of attraction of one of the two stable equilibria.

The equilibrium analysis of this model is due to Finney (1952). Our proof of convergence to the stable equilibria follows Karlin and Feldman (1968).

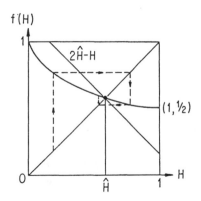

Fig. 5.4. The iteration of (5.82).

4. Zygote Elimination with Three Distinct Phenotypes

The fertile crosses are the same as in the pollen elimination model above, but now we write our equations by specializing (5.72):

$$XP' = PH, \tag{5.84a}$$
$$XH' = PH + 2PR + HR, \tag{5.84b}$$
$$XR' = HR, \tag{5.84c}$$

with

$$X = 1 - P^2 - H^2 - R^2. \tag{5.85}$$

If $R = 0$, (5.84c) informs us that $R' = 0$, so we obtain from (5.84a) and (5.84b) $P' = H' = \frac{1}{2}$. Thus, with $R(0) = 0$, the population reaches the equilibrium $P = H = \frac{1}{2}$ in one generation. The initial condition $H(0) = 0$ leads to extinction in two generations. Hence, we assume that $R(0) > 0$ and $H(0) > 0$, whence $R(t) > 0$ and $H(t) > 0$. Dividing (5.84a) by (5.84c) immediately shows that

$$P(t)/R(t) = P(0)/R(0) \equiv k \ . \tag{5.86}$$

We derive an equation for $x = R/H$ by inserting (5.86) into the ratio of (5.84c) to (5.84b), with the result

$$x' = (1 + k + 2kx)^{-1}, \qquad 0 < x < \infty \ . \tag{5.87}$$

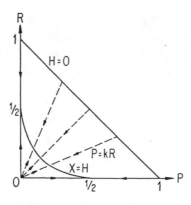

Fig. 5.5. The iteration of (5.84).

From Sect. 2.2 we see easily that (5.87) is Case 2a of the linear fractional transformation. Then (2.28) and (2.31) imply global oscillating convergence to

$$\hat{x} = (4k)^{-1}[-(1 + k) + (1 + 10k + k^2)^{1/2}]$$

at the geometric rate μ^t, where

$$\mu = -(4k)^{-1}[1 + 6k + k^2 - (1 + k)(1 + 10k + k^2)^{1/2}] \ .$$

Combining (5.86) with the definition of x, we get

$$\hat{R} = \hat{x}/[1 + (1 + k)\hat{x}] \ .$$

We note from (5.84c) that the equilibria lie on the curve $X = H$, which, with the aid of (5.85), is revealed as the ellipse

$$2P^2 + 2PR + 2R^2 - 3P - 3R + 1 = 0 \ .$$

We owe these results, displayed in Fig. 5.5, to Finney (1952) and Karlin and Feldman (1968). Solutions were also presented by Cannings (1968) and Falk and Li (1969).

5.7 Problems

5.1 Consider a diallelic plant population, with the usual notation, viz., $p = p_1, q = 1 - p$, $P = P_{11}$, $Q = P_{12}$, and $R = P_{22}$. Assume that all AA plants in the infinitely large population are fertilized at random and the allele a causes selfing, so that only a fraction α of Aa plants and a fraction γ of aa plants is pollinated at random. To allow for the possibility that selfing plants make a smaller contribution of randomly fertilizing pollen than nonselfers, let κ and λ be the respective factors for Aa and aa selfers by which this contribution is reduced compared with cross-pollinators. Derive the recursion relations for P, Q, and R. Show that

$$2T\Delta q = P(\mu Q + \nu R) + (\alpha\nu - \gamma\mu)QR \; ,$$

where $\mu = (1 - \alpha)\kappa$, $\nu = (1 - \gamma)\lambda$, and

$$T = P + 2(\alpha + \mu)Q + (\gamma + \nu)R \; .$$

Deduce that if the selfing gene, a, is initially rare, it is established in the population (i.e., q increases) if and only if at least one of the selfing genotypes contributes to cross-fertilization. Notice that if $\alpha\nu > \gamma\mu$, the cross-pollinating allele is eventually lost (Nagylaki, 1976a).

5.2 Solve the difference equations

(a) $z' - \lambda z = \alpha t$, $\quad z(0) = z_0$;

(b) $z'' - \lambda z = \mu^t$, $\quad z(0) = z_0$, $\quad z(1) = z_1$.

5.3 Show that the unique equilibrium of (5.37) satisfies $\hat{P} > p^2$ for $0 < p < 1$ and $0 \le \alpha, \beta < 1$, unless $\alpha = \beta = 0$.

5.4 Derive (5.55a).

5.5 Derive (5.59), (5.67), and (5.70).

5.6 Let P, $2Q$, and R be the respective frequencies of AA, Aa, and aa in a monoecious, diallelic, single-locus model. Homozygotes mate only with homozygotes, but mate at random within the homozygous population, and heterozygotes mate only with heterozygotes. Homozygous crosses have fertility one, but heterozygous ones have fertility $1 + s$, with $s > 0$.

(a) Write the recursion relations for P, Q, and R.
(b) Locate all the equilibria.
(c) Prove that the Hardy-Weinberg equilibrium $P = Q = R = \frac{1}{4}$ is globally asymptotically stable.

5.7 Take $s = 0$ in Problem 5.6.

(a) Write the recursion relations for P, Q, and R.
(b) Show that the frequency, p, of A in the entire population is conserved (i.e., constant in time).
(c) Demonstrate that for each p, Q converges globally to the unique equilibrium $(q = 1 - p)$

$$\hat{Q}(p) = \tfrac{1}{8}[3 - (9 - 32pq)^{1/2}] \ .$$

(d) Establish that the convergence is oscillatory unless $p = \tfrac{1}{2}$.
(e) Prove that at equilibrium there is a deficiency of heterozygotes relative to panmixia with allelic frequency p: $\hat{Q}(p) \leq pq$, in which equality holds if and only if $p = \tfrac{1}{2}$.

5.8 Let P, $2Q$, and R be the respective frequencies of AA, Aa, and aa in a monoecious, diallelic, single-locus model. Mating is random, but homozygote $\times$ heterozygote crosses are sterile.

(a) Write the recursion relations for P, Q, and R.
(b) Locate all the equilibria.
(c) Show that after one generation there cannot be a deficiency of homozygotes compared with Hardy-Weinberg proportions.
(d) Prove that
   (i) $P(0) = R(0)$ implies $P(t) = Q(t) = R(t) = \tfrac{1}{4}$ for $t \geq 1$;
   (ii) $P(0) > R(0)$ implies $P(t) \to 1$ as $t \to \infty$;
   (iii) $P(0) < R(0)$ implies $R(t) \to 1$ as $t \to \infty$.

5.9 The continuous mapping $x' = f(x)$ of the interval I: $[a, b]$ into itself is called a *contraction* if the *Lipschitz condition*

$$|f(x_1) - f(x_2)| \leq L|x_1 - x_2| \tag{5.88}$$

is satisfied for all x_1 and x_2 in I with *Lipschitz constant* $L < 1$.

(a) Show that $x(t)$ converges globally to the unique equilibrium $\hat{x}$ that satisfies $\hat{x} = f(\hat{x})$.
(b) Establish that

$$|x(t) - \hat{x}| \leq L^t(1 - L)^{-1}|x(1) - x(0)| \ .$$

(c) Prove that if $f(x)$ is differentiable, then (5.88) holds if and only if

$$\left| \frac{df}{dx}(x) \right| \leq L$$

for all x in I.

5.10 Assume that the continuously differentiable mapping $x' = f(x)$ of $[a, b]$ into itself has the following properties.

(i) There exists a fixed point $\hat{x}$ in (a, b) such that $\text{sgn}\,[f(x) - x] = -\text{sgn}\,(x - \hat{x})$.

(ii) $\dfrac{df}{dx}(x) > -1$ for all x in (a, b).

(iii) There is no x in (a, b) such that $f(x) = a$ or $f(x) = b$.

Use (5.83) on $[a, b]$ to show that if $a < x(0) < b$, then $x(t) \to \hat{x}$ as $t \to \infty$.

5.11 Suppose that the mapping $x' = f(x)$ of $[a, b]$ into itself has a nonpositive second derivative on $[a, b]$ and there exists an isolated fixed point $\hat{x}$ in (a, b). Prove that assumption (i) of Problem 5.10 is satisfied. Hence, if (ii) and (iii) also hold, $a < x(0) < b$ implies $x(t) \to \hat{x}$.

5.12 Consider the mapping $x' = \lambda(x)x$ near $x = 0$. Let $\lambda_0 = \lambda(0)$ and $\lambda(x)/\lambda_0 = 1 + \mu(x)$; assume that $\lambda(x)$ is continuous, $\lambda_0 \neq 0$, $-1 < \lambda_0 < 1$, and $\mu(x) \doteq O(|\ln|x||^{-\alpha})$ as $x \to 0$, for some $\alpha > 1$. Show that $x(t) \sim c\lambda_0^t$ as $t \to \infty$, for some $c \neq 0$. This result includes the usual case: $\mu(x) = O(x)$ as $x \to 0$.

5.13 To calculate slow rates of convergence, examine $\Delta x = -f(x)$ near $x = 0$. Posit that $f(x) \neq 0$ and $\text{sgn}\,f(x) = \text{sgn}\,x$ for $x \neq 0$, and df/dx is continuous and equal to zero at $x = 0$. The last condition excludes the geometric convergence of Problem 5.12.

(a) Prove that $x(t) \to 0$ monotonically as $t \to \infty$, so that it suffices to consider $x > 0$.

(b) Demonstrate that

$$\int_{x(t)}^{\xi} \frac{dy}{f(y)} \sim t \qquad (5.89)$$

as $t \to \infty$, where ξ is a constant. This allows derivation of the asymptotic behavior of $x(t)$. Observe that (5.89) establishes that, under our hypotheses, the asymptotic form of $x(t)$ may be derived from the approximation $dx/dt \approx -f(x)$.

(c) Show that if $f(x) \sim g(x)$ as $x \to 0+$, then f may be replaced by g in (5.89). Thus, the asymptotic behavior of $x(t)$ as $t \to \infty$ depends only on that of $f(x)$ as $x \to 0+$. This greatly facilitates the calculation of the former.

(d) The following examples indicate that the faster $f(x) \to 0$ as $x \to 0$, the slower $x(t) \to 0$ as $t \to \infty$. Assuming that $b, c, \alpha > 0$ and $\beta > 1$, prove that the given behavior of $f(x)$ as $x \to 0+$ implies the stated behavior of $x(t)$ as $t \to \infty$:

$$f(x) \sim cx(-\ln x)^{-\alpha}, \qquad \ln x(t) \sim -[c(\alpha+1)t]^{1/(\alpha+1)}\ ; \quad (5.90a)$$

$$f(x) \sim cx^{\beta}, \qquad x(t) \sim [c(\beta-1)t]^{-1/(\beta-1)}\ ; \quad (5.90b)$$

$$f(x) \sim ce^{-bx^{-\alpha}}, \qquad x(t) \sim (b/\ln t)^{1/\alpha}\ . \quad (5.90c)$$

6. Migration and Selection

Since many natural populations are divided into subpopulations, it is important to investigate the effects of the joint action of selection and migration. We shall suppose that the population occupies a number of distinct niches, each of which has its own selection pattern. The niches may be defined by any pertinent set of environmental variables. If a population is distributed in clusters, this scheme provides a reasonable model for geographical variation. The study of the spatial variation of gene frequencies, however, requires more advanced analysis; see Nagylaki (1989c) and references therein. In this chapter, we shall confine ourselves to the easier problem of the maintenance of genetic diversity. For time-dependent environments, we examined this question in Sect. 4.5 by introducing the concept of protected polymorphism. We shall use the same approach here to expound multi-niche selection.

We shall restrict our treatment to a single autosomal locus and assume that the sexes need not be distinguished. Only genotype-independent migration will be treated. Selection operates entirely through viability differences. Most of our examples will concern the diallelic case. In Sect. 6.1, we shall give a complete analysis of the island model. The general formulation and analysis of Sect. 6.2 will be applied to the Levene (1953) model and to the two-niche situation in Sects. 6.3 and 6.4.

6.1 The Island Model

We designate the frequencies of the allele A_i in zygotes and adults on our figurative island by p_i and $p_i^\star$, respectively. The corresponding population sizes are N and $N^\star$. We assume that a proportion α of adults is removed from the island either by mortality or emigration and a fraction β of migrants with constant allelic frequencies $\overline{p}_i$ is added. If the migration mechanism is exchange with a metaphoric continent, its population must exceed that of the island sufficiently to allow us to neglect the change in $\overline{p}_i$. Clearly, the number of A_i genes in zygotes in the next generation is given by

$$n_i' = [(1 - \alpha)p_i^\star + \beta\,\overline{p}_i](2N^\star) \ ,$$

whence the number of zygotes in the next generation reads

$$N' = \tfrac{1}{2} \sum_i n'_i = (1 - \alpha + \beta)N^\star \ .$$

Therefore,

$$p'_i = n'_i/(2N') = (1 - m)p^*_i + m\bar{p}_i \ , \tag{6.1}$$

where

$$m = \beta/(1 - \alpha + \beta) \ .$$

Evidently, m is the fraction of zygotes with immigrant parents. If the influx and efflux of individuals are equal, then $\alpha = \beta = m$, the proportion of adults replaced per generation. We owe this special case of the model to Haldane (1930) and Wright (1931). Since efflux does not alter gene frequencies, the fact that with the above reinterpretation their formulation always applies is hardly surprising.

As explained in Sect. 2.3, (6.1) is a particular case of the mutation-selection model (4.95), (4.97).

To complete the model, we posit random mating on the island after migration. The only effect of reproduction here is to return the population to Hardy-Weinberg proportions. Then

$$p^*_i = p_i w_i/\bar{w} \ , \tag{6.2}$$

the fitnesses being specified by the Hardy-Weinberg formulae (4.25b). It is clear from (6.1) that no allele carried to the island by immigrants can be lost. Therefore, we shall focus our attention on the diallelic case with $\bar{p} = 0$, i.e., the immigrants are all aa.

Combining (6.1) and (6.2), we obtain the transformation

$$p' = (1 - m)pw_1/\bar{w} \equiv f(p) \ . \tag{6.3}$$

Since $f(1) = 1 - m < 1$ for $m > 0$, therefore, in agreement with the remark in the previous paragraph, a cannot be lost. As $p \to 0$, (6.3) yields

$$f(p) \sim [(1 - m)w_{12}/w_{22}]p \ . \tag{6.4}$$

The allele A will be protected if the above bracket exceeds unity, i.e., if

$$m < 1 - \frac{w_{22}}{w_{12}} \ . \tag{6.5}$$

Obviously, $p = 0$ is an equilibrium. If there is no other equilibrium, then $f(1) < 1$ implies that $p' = f(p) < p$ for $0 < p \leq 1$, so A will be lost.

We shall now present a complete analysis of (6.3). The equilibria with $p \neq 0$ satisfy the quadratic

$$\bar{w} = (1 - m)w_1 \ . \tag{6.6}$$

Since we are interested in the maintenance of genetic variability by migration and selection, we suppose that AA is favored over the immigrant genotype,

aa, and, for simplicity, exclude overdominance and underdominance in the remainder of this section. By suitable scaling, we can parametrize the fitnesses as

$$w_{11} = 1 + s, \qquad w_{12} = 1 + hs, \qquad w_{22} = 1 - s , \qquad (6.7)$$

where s $(0 < s \le 1)$ represents the intensity of selection and h $(-1 \le h \le 1)$ determines the degree of dominance. We take $0 < m < 1$.

From (4.25b) and (6.7) we derive

$$w_1 = 1 + hs + s(1-h)p , \qquad (6.8a)$$
$$\overline{w} = 1 - s + 2s(1+h)p - 2hsp^2 . \qquad (6.8b)$$

Substituting (6.8) into (6.6) produces the quadratic *equilibrium* equation

$$2hp^2 - [1 + 3h + m(1-h)]p + [1 + h(1-m) - \nu] = 0 ,$$

where $\nu = m/s \ge m$. Therefore, the equilibria are

$$p_\pm = (4h)^{-1}\{1 + 3h + m(1-h) \pm [(1-h)^2(1+m)^2 + 8h(\nu+m)]^{1/2}\} . \quad (6.9)$$

As $h \to 0$, (6.9) gives the correct limit,

$$p_- = (1-\nu)/(1+m), \qquad h = 0 ,$$

for the case without dominance. The equilibria depend on three parameters, h, m, and ν, the selection intensity s being absorbed into the migration-selection ratio ν. Of course, we require that p_+ and p_- be real and $0 \le p_\pm \le 1$. To find out what parameter ranges yield acceptable equilibria, we impose these conditions on (6.9). After some elementary but tedious calculation, the following results emerge (Problem 6.1).

Define $\mu = \nu^{-1} = s/m \le m^{-1}$, the strength of selection relative to that of migration, and

$$h_0 = -(1+m)/(3-m) , \qquad (6.10)$$
$$\mu_1 = [1 + h(1-m)]^{-1} , \qquad (6.11a)$$
$$\mu_2 = -8h[(1-h)^2(1+m)^2 + 8hm]^{-1} . \qquad (6.11b)$$

Observe that as m increases from 0 to 1, h_0 decreases from $-\frac{1}{3}$ to -1. Hence, for $0 < m < 1$, if there is no dominance ($h = 0$) or A is dominant ($h = 1$), then $h > h_0$; if A is recessive ($h = -1$), then $h < h_0$. We fix h and m and study p_+ and p_- as functions of μ; we display the results in Fig. 6.1. We remark that

$$p_-(\mu_2) = (4h)^{-1}[1 + 3h + m(1-h)] , \qquad (6.12a)$$
$$p_-(m^{-1}) = (4h)^{-1}\{1 + 3h + m(1-h)$$
$$- [(1-h)^2(1+m)^2 + 16hm]^{1/2}\} . \qquad (6.12b)$$

For weak selection and migration, $s, m \ll 1$, the correct approximation is obtained in (6.9) to (6.12) by setting $m = 0$ wherever it appears explicitly.

It remains to justify the *stability* behavior indicated in Fig. 6.1. The mapping (6.3) is equivalent to that for mutation and selection with mutation rates $u_{12} = m$ and $u_{21} = 0$ and fitnesses w_{ij}. Hence, nonoscillatory convergence to some equilibrium point for all initial conditions is guaranteed by Problem 4.6. There are three cases.

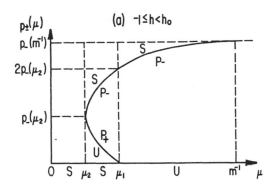

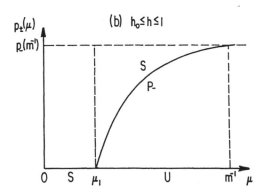

Fig. 6.1. The equilibrium and stability structure of (6.3).

1. $\mu \le \mu_2$ and $-1 \le h < h_0$, or $\mu \le \mu_1$ and $h_0 \le h \le 1$

By the comment below (6.5), since there is no internal equilibrium in these two situations, therefore $p(t) \to 0$ as $t \to \infty$, i.e., A is ultimately lost. [In the degenerate case $\mu = \mu_2$ and $-1 \le h < h_0$, $p_+(\mu_2) = p_-(\mu_2)$ is an internal equilibrium, but if $p(0) < p_\pm(\mu_2)$, then $p(t) \to 0$; if $p(0) > p_\pm(\mu_2)$, then $p(t) \to p_\pm(\mu_2)$ and a slight perturbation to $p < p_\pm(\mu_2)$ will cause loss of A.]

From (6.4), the ultimate rate of loss of A reads

$$[(1 - m)w_{12}/w_{22}]^t \ . \tag{6.13}$$

2. $\mu_2 < \mu \leq \mu_1$ and $-1 \leq h < h_0$

Since there are three equilibria, 0, p_+, and p_-, with $0 < p_+ < p_-$, clearly we have $f(p) > p$ if and only if $p_+ < p < p_-$. Then Problem 4.6 informs us that

$$p(t) \to 0 \quad \text{if} \quad p(0) < p_+ \ , \tag{6.14}$$

$$p(t) \to p_- \quad \text{if} \quad p(0) > p_+ \ . \tag{6.15}$$

The rate of convergence in (6.14) is, of course, again (6.13). For (6.15), we apply (5.64), Problem 4.5, and (6.6) to deduce the rate λ^t, where

$$\lambda = \frac{df}{dp}(p_-) \tag{6.16}$$

$$= (1 - m)[\overline{w}(p_-)]^{-2}[q_- w_{22} w_1(p_-) + p_- w_{11} w_2(p_-)] \tag{6.17}$$

$$= [w_{22} q_-^2 + w_{11} p_-(q_- - m)]/[q_- \overline{w}(p_-)] \ . \tag{6.18}$$

3. $\mu > \mu_1$

Since 0 and p_- are the only equilibria, evidently $f(p) > p$ if and only if $0 < p < p_-$, whence, by Problem 4.6, $p(t) \to p_-$. The asymptotic rate of approach to equilibrium is again given by (6.18).

For the much easier analysis with weak migration and selection, see Haldane (1930) and Nagylaki (1975b).

Several qualitative conclusions can be drawn from Fig. 6.1. For the existence of a polymorphism, the intensity of selection compared with that of migration, μ, must exceed a critical value (μ_1 or μ_2), which is usually of the order of unity. For most parameter values, A is absent or has a high frequency; it is rare only if $h_0 \leq h \leq 1$ and μ slightly exceeds μ_1. If $-1 \leq h < h_0$, i.e., A is partially or completely recessive, as conditions for maintaining A become unfavorable in the sense that $\mu \to \mu_2+$, the equilibrium gene frequency $p_-(\mu) \to p_-(\mu_2) > 0$. One might naively have expected A to disappear, $p_-(\mu) \to 0$, as indeed it does for $h_0 \leq h \leq 1$. In sum, the comments of this paragraph suggest that, unlike mutation and selection, migration and selection may not maintain alleles at low frequencies at many loci, thereby possibly limiting response to environmental change.

6.2 General Analysis

We assume that the population occupies a number of distinct niches, each of which has its own selection scheme. To describe the migration among these colonies, we employ the forward and backward migration patterns, first introduced by Malécot (1948) for the study of migration and random drift in the absence of selection. The *forward* and *backward migration matrices* (Malécot, 1950; Bodmer and Cavalli-Sforza, 1968), $\tilde{M}$ and M, were first applied to the migration-selection problem by Bulmer (1972). Their elements, $\tilde{m}_{ij}$ and m_{ij}, may be functions of our sundry variables and parameters. We suppose that adults migrate (we discuss alternative hypotheses below) and define $\tilde{m}_{ij}$ and m_{ji} as follows. The probability that an adult in niche i migrates to niche j is $\tilde{m}_{ij}$. Since every niche is assumed to have a very large population, $\tilde{m}_{ij}$ represents the fraction of adults that migrate from niche i to niche j in each generation. We designate by m_{ji} the probability that an adult in niche j migrated from niche i. The migration matrices satisfy the normalization conditions

$$\sum_j \tilde{m}_{ij} = 1, \qquad \sum_i m_{ji} = 1 . \tag{6.19}$$

The life cycle is selection (possibly including population regulation), migration, population regulation (if present at this stage), and reproduction. If it is the zygotic rather than the mating adult population that is regulated, we may interchange the last two steps without modifying our equations. Thus, the population may be controlled once, twice, or not at all in one generation. The respective proportions of zygotes, pre-migration adults, post-migration adults, and post-regulation adults in niche i are c_i, $c_i^\star$, $c_i^{\star\star}$, and c_i'. We denote the frequency of the allele A_j in niche i at the corresponding stages by $p_{i,j}$, $p_{i,j}^\star$, $p_{i,j}'$, and $p_{i,j}'$, respectively. We summarize this information in the following scheme.

Zygote $\longrightarrow$ Adult $\longrightarrow$ Adult $\longrightarrow$ Adult $\longrightarrow$ Zygote

selection   migration   regulation   reproduction

$c_i p_{i,j}$ $c_i^\star p_{i,j}^\star$ $c_i^{\star\star} p_{i,j}'$ $c_i' p_{i,j}'$ $c_i' p_{i,j}'$

In this formulation, reproduction serves only to return each subpopulation to Hardy-Weinberg proportions through random mating in each niche.

Since no individuals are lost in migration, we have

$$c_j^{\star\star} = \sum_i c_i^\star \tilde{m}_{ij} , \tag{6.20a}$$

$$c_i^\star = \sum_j c_j^{\star\star} m_{ji} . \tag{6.20b}$$

Expressing the joint probability that an adult is in niche i and migrates to niche j in terms of the prospective and retrospective conditional probabilities, $\tilde{m}_{ij}$ and m_{ji}, yields

$$c_i^\star \tilde{m}_{ij} = c_j^{\star\star} m_{ji} . \tag{6.21}$$

The appropriate summations and use of (6.19) in (6.21) recapture (6.20). Inserting (6.20a) into (6.21), we deduce the desired connection between the forward and backward migration matrices:

$$m_{ji} = c_i^\star \tilde{m}_{ij} \bigg/ \sum_k c_k^\star \tilde{m}_{kj} . \tag{6.22}$$

We designate the fitness of $A_j A_k$ individuals in niche i by $w_{i,jk}$. Therefore,

$$p_{i,j}^\star = p_{i,j} w_{i,j} / \overline{w}_i \ , \tag{6.23a}$$

where the allelic fitnesses and the mean fitness in niche i read

$$w_{i,j} = \sum_k w_{i,jk} p_{i,k} \ , \tag{6.23b}$$

$$\overline{w}_i = \sum_{jk} w_{i,jk} p_{i,j} p_{i,k} \ . \tag{6.23c}$$

The gene frequencies in the next generation may be calculated directly from the backward migration matrix:

$$p_{i,j}' = \sum_k m_{ik} p_{k,j}^\star \ . \tag{6.24}$$

Observe that if we consider the flow of genes, it is the forward migration matrix that enters:

$$c_i^{\star\star} p_{i,j}' = \sum_k c_k^\star p_{k,j}^\star \tilde{m}_{ki} \ . \tag{6.25}$$

Since (6.21) immediately reduces (6.25) to (6.24), we shall employ the simpler equation (6.24).

If we know the fitnesses, $w_{i,jk}$, and the distribution of the immigrants into each panmictic subpopulation (or *deme*), M, then (6.23) and (6.24) completely determine the model. But if, instead of M, we are given the distribution of the emigrants from each deme, $\tilde{M}$, to have a well-posed problem in (6.23) and (6.24), we must compute M from (6.22). This requires an *Ansatz* for $\mathbf{c}^\star$ in terms of $\mathbf{c}$ and a hypothesis for the variation, if any, of $\mathbf{c}$.

One must specify the adult distribution before migration, $\mathbf{c}^\star$, in accordance with the nature of selection. The two possible extreme assumptions were lucidly described by Dempster (1955). If the fraction of adults in every niche is fixed, a mode now usually called *soft selection* (Wallace, 1968), we have $c_i^\star = c_i$. As Dempster (1955) observed, this should be a good approximation if the population is regulated within each niche. If it is the total population size that is controlled, Dempster pointed out that it is better to suppose that the fraction of zygotes in each niche is prescribed and the fraction of adults is proportional to the mean fitness in the niche:

$$c_i^\star = c_i \overline{w}_i / \overline{w} \ , \tag{6.26}$$

where

$$\overline{w} = \sum_i c_i \overline{w}_i \tag{6.27}$$

is the mean fitness of the population. Wallace (1968) termed (6.26) *hard selection*. Since plants compete for resources locally, soft selection must apply to them. For animals, the choice of scheme depends on the situation (Wallace, 1968).

Let us consider now the influence of the order of effects in our model. If we replace adult migration by gametic dispersion (as may be reasonable for some lower vascular plants) and denote the allelic frequencies after dispersion by $p_{i,j}$, no change in our formulation is required. This simple observation is disguised if allelic frequencies before dispersion are used (Bulmer, 1972), because, in contrast to our definition, they are not the same as the gene frequencies in zygotes.

If zygotes rather than gametes disperse, new formulation and analysis are required. We shall discuss this case at the end of this section.

For soft selection, the diploid model includes as a special case the haploid one with the same forward migration matrix: the substitution $w_{i,jk} = b_{i,j}b_{i,k}$ reduces (6.23) and (6.25) with $c_i^* = c_i$ to a haploid scheme with fitnesses $b_{i,j}$. For hard selection, (6.26) precludes this simplification.

Let us now derive sufficient conditions for the existence of a protected polymorphism in the diallelic case. We write p_i $(= p_{i,1})$ for the frequency of A in niche i and seek sufficient conditions for the instability of the trivial equilibria $p_i = 0$ and $p_i = 1$. These guarantee that both alleles will increase in frequency when rare. We define the ratios of the homozygote fitnesses in niche i to that of the heterozygotes by

$$u_i = w_{i,11}/w_{i,12}\,, \qquad v_i = w_{i,22}/w_{i,12} \tag{6.28}$$

and posit that u_i and v_i are constant. With frequency-dependent selection, $u_i = u_i(p_i)$ and $v_i = v_i(p_i)$, the protection conditions would involve only $u_i(1)$ and $v_i(0)$.

Soft Selection

We commence with soft selection, $c_i^* = c_i$. Then (6.22) becomes

$$m_{ji} = c_i \tilde{m}_{ij} \bigg/ \sum_k c_k \tilde{m}_{kj}\ . \tag{6.29}$$

Suppose that M is constant. This is equivalent to constant $\tilde{M}$ if $\mathbf{c}$ is constant. However, without post-migration population regulation, $\mathbf{c}$ will generally depend on time. Indeed, in that case, from (6.20a) we have

$$\mathbf{c}' = \mathbf{c}^{**} = L\mathbf{c}\ ,$$

whence

$$\mathbf{c}(t) = L^t \mathbf{c}(0)\ ,$$

where L is the transpose of $\tilde{M}$. Therefore, our results are more likely to apply with population control after migration: $\mathbf{c}' = \mathbf{c}$.

If a migration pattern does not change the niche proportions, we shall call it *conservative*. For such a scheme, (6.21) yields $c_i \tilde{m}_{ij} = c_j m_{ji}$, whence

$$c_j = \sum_i c_i \tilde{m}_{ij}\,, \qquad c_i = \sum_j c_j m_{ji}\ .$$

If the deme sizes are all the same, $c_i = c$, then $m_{ij} = \tilde{m}_{ji}$, so constancy of the forward migration matrix always implies that of the backward one. Conservative migration has two interesting special cases: reciprocal migration and doubly stochastic migration with equal deme sizes.

Let us call dispersion *reciprocal* if the number of individuals that migrate from deme i to deme j equals the number that migrate from deme j to deme i: $c_i \tilde{m}_{ij} = c_j \tilde{m}_{ji}$. With this type of exchange for all pairs of niches, (6.19) and (6.20a) immediately demonstrate the invariability of the niche proportions: $c_i^{**} = c_i$. From (6.21) we infer the identity of the forward and backward migration matrices: $m_{ij} = c_j \tilde{m}_{ji}/c_i = \tilde{m}_{ij}$. Hence, constant $\tilde{M}$ is always equivalent to constant M. Note that M and $\tilde{M}$ are symmetric if the demes are of equal size. An example of reciprocity is the random outbreeding and site homing model (Deakin, 1966; Maynard Smith, 1966, 1970; Christiansen, 1974, 1975), of which the Levene (1953) model, discussed in the next section, is a special case.

A migration scheme is *doubly stochastic* if

$$\sum_i \tilde{m}_{ij} = 1 \ .$$

With equal colony sizes, (6.20a) immediately shows that such a pattern is conservative. Therefore, $m_{ij} = \tilde{m}_{ji}$, so M is also doubly stochastic. This situation will arise in problems with a natural periodicity, such as a set of demes in a circular or torus-like arrangement. Then we posit equal niche proportions and *homogeneous* migration: $\tilde{m}_{ij} = \tilde{m}_{j-i}$, i.e., dependence only on displacement, rather than on both initial and final positions. Since $m_{ij} = \tilde{m}_{ji} = \tilde{m}_{i-j}$, the backward migration pattern also is homogeneous. If migration is symmetric, $\tilde{m}_{ij} = \tilde{m}_{ji}$, and the deme sizes are equal, then dispersion is both reciprocal and doubly stochastic.

To derive a sufficient condition for the protection of A, we linearize (6.23) near $\mathbf{p} = \mathbf{0}$. Defining the diagonal matrix D by $d_{ii} = v_i^{-1}$ and putting $Q = MD$, from (6.23), (6.24), and (6.28) we obtain the linearized recursion relation

$$\mathbf{p}' = Q\mathbf{p}, \qquad \text{with} \qquad q_{ij} = m_{ij}/v_j \ . \tag{6.30}$$

Recalling (5.15), we conclude that if the absolute value of at least one eigenvalue of Q exceeds unity, then $\mathbf{p} = \mathbf{0}$ is unstable (Bulmer, 1972).

It is obvious from (6.28) and (6.30) that universal heterozygote lethality ($w_{i,12} = 0$ for all i) precludes protection. We assume that $w_{i,12} > 0$ for all i, in which case, if $0 < p_j < 1$ for some j, then A cannot be lost by a jump of the exact system (6.23), (6.24) to $\mathbf{p} = \mathbf{0}$. Now, Q may have some eigenvalues less than one in absolute value. The corresponding eigenspace extends to a stable manifold of (6.23), (6.24), from which $\mathbf{p}(t)$ would converge to $\mathbf{0}$, were it not for the biological certainty of small perturbations that would move the population off this manifold. We must still ascertain, however, that the population does not repeatedly approach arbitrarily closely the stable eigenspace near $\mathbf{p} = \mathbf{0}$. For if it did, say by virtue of suitable cycling, random drift would eventually cause loss of A.

We must require that the descendants of individuals in every niche be able eventually to reach every other niche. Therefore, we suppose henceforth that there do not exist complementary sets of integers I and J such that $m_{ij} = 0$ for i in I and j in J, i.e., the matrix M is *irreducible*. Then Q is irreducible. Since Q is clearly nonnegative ($q_{ij} \geq 0$ for all i and j), Frobenius' theorem (Gantmacher, 1959, Vol. II, p. 53) informs us that it has a real, positive, nondegenerate (or simple) eigenvalue, λ_0 (called the *maximal* eigenvalue), not exceeded by any other eigenvalue in modulus, to which corresponds a positive eigenvector (i.e., all components of this *maximal* eigenvector may be chosen to be positive). By employing the general decomposition (5.13) and the fact that if a subspace is invariant with respect to an operator, then its orthogonal complement is invariant with respect to the adjoint operator (Gantmacher, 1959, Vol. I, p. 266), one can prove that the positive maximal left eigenvector of Q is orthogonal to every vector in the eigenspace complementary to the maximal (right) eigenvector. Therefore, there are no nonnegative vectors in that space, and consequently $\lambda_0 > 1$ does imply protection of A.

We see at once from (6.28) that (6.30) does not apply if aa is lethal in any deme: $w_{i,22} = 0$ for some i. But in such a deme, as $p_i \to 0$, $p_i^\star \to \frac{1}{2}$ from (6.23), as is obvious, and hence (6.24) shows that $p_j' \not\to 0$ for some j as $\mathbf{p} \to \mathbf{0}$. Therefore, A is protected.

The maximal eigenvalue λ_0 satisfies the inequality

$$\min_i \sum_j q_{ij} \leq \lambda_0 \leq \max_i \sum_j q_{ij} \;, \tag{6.31}$$

with equality if and only if all the row sums are the same (Gantmacher, 1959, Vol. II, p. 63). Suppose aa is at least as fit as Aa in every niche and more fit in at least one: $v_i \geq 1$ for all i and $v_j > 1$ for some j. Evidently, $q_{ij} \leq m_{ij}$; the irreducibility of M implies that there is no j such that $m_{ij} = 0$ for all i. Therefore, the row sums in (6.31) are not all equal and (6.19) and (6.31) yield

$$\lambda_0 < \sum_j m_{ij} = 1 \;.$$

As expected, A is not protected. If Aa is at least as fit as aa in all niches and more fit in at least one, the above reasoning tells us that $\lambda_0 > 1$, i.e., A is protected. These results were proved by a different method by Bulmer (1972).

Consider (6.23) and (6.24) in the absence of migration, $m_{ij} = \delta_{ij}$. It is easy to locate all the equilibria of this system. Karlin and McGregor (1972, 1972a) have shown that if one is not an eigenvalue of Q, then for sufficiently weak migration, in the neighborhood of each stable equilibrium of the uncoupled system, there exists exactly one equilibrium of the weakly coupled one, and it is stable. In the neighborhood of each unstable equilibrium of the system without migration, there exists at most one equilibrium of the system with weak migration; if it exists, such an equilibrium is unstable. Now, if Aa is favored over aa in at least one niche (it may be less fit than aa in the other niches), i.e., $v_i < 1$ for at least one i, then in the absence of migration, the equilibrium $\mathbf{p} = \mathbf{0}$ is unstable. By the above principle, for sufficiently weak migration, this equilibrium is still unstable. Therefore, A will remain in the population. Although still intuitively reasonable, this result is much less obvious than the two in the previous paragraph. For weak migration, the latter also follow directly from the general principle.

For a protected polymorphism, we stipulate also that a be preserved. As $p_i \to 1$, we obtain instead of m_{ij}/v_j the matrix elements m_{ij}/u_j and require again that the maximal eigenvalue exceed unity.

If there is complete dominance, the eigenvalue condition cannot be satisfied: e.g., if a is dominant, $v_i = 1$, so $q_{ij} = m_{ij}$ and (6.31) gives $\lambda_0 = 1$. We shall study this important case after investigating hard selection.

Christiansen (1974) has derived a sufficient condition for the protection of A that is equivalent to $\lambda_0 > 1$, but is easier to apply. Let $I^{(n)}$ and $Q^{(n)}$ respectively designate the $n \times n$ unit matrix and the square matrix formed from the first n rows and columns of Q. We require

$$\det\left(I^{(n)} - Q^{(n)}\right) < 0 \tag{6.32}$$

for some n, where $n = 1, 2, \ldots, N$ and N is the number of niches.

For further analysis, see Karlin (1976) and Karlin and Richter-Dyn (1976).

Hard Selection

We turn now to the hard-selection model. Substituting (6.26) into (6.22), we find

$$m_{ji} = c_i \overline{w}_i \tilde{m}_{ij} \bigg/ \sum_k c_k \overline{w}_k \tilde{m}_{kj} \ . \tag{6.33}$$

Since $\mathbf{c}^*$ depends on gene frequency, M is now quite unlikely to remain constant. It is much more probable that $\tilde{M}$ is constant. If we suppose that the population is regulated after migration, $\mathbf{c}' = \mathbf{c}$, then we can evaluate M from (6.33). As $\mathbf{p} \to 0$, (6.23), (6.24), and (6.33) yield the linearized matrix difference equation

$$\mathbf{p}' = R\mathbf{p}, \qquad \text{where} \qquad r_{ij} = c_j w_{j,12} \tilde{m}_{ji} \bigg/ \sum_k c_k w_{k,22} \tilde{m}_{ki} \ . \tag{6.34}$$

The allele A will be protected if the maximal eigenvalue of R exceeds unity (Christiansen, 1975). To protect a, we require the same condition with $w_{k,22}$ replaced by $w_{k,11}$ in (6.34).

We can compare hard selection with soft selection by assuming that the fitnesses $w_{i,jk}$, the forward migration matrices $\tilde{M}$, and the zygotic distributions $\mathbf{c}$ are the same. Since we no longer require (6.33) here, we employ (6.29) to define the constant backward migration matrix M for soft selection and rewrite (6.34) in the form $R = D^{(1)} M D^{(2)}$, where the diagonal matrices $D^{(1)}$ and $D^{(2)}$ have the elements

$$d_{ii}^{(1)} = \sum_k c_k \tilde{m}_{ki} \bigg/ \sum_k c_k w_{k,22} \tilde{m}_{ki} \, , \qquad d_{ii}^{(2)} = w_{i,12} \ . \tag{6.35}$$

But for any two matrices B and C, BC and CB have the same eigenvalues (see Problem 6.3). Consequently, the eigenvalues of R are identical to those of $S = MD^{(2)}D^{(1)}$. Since (6.35) gives $s_{ij} = m_{ij}/\nu_j$, where

$$\nu_j = \left(\sum_k c_k w_{k,22} \tilde{m}_{kj} \right) \left(w_{j,12} \sum_k c_k \tilde{m}_{kj} \right)^{-1} , \qquad (6.36)$$

we obtain for hard selection just the soft-selection matrix (6.30) with v_j replaced by ν_j. Recalling (6.28), we may rephrase this in terms of the substitution

$$w_{j,22} \to \sum_k c_k w_{k,22} \tilde{m}_{kj} \Big/ \sum_k c_k \tilde{m}_{kj} .$$

We shall see in Sect. 6.3 that even in the special case of the Levene (1953) model, no general statement can be made regarding the relative stringency of the protection conditions for soft and hard selection.

Our discussion for soft selection of the case of recessive lethal a applies unaltered here.

If migration is sufficiently weak, the principle of Karlin and McGregor (1972, 1972a) described above demonstrates that A is protected if Aa is fitter than aa in at least one niche, and that A is not protected if Aa is less fit than aa in every niche.

We can derive the other results that correspond to those we obtained for soft selection by rewriting r_{ij} in (6.34) as

$$r_{ij} = m_{ij}^\star/v_j , \qquad \text{where} \qquad m_{ij}^\star = c_j w_{j,22} \tilde{m}_{ji} \Big/ \sum_k c_k w_{k,22} \tilde{m}_{ki} . \qquad (6.37)$$

Naturally, $M^\star$ is the backward migration matrix computed from (6.33) in the limit $\mathbf{p} \to \mathbf{0}$. Comparing (6.30) with (6.37) and noting that

$$\sum_j m_{ij}^\star = 1 ,$$

we deduce, precisely as for soft selection:

(i) if aa is at least as fit as Aa in every niche and more fit in at least one, A is not protected;

(ii) if aa and Aa are interchanged in (i), A is protected;

(iii) if a is dominant, the maximal eigenvalue of R is one, so protection must be analyzed *ab initio*.

The most interesting feature of hard selection, obvious by inspection of (6.34), (6.36), or (6.37), is that, in contrast to soft selection, the model (including the protection conditions) can no longer be expressed in terms of the fitness ratios (6.28). Mathematically, this happens because the hard-selection backward migration matrix, (6.33), depends on the fitnesses, whereas the one for soft selection, (6.29), does not. The biological reason is that with hard selection the fitnesses affect the deme sizes, and since the gene frequencies in the demes are different, subpopulation numbers influence gene frequencies through migration. If (and only if) the heterozygote fitnesses are the same in all the niches, (6.34) simplifies to the result of Christiansen (1975):

$$r_{ij} = c_j \tilde{m}_{ji} \Big/ \sum_k c_k v_k \tilde{m}_{ki} \ . \tag{6.38}$$

A Recessive

We have $w_{i,12} = w_{i,22}$ for all i. For soft selection, a will be protected if the maximal eigenvalue of the matrix with elements m_{ij}/u_j exceeds one. For hard selection, the pertinent matrix elements are

$$c_j w_{j,12} \tilde{m}_{ji} \Big/ \sum_k c_k w_{k,11} \tilde{m}_{ki} \ .$$

We wish to deduce sufficient conditions for protection of A. We must now compute the effect of selection to second order as $\mathbf{p} \to \mathbf{0}$: (6.23) and (6.28) yield

$$p_i^\star = p_i + (u_i - 1)p_i^2 + O(p_i^3) \ . \tag{6.39}$$

Soft and hard selection require separate analyses.

With *soft selection*, we assume that M is constant. In view of (6.19) and (6.31), M has the maximal eigenvalue one, with the maximal eigenvector

$$\mathbf{V} = \begin{pmatrix} 1 \\ 1 \\ \vdots \\ 1 \end{pmatrix} \ .$$

Let us suppose that all the other eigenvalues are (strictly) less than one in absolute value. This restriction is equivalent to aperiodicity of the backward migration matrix (Gantmacher, 1959, Vol. II, p. 88); it holds if and only if some power of M is positive (Gantmacher, 1959, Vol. II, p. 80). Since we have already assumed that M is irreducible, the extremely weak condition $m_{ii} > 0$ for some i suffices for aperiodicity (Feller, 1968, Vol. I, p. 426). Thus, we require only that individuals remain in some colony with positive probability. Substituting (6.39) into (6.24) produces

$$\mathbf{p}' = M\mathbf{p} + \mathbf{f}(\mathbf{p}) + O(|\mathbf{p}|^3) \ , \tag{6.40a}$$

where

$$f_i(\mathbf{p}) = \sum_j m_{ij}(u_j - 1)p_j^2 \ . \tag{6.40b}$$

For $|\mathbf{p}| \ll 1$ and long times, we may invoke (5.15) to infer from (6.40) that $\mathbf{p}(t) \approx x(t)\mathbf{V}$, where $x(t)$ is the component of $\mathbf{p}(t)$ along $\mathbf{V}$. The component of $\mathbf{p}(t)$ in the eigenspace of M complementary to $\mathbf{V}$ decreases geometrically, controlled by the dominant linear part of (6.40a). Since this observation holds to first order, we may replace $\mathbf{p}$ by $x\mathbf{V}$ in the higher-order terms in (6.40a):

$$\mathbf{p}' = M\mathbf{p} + \mathbf{f}(x\mathbf{V}) + O(x^3) \ . \tag{6.41}$$

By Frobenius' theorem (Gantmacher, 1959, Vol. II, p. 53), we can choose $\mathbf{U}$, the left eigenvector of M corresponding to the eigenvalue one, to be positive, and we take also $\mathbf{U}^T\mathbf{V} = 1$, where the superscript T signifies transposition. The allele A will be protected if $x(t) = \mathbf{U}^T\mathbf{p}(t)$, the component of $\mathbf{p}$ along $\mathbf{V}$, increases (note that $\mathbf{U}$ is orthogonal to the complement of $\mathbf{V}$). Taking the scalar product of (6.41) with $\mathbf{U}$ and using the fact that $\mathbf{U}^T M = \mathbf{U}^T$, we find

$$x' = x + \mathbf{U}^T\mathbf{f}(x\mathbf{V}) + O(x^3) \ . \tag{6.42}$$

The positivity of x allows us to derive from (6.42) the sufficient condition $\mathbf{U}^T\mathbf{f}(x\mathbf{V}) > 0$ for protection of A. With the aid of (6.19), (6.40b), $\mathbf{U}^T M = \mathbf{U}^T$, and the normalization

$$\sum_i U_i - 1 \tag{6.43}$$

of $\mathbf{U}$, we obtain the explicit criterion

$$\mathbf{U}^T\mathbf{u} = \sum_i U_i u_i > 1 \ . \tag{6.44}$$

This condition depends only on the fitness ratios u_i and the maximal left eigenvector $\mathbf{U}$ of the backward migration matrix M. The deme sizes c_i affect (6.44) only to the extent that they influence $\mathbf{U}$.

From (6.43) and (6.44) we can easily see that if AA is at least as fit as Aa in every niche and more fit in at least one, then A is protected. If AA and Aa are interchanged in the previous statement, A is not protected.

The criterion (6.44) was derived independently by Karlin (1977a) and Nagylaki (1977a). Karlin's proof does not assume that the absolute values of all eigenvalues other than the maximal one are less than 1. Karlin (1977a) has also shown that the inequalities $\mathbf{U}^T\mathbf{u} < 1$ and $\hat{\lambda}_0 < 1$, where $\hat{\lambda}_0$ is the maximal eigenvalue of the matrix with elements m_{ij}/u_j, cannot hold simultaneously. Hence, excluding the special cases $\mathbf{U}^T\mathbf{u} = 1$ and $\hat{\lambda}_0 = 1$, at least one of the two alleles must be protected. This is, of course, not so without complete dominance, for then different fitness ratios influence the protection of the two alleles. Underdominance in a panmictic population is the most trivial counterexample.

Christiansen (1974) obtained the condition

$$\mathbf{c}^T M\mathbf{u} > 1 \tag{6.45}$$

as the criterion for protection with a constant zygotic distribution $\mathbf{c}$. This condition agrees with (6.44) if and only if $\mathbf{c}^T M = \mathbf{U}^T$, which is true if $\mathbf{c} = \mathbf{U}$, and should $\det M = 0$, may be fortuitously valid for particular vectors $\mathbf{c}$.

From (6.20b) we conclude at once that $\mathbf{c}^T M = \mathbf{c}^T$ if and only if migration is conservative; this reduces (6.45) to the arithmetic-mean condition

$$\mathbf{c}^T \mathbf{u} = \sum_i c_i u_i > 1 \tag{6.46a}$$

for protecting A, but does not always hold. For example, if M is symmetric, evidently $\mathbf{U} = N^{-1}\mathbf{V}$, so (6.44) becomes

$$N^{-1} \sum_i u_i > 1 \ . \tag{6.46b}$$

But symmetric M corresponds to conservative migration if and only if the N subpopulations are of equal size, and it is only then that (6.46a) and (6.46b) agree.

Christiansen's condition (6.45) is generally false because he replaced $\mathbf{p}(t)$ by $x(t)\mathbf{V}$ in the linear term $(M\mathbf{p})$ on the right-hand side of (6.40a). The resulting equation is correct only to first order in $\mathbf{p}$, and protection is determined by the quadratic terms.

For conservative migration, we can see that protection of A is most likely if AA is strongly favored in large demes rather than in small ones. Without loss of generality, assume that $c_1 \leq c_2 \leq \ldots \leq c_N$. Then, as pointed out by Karlin (1977a), the sum in (6.46a) is a maximum for given values of u_i if $u_1 \leq u_2 \leq \ldots \leq u_N$. This follows from the trivial observation that $a_1 \leq a_2$ and $b_1 \leq b_2$ jointly imply $a_1 b_1 + a_2 b_2 \geq a_1 b_2 + a_2 b_1$. Thus, the probability of protecting A is greatest if the deme sizes and fitness ratios are in the same order.

Let us consider now the case of *hard selection*. As before, we suppose that $\tilde{M}$ and $\mathbf{c}$ are constant. The selection equation (6.39) still applies, of course, but we must now calculate M from (6.33). From (6.23c) we find

$$\overline{w}_i = w_{i,22}[1 + (u_i - 1)p_i^2] \ ,$$

whence (6.33) immediately informs us that

$$M = M^\star + O(|\mathbf{p}|^2) \ , \tag{6.47}$$

where $M^\star$ is the backward migration matrix in (6.37). Inserting (6.39) and (6.47) into (6.24) gives (6.40) with M replaced by $M^\star$. From (6.44) we have the protection condition (Nagylaki, 1977a)

$$\mathbf{Y}^T \mathbf{u} = \sum_i Y_i u_i > 1 \ , \tag{6.48}$$

where $\mathbf{Y}$ is the maximal left eigenvector of $M^\star$ and

$$\sum_i Y_i = 1 \ . \tag{6.49}$$

As expected, the obvious conclusions for AA favored over Aa and *vice versa* follow as for soft selection. Again, the hard-selection result depends on fitnesses, not just on fitness ratios. We shall see that even in the Levene model, one can draw no general conclusion concerning the relative stringency of (6.44) and (6.48). If M^* is symmetric, (6.48) simplifies to (6.46b). If the dominant fitnesses, $w_{i,22}$, are the same in all niches, then (6.29) and (6.37) inform us that $M^* = M$, so $\mathbf{Y} = \mathbf{U}$ and the soft- and hard-selection requirements (6.44) and (6.48) are identical.

Juvenile Dispersion

For some marine organisms and for plants with seeds carried by wind, water, or animals, migration is best modeled by zygotic (or, at least, juvenile) dispersion. If the frequency of A_j in niche i immediately after meiosis is $p_{i,j}$, which gives zygotic frequencies $p_{i,j}p_{i,k}$ just after conception, the post-migration zygotic proportions read

$$P_{i,jk} = \sum_l m_{il} p_{l,j} p_{l,k} \ , \tag{6.50}$$

where M is now the zygotic backward migration matrix. Including selection yields the recursion relations

$$p'_{i,j} = \sum_k w_{i,jk} P_{i,jk} \Big/ \sum_{kl} w_{i,kl} P_{i,kl} \ . \tag{6.51}$$

The model described by (6.50) and (6.51) is quite different from the one specified by (6.23) and (6.24).

With the zygotic distribution $\mathbf{c}$ regulated to constancy, we may assume that M is constant regardless of the nature of selection. Then it is easy to derive the criteria for a protected polymorphism in the diallelic case. For protection of A, we linearize (6.50) and (6.51) as $\mathbf{p} \to \mathbf{0}$, where p_i represents the frequency of A in gametes in niche i. An easy calculation yields

$$\mathbf{p}' = T\mathbf{p}, \quad \text{with} \quad t_{ij} = m_{ij}/v_i \ . \tag{6.52}$$

Since $T = DM$ has the same eigenvalues as MD (see Problem 6.3), comparing (6.52) with (6.30), we see at once that the sufficient conditions for protection are the same as for adult migration with soft selection (Christiansen, 1975). Furthermore, the discussion between (6.30) and (6.32) applies without essential change.

It remains only to consider the case of *recessive* A. Expanding (6.50) and (6.51) to second order in $|\mathbf{p}|$, we find

$$\mathbf{p}' = M\mathbf{p} + \mathbf{g}(\mathbf{p}) + O(|\mathbf{p}|^3) \ , \tag{6.53a}$$

where

$$g_i(\mathbf{p}) = (u_i - 1) \sum_j m_{ij} p_j^2 \ . \tag{6.53b}$$

Obviously, (6.40a) and (6.53a) have the same form. Therefore, A will be protected if $\mathbf{U}^T \mathbf{g}(x\mathbf{V}) > 0$, where $\mathbf{U}$, the maximal left eigenvector of M, is again normalized as in (6.43). By a trivial reduction that uses (6.19) and (6.43), we obtain again the criterion (6.44). Thus, in all cases the sufficient conditions for protection with juvenile dispersion are the same as with adult migration (Nagylaki, 1977a).

Karlin and Kenett (1977) discuss the effects of simultaneous juvenile and adult migration and of mating areas distinct from habitat sites. Strobeck (1974) had been the first to introduce mating groups as a generalization of the Levene (1953) model. For migration and selection in dioecious populations, investigated by Nagylaki (1979a), refer to Problems 7.6, 7.7, and 7.8. Moody (1979) studies X-linked loci. Because of the possibility of selfing and the fact that both pollen and seeds (rather than sperm and eggs, zygotes, or adults) disperse, most plant populations require modifications of our basic model. The interested reader may consult Nagylaki (1976b) and, for a much more general investigation, Moody (1979).

Karlin (1982) presents an extensive review of the migration-selection problem and treats many examples.

6.3 The Levene Model

Since we are interested primarily in the sufficient conditions for a protected polymorphism, the result at the end of the last section permits us to restrict our discussion to adult migration. The basic assumption of the Levene (1953) model is that migration is random. More precisely, individuals disperse independently of their niche of origin : $\tilde{m}_{ij} = \mu_j$, for some constants μ_j. We treat soft and hard selection separately.

For *soft selection*, (6.22) directly yields $m_{ji} = c_i$. If the demes are regulated to constant proportions, this is all we need. But the proportions remain constant even without regulation, for (6.20a) gives $c'_j = c^{**}_j = \mu_j$. Thus, we have the standard interpretation, $c_i = \mu_i$.

We proceed to demonstrate that the geometric mean, w^*, of the average fitnesses in the various niches is nondecreasing, the change in w^* being zero only at equilibrium. Such a property has been proved only in the Levene model. From (6.24) we observe that, not surprisingly, after one generation the gene frequencies in zygotes are the same in all the niches. Therefore, we may simplify our notation by letting p_i represent the frequency of A_i in zygotes. Inserting (6.23a) into (6.24), we find

$$p'_i = p_i \sum_j c_j w_{j,i}/\overline{w}_j \ . \tag{6.54}$$

Since (6.23b) and (6.23c) may be rewritten as

$$w_{i,j} = \sum_k w_{i,jk} p_k \ , \tag{6.55a}$$

$$\overline{w}_i = \sum_{jk} w_{i,jk} p_j p_k \ , \tag{6.55b}$$

we can express (6.54) in the form

$$p'_i = \tfrac{1}{2} p_i \sum_j \frac{c_j}{\overline{w}_j} \frac{\partial \overline{w}_j}{\partial p_i} = \tfrac{1}{2} p_i \sum_j c_j \frac{\partial}{\partial p_i} \ln \overline{w}_j = \tfrac{1}{2} p_i \frac{\partial}{\partial p_i} \ln w^* \ , \tag{6.56}$$

where

$$w^* = \prod_i \overline{w}_i^{c_i} \ . \tag{6.57}$$

We might hope that w^* is nondecreasing because (6.56) has the form of multiallelic selection in a single panmictic population. Proof is required, however: for random mating in a single population, the mean fitness is a homogeneous quadratic in $\mathbf{p}$, which w^* manifestly is not.

From (6.55), (6.56), and (6.57) we can easily verify that

$$\sum_i p'_i = 1 \ .$$

Making this explicit in (6.56) leads to

$$p'_i = p_i \frac{\partial w^*}{\partial p_i} \bigg/ \sum_j p_i \frac{\partial w^*}{\partial p_j} \ . \tag{6.58}$$

We can approximate (6.57), and hence the orbit $\mathbf{p}(t)$, as closely as desired by the replacement

$$w^* \to W = \prod_i \overline{w}_i^{\gamma_i} \ ,$$

with positive, rational γ_i, provided we choose γ_i sufficiently close to c_i. But there exists a positive integer l such that $F \equiv W^l$ is a homogeneous polynomial in $\mathbf{p}$ with nonnegative coefficients. So, with arbitrarily high accuracy, (6.58) becomes

$$p'_i = p_i \frac{\partial F}{\partial p_i} \bigg/ \sum_j p_j \frac{\partial F}{\partial p_j} \ .$$

Now it follows directly from a result of Baum and Eagon (1967) that $F(\mathbf{p})$ is nondecreasing along trajectories and the change in F is zero only at equilibrium. The same obviously holds for W, and consequently for $w^*(\mathbf{p})$.

The above argument deviates from that of Cannings (1971) only in replacing w^* by F to meet the restrictions of Baum and Eagon's theorem. Earlier, Li (1955a) had proposed (6.57) for two alleles. The existence of the nondecreasing function w^* implies that the system (6.54) must converge to some equilibrium point (or surface) for all initial conditions. In particular, for the diallelic case, discussed below, if there is a protected polymorphism, the gene frequency must converge to some stable internal equilibrium point.

Passing to two alleles, we can easily deduce the criteria for protection directly from (6.54). But even this simple calculation is unnecessary. Notice that $q_{ij} = c_j/v_j$, independent of i, in (6.31). Therefore, the left- and right-hand sides of (6.31) are equal and

$$\lambda_0 = \sum_i c_i/v_i \ .$$

So, A will be protected from disappearance if the harmonic mean of the fitness ratios v_i is less than one (Levene, 1953):

$$\tilde{v} \equiv \left(\sum_i c_i/v_i\right)^{-1} < 1 \ . \tag{6.59}$$

Clearly, a will be protected if (6.59) holds with v_i replaced by u_i.

If A is recessive ($v_i = 1$), we can again derive the sufficient condition for its protection by expanding (6.54). However, observing that $\mathbf{c}$ is the maximal left eigenvector of M, we see at once from (6.44) that the arithmetic mean of the fitness ratios u_i must exceed unity (Prout, 1968):

$$\overline{u} \equiv \sum_i c_i u_i > 1 \ . \tag{6.60}$$

Thus, a sufficient condition for a protected polymorphism reads

$$\tilde{u} < 1 < \overline{u} \ . \tag{6.61}$$

Equation (6.60) requires superiority of AA over Aa in the mean of the fitness ratios $w_{i,11}/w_{i,12}$. Now, $\overline{u} \geq \tilde{u}$, with equality if and only if all the u_i are the same (Problem 6.8). Therefore, fitness ratios u_i certainly exist for which (6.61) applies. Furthermore, (6.59) can be satisfied without heterozygote advantage in the mean of the v_i ($\overline{v} < 1$). Such superiority does imply (6.59).

Observe that A is protected if aa is sufficiently harmful in some deme: (6.59) holds if there exists some j such that $v_j < c_j$. With complete dominance, there is a protected polymorphism if the recessive is sufficiently beneficial in some niche and sufficiently deleterious in another: (6.61) is valid if $u_j > 1/c_j$ and $u_k < c_k$ for some j and k such that $j \neq k$.

By the discussion relating to (6.46a), for given niche proportions c_i and fitness ratios v_i, $\tilde{v}$ is minimized if the deme sizes and the reciprocal fitness ratios, $1/v_i$, are in the same order. Hence, (6.59) shows that A is most likely to be preserved if Aa is most advantageous relative to aa in the larger subpopulations.

We continue with *hard selection*. Suppose that the zygotic distribution $\mathbf{c}$ is regulated to constancy. With $\tilde{m}_{ij} = \mu_j$, from (6.33) and (6.27) we find

$$m_{ji} = c_i \overline{w}_i/\overline{w} \ . \tag{6.62}$$

Therefore, as for soft selection, (6.24) informs us that after one generation the allelic frequencies in zygotes in all the niches are equal. Then we may again designate the frequency of A_i in zygotes by p_i; $\overline{w}_i$ and $\overline{w}$ are given by (6.55b) and (6.27), respectively. Inserting (6.23a) and (6.62) into (6.24), we obtain

$$p_i' = p_i \overline{w}^{-1} \sum_k c_k w_{k,i} \ . \tag{6.63}$$

The arithmetic-mean fitnesses of A_iA_j, A_i, and the entire population read

$$z_{ij} = \sum_k c_k w_{k,ij} \ , \tag{6.64a}$$

$$z_i = \sum_j z_{ij} p_j \ , \tag{6.64b}$$

$$\bar{z} = \sum_{ij} z_{ij} p_i p_j \ . \tag{6.64c}$$

Employing (6.55) and (6.64a) in (6.64b) and (6.64c) yields

$$z_i = \sum_k c_k w_{k,i}, \qquad \bar{z} = \bar{w} \ ,$$

with the aid of which (6.63) simplifies to

$$p_i' = p_i z_i / \bar{z} \ , \tag{6.65}$$

the classical selection equation for a single random mating population with fitnesses z_{ij}. For two alleles, this result is due to Dempster (1955).

It follows from Sect. 4.3 that the mean fitness $\bar{z}$ is nondecreasing and the population converges to some equilibrium point or surface. With two alleles, Sect. 4.2 tells us that A is protected if and only if either

$$z_{12} > z_{22} \tag{6.66a}$$

or

$$z_{11} > z_{12} = z_{22} \ . \tag{6.66b}$$

Of course, (6.66b) means that A is recessive. There is a protected polymorphism if and only if arithmetic-mean overdominance applies:

$$z_{12} > z_{11}, z_{22} \ . \tag{6.67}$$

Let us compare the stringency of the conditions for a protected polymorphism with soft and hard selection. If there is complete dominance, the soft-selection criterion is (6.61), whereas (6.67) rules out protection for hard selection. In the absence of complete dominance, we desire to compare the requirements for protecting A, given by (6.59) for soft and (6.66a) for hard selection. From (6.28) and (6.64a) it is easy to see that if aa has the same fitness in every niche, then (6.59) and (6.66a) are the same. If the heterozygote fitness is the same in every niche, (6.66a) becomes

$$\bar{v} = \sum_i c_i v_i < 1 \ . \tag{6.68}$$

But the arithmetic mean exceeds the harmonic mean unless all the fitness ratios are equal, in which case $\bar{v} = \tilde{v}$ (Problem 6.8). Hence, protection of A with hard selection implies its protection with soft selection, but not *vice versa*.

It is not true, however, that the hard-selection criterion (6.66a) is always more restrictive than the soft-selection criterion (6.59). Glancing at (6.59) and (6.66a), we see that the soft-selection condition is more stringent than the hard one if and only if

$$\tilde{v} > z_{22}/z_{12} \; ,$$

which we rearrange in the form

$$\sum_i c_i w_{i,12} > \left(\sum_i c_i w_{i,22} \right) \left(\sum_i c_i w_{i,12}/w_{i,22} \right) \; . \tag{6.69}$$

To give an example of protection with hard selection that does not imply protection with soft selection, we take $w_{i,12} = w_{i,22}^2$. Then (6.69) reduces to

$$\sum_i c_i w_{i,22}^2 > \left(\sum_i c_i w_{i,22} \right)^2 \; . \tag{6.70}$$

Choosing $\mu = 2$ in the special case (4.47) of Jensen's inequality informs us that (6.70) holds as long as not all the fitnesses $w_{i,22}$ are the same.

For further results, see Karlin (1977).

6.4 Two Diallelic Niches

As in Sect. 6.3, we shall treat only adult migration. Since the hard-selection protection criteria may be obtained from the soft-selection ones by the replacement $M \to M^\star$, in which the limiting backward migration matrix $M^\star$ is given by (6.37), we shall therefore restrict ourselves to soft selection. We use the notation of Sect. 6.2, setting $v_i = 1 - s_i$, and seek sufficient conditions for protection of A.

We shall consider the case of recessive A, $s_1 = s_2 = 0$, separately below. After (6.31), we demonstrated that if $s_i \leq 0$ for all i and $s_j < 0$ for some j, then A is not protected. We proved also that $s_i \geq 0$ for all i and $s_j > 0$ for some j together ensure preservation of A in the population. Therefore, we may assume that $s_1 s_2 < 0$, i.e., neither selection coefficient vanishes and s_1 and s_2 have the opposite sign. Putting

$$M = \begin{pmatrix} 1 - m_1 & m_1 \\ m_2 & 1 - m_2 \end{pmatrix} \; , \tag{6.71}$$

from (6.30) and (6.32) we find that A is protected if either

$$m_1 < s_1 \tag{6.72a}$$

or

$$\frac{m_1}{s_1} + \frac{m_2}{s_2} < 1 \; . \tag{6.72b}$$

Now, (6.72a) fails if $s_1 < 0$, although (6.72b) may not. If $s_1 > 0$ and $s_2 < 0$, (6.72a) is more severe than (6.72b). Therefore, we may confine ourselves to the weaker requirement (6.72b). The region of protection of A is displayed in Fig. 6.2 for fixed m_i.

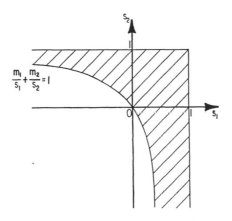

Fig. 6.2. The region of protection of A (hatched).

The backward migration matrix for random outbreeding and site homing (Deakin, 1966; Maynard Smith, 1966, 1970; Christansen, 1974) with an arbitrary number of niches has the elements

$$m_{ii} = 1 - k + kc_i \ ,$$
$$m_{ij} = kc_j, \qquad i \neq j \ . \tag{6.73}$$

It is easy to verify that $M = \tilde{M}$ and migration leaves the population distribution, $\mathbf{c}$, invariant. The constant k $(0 \leq k \leq 1)$ represents the proportion of outbreeding. Observe that $k = 0$ corresponds to no migration, whereas $k = 1$ gives the Levene model, which includes no homing tendency.

The general two-niche model has three independent parameters: m_1, m_2, and, say, c_1. Given m_1 and m_2, however, c_1 does not affect the gene frequencies. We may study the influence of the overall rate of migration by setting

$$m_1 = \kappa\gamma_2, \qquad m_2 = \kappa\gamma_1 \ , \tag{6.74a}$$

with $\gamma_1 + \gamma_2 = 1$ and $0 \leq \kappa\gamma_i \leq 1$, fixing γ_i, and varying κ. For two niches, (6.73) becomes

$$m_1 = kc_2, \qquad m_2 = kc_1 \ , \tag{6.74b}$$

which is the special case of (6.74a) with $\kappa = k \leq 1$ and $\gamma_i = c_i$.

The particular case of (6.72b) for random outbreeding and site homing was derived by Maynard Smith (1970), whose h is $2(1 - k)$. We owe (6.72b) to Bulmer (1972).

Figure 6.2 shows that if aa is sufficiently deleterious relative to Aa, then A is protected. Explicitly, it suffices to have either $s_1 > m_1$ or $s_2 > m_2$. From (6.72b) and (6.74a) we see that as the amount of dispersion, measured by κ, decreases, the region of protected polymorphism in Fig. 6.2 expands. As $\kappa \to 0$, the hyperbolic boundary degenerates into the lines $s_2 = 0$, $s_1 \leq 0$; $s_1 = 0$, $s_2 \leq 0$. Hence, if $s_i > 0$ for some i, then A is protected for sufficiently weak migration. This illustrates a general result proved in Sect. 6.2.

We proceed to discuss the case of *recessive* A. With $v_i = 1$ and $u_i = 1 - \sigma_i$, according to (6.72b), a is protected if

$$\frac{m_1}{\sigma_1} + \frac{m_2}{\sigma_2} < 1 \ . \tag{6.75}$$

The normalized maximal left eigenvector of (6.71) reads

$$\mathbf{U} = \begin{pmatrix} m_2/(m_1 + m_2) \\ \\ m_1/(m_1 + m_2) \end{pmatrix} \ . \tag{6.76}$$

Therefore, the criterion (6.44) for protecting A becomes

$$m_2 \sigma_1 + m_1 \sigma_2 < 0 \ . \tag{6.77}$$

We exhibit the region of protected polymorphism with fixed m_i in Fig. 6.3. Again, as the amount of migration increases, the region of protected polymorphism shrinks.

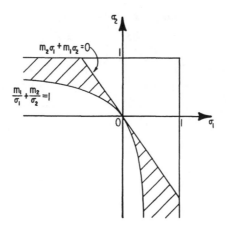

Fig. 6.3. The region of protected polymorphism with complete dominance (hatched).

For random outbreeding and site homing, inserting (6.74b) into (6.75) and (6.77) yields

$$k \left(\frac{c_2}{\sigma_1} + \frac{c_1}{\sigma_2} \right) < 1 \; , \tag{6.78a}$$

$$c_1 \sigma_1 + c_2 \sigma_2 < 0 \; . \tag{6.78b}$$

The condition (6.78b), which requires a negative mean selection coefficient, is due to Maynard Smith (1970).

6.5 Problems

6.1 Derive the equilibrium structure exhibited in Fig. 6.1.

6.2 Give an independent proof of global nonoscillating convergence for the island model without dominance.

6.3 To show that the matrices AB and BA have the same eigenvalues, note first that $\det(AB) = \det(BA)$. Therefore, zero is an eigenvalue of AB if and only if it is an eigenvalue of BA. Next, verify the identity

$$\lambda(\lambda I - BA)^{-1} = I + B(\lambda I - AB)^{-1} A \; ,$$

where I is the identity matrix. Hence, for $\lambda \neq 0$, $(\lambda I - BA)^{-1}$ exists if and only if $(\lambda I - AB)^{-1}$ exists. So, $\det(\lambda I - BA) = 0$ if and only if $\det(\lambda I - AB) = 0$, which proves the identity of the nonzero eigenvalues.

6.4 Demonstrate that migration does not change the allelic frequencies in the total (subdivided) population.

6.5 Assume that there is no selection and the niche proportions are constant: $c_i' = c_i$. Suppose that the constant backward migration matrix, M, is irreducible and let $\mathbf{U}$ designate its unique, positive maximal left eigenvector satisfying (6.43). Define the weighted average frequency of A_j in the entire population as

$$P_j(t) = \sum_i U_i p_{i,j}(t) \; .$$

Note that P_j is the actual frequency of A_j if and only if migration is conservative.

(a) Show that $P_j(t)$ is constant, and hence that no allele can be lost.
(b) Prove that if M is aperiodic as well as irreducible, then the frequency of A_j is asymptotically the same in all demes: $p_{i,j}(t) \rightarrow P_j(0)$ as $t \rightarrow \infty$.

6.6 Sometimes migrating individuals have higher mortality. With the possibility of death *en route*, $\tilde{m}_{ij}$ is reduced for $i \neq j$, and (6.19) must be generalized to

$$\sum_j \tilde{m}_{ij} \leq 1, \qquad \sum_i m_{ji} = 1 .$$

Observe that (6.20), (6.21), and (6.25) hold if we replace the niche proportions by the numbers in the various demes. Since (6.22) and equations not involving the forward migration matrix remain valid, so do the analyses in this chapter.

Mortality during dispersion reduces the effective amount of migration. Infer from (6.22) that decreasing $\tilde{m}_{ij}$ for $i \neq j$ without changing $\tilde{m}_{ii}$ increases m_{ii}.

6.7 Deduce (6.59) and (6.60) from (6.54).

6.8 To establish that the arithmetic mean, $\bar{u}$, exceeds the harmonic mean, $\tilde{u}$, unless all the u_i are the same, use Schwarz's inequality to obtain

$$1 = \left(\sum_i c_i \right)^2 = \left(\sum_i \sqrt{c_i/u_i}\, \sqrt{c_i u_i} \right)^2$$

$$\leq \left(\sum_i c_i/u_i \right) \left(\sum_i c_i u_i \right) = \bar{u}/\tilde{u} .$$

Equality holds if and only if $\sqrt{c_i/u_i} = \beta\sqrt{c_i u_i}$, for some constant β. This requires that all the u_i be the same. For an alternative proof, take $g(x) = 1/x$ in (4.46).

6.9 Derive (6.72).

6.10 Apply Problems 4.5 and 4.6 to establish that the gene frequency always converges to some equilibrium point or surface for the diallelic Levene model with soft selection.

7. X-Linkage

As discussed at the beginning of Sect. 3.2, X-linked loci are of particular interest and usefulness in human, mouse, and *Drosophila* genetics. In species with arrhenotokous parthenogenesis, all loci are effectively X-linked. We shall confine ourselves to discrete, nonoverlapping generations; for the continuous-time theory, the reader may refer to Nagylaki (1975c). In Sect. 7.1 we present the general formulation for selection and mutation with multiple alleles. We discuss the dynamics of selection with two alleles in Sect. 7.2 and treat multiallelic mutation-selection balance in Sect. 7.3. A weak-selection analysis of multiallelic selection is carried out in Sect. 7.4. The formulation and analysis of selection at an autosomal locus in a dioecious population resembles the theory for X-linkage; see Problem 7.4 and Nagylaki (1979b).

7.1 Formulation for Multiallelic Selection and Mutation

We follow the notation and reasoning in Sects. 3.2, 4.1, and 4.9. Let p_i and Q_{ij} represent the respective frequencies of A_i males and ordered $A_i A_j$ females among zygotes. Thus, the frequency of A_i in female zygotes is

$$q_i = \sum_j Q_{ij} \ . \tag{7.1}$$

We designate by u_i and v_{ij} the respective probabilities that A_i males and $A_i A_j$ females survive to reproductive age. The mean viability, v_i, of a female who carries A_i is given by

$$q_i v_i = \sum_j v_{ij} Q_{ij} \ , \tag{7.2}$$

and the mean viabilities of males and females read

$$\bar{u} = \sum_i u_i p_i \ , \qquad \bar{v} = \sum_{ij} v_{ij} Q_{ij} \ . \tag{7.3}$$

Employing a tilde to indicate adult frequencies, we have

$$\tilde{p}_i = p_i u_i / \bar{u} \ , \tag{7.4a}$$

$$\tilde{Q}_{ij} = Q_{ij} v_{ij} / \bar{v} \ , \qquad \tilde{q}_i = q_i v_i / \bar{v} \ . \tag{7.4b}$$

We designate the frequency of the mating of an A_i male with an (ordered) $A_j A_k$ female by $X_{i,jk}$. Hence,

$$\sum_{ijk} X_{i,jk} = 1 \ . \tag{7.5}$$

Let $f_{i,jk}$ and $g_{i,jk}$ be the average numbers of males and females from an $A_i \times A_j A_k$ union. Although these fertilities will almost invariably be the same, we need not impose this restriction at this point. To incorporate mutation, suppose the probabilities that an A_i allele in a zygote appears as A_j in a gamete are R_{ij} and S_{ij} in males and females, respectively. Since there is some evidence that the mutation rates in man (Vogel, 1977) and mice (Searle, 1974) may be different in the two sexes, we shall not assume that $R_{ij} = S_{ij}$.

Denoting the respective numbers of A_i males and ordered $A_i A_j$ females in zygotes by $n_{1,i}$ and $n_{2,ij}$, we obtain

$$n'_{1,i} = M \sum_{klm} X_{k,lm} f_{k,lm} S_{li} \ ,$$

$$n'_{2,ij} = \tfrac{1}{2} M \sum_{klm} X_{k,lm} g_{k,lm} (R_{ki} S_{lj} + R_{kj} S_{li}) \ ,$$

where M is the number of matings. If N_1 and N_2 represent the respective numbers of male and female zygotes, the identities

$$\sum_{j} R_{ij} = 1, \qquad \sum_{j} S_{ij} = 1 \tag{7.6}$$

immediately lead to

$$N'_1 = M\overline{f}, \qquad N'_2 = M\overline{g} \ , \tag{7.7a}$$

where

$$\overline{f} = \sum_{klm} X_{k,lm} f_{k,lm} \ , \qquad \overline{g} = \sum_{klm} X_{k,lm} g_{k,lm} \ . \tag{7.7b}$$

Hence, the genotypic frequencies in zygotes in the next generation are

$$p'_i = \overline{f}^{-1} \sum_{klm} X_{k,lm} f_{k,lm} S_{li} \ , \tag{7.8a}$$

$$Q'_{ij} = \tfrac{1}{2} \overline{g}^{-1} \sum_{klm} X_{k,lm} g_{k,lm} (R_{ki} S_{lj} + R_{kj} S_{li}) \ . \tag{7.8b}$$

In general, (7.8b) cannot be simplified further. To elucidate the underlying biology of the model, we consider gene frequencies. From (7.8b), with the aid of (7.1) and (7.6), we get

$$q'_i = \tfrac{1}{2} \overline{g}^{-1} \sum_{klm} X_{k,lm} g_{k,lm} (R_{ki} + S_{li}) \ . \tag{7.9}$$

Designating the numbers of adult males and females by $\tilde{N}_1$ and $\tilde{N}_2$, we find that the average numbers of females born to an A_i male (a_i), males born to an $A_i A_j$ female (b_{ij}), and females born to an $A_i A_j$ female (c_{ij}) are given by

$$\tilde{N}_1 \tilde{p}_i a_i = M \sum_{kl} X_{i,kl} g_{i,kl} \ , \tag{7.10a}$$

$$\tilde{N}_2 \tilde{Q}_{ij} b_{ij} = M \sum_{k} X_{k,ij} f_{k,ij} \ , \tag{7.10b}$$

$$\tilde{N}_2 \tilde{Q}_{ij} c_{ij} = M \sum_{k} X_{k,ij} g_{k,ij} \ . \tag{7.10c}$$

The average numbers of males (b_i) and females (c_i) born to a female who carries A_i satisfy

$$\tilde{q}_i b_i = \sum_{j} \tilde{Q}_{ij} b_{ij} \ , \qquad \tilde{q}_i c_i = \sum_{j} \tilde{Q}_{ij} c_{ij} \ , \tag{7.11a}$$

whence

$$\tilde{N}_2 \tilde{q}_i b_i = M \sum_{jk} X_{k,ij} f_{k,ij} \ , \tag{7.11b}$$

$$\tilde{N}_2 \tilde{q}_i c_i = M \sum_{jk} X_{k,ij} g_{k,ij} \ . \tag{7.11c}$$

For the mean fertilities per individual,

$$\bar{a} = \sum_{i} \tilde{p}_i a_i \ , \qquad \bar{b} = \sum_{i} \tilde{q}_i b_i \ , \qquad \bar{c} = \sum_{i} \tilde{q}_i c_i \ , \tag{7.12a}$$

we obtain from (7.7b), (7.10a), (7.11), and (7.12a)

$$\tilde{N}_1 \bar{a} = M \bar{g} = \tilde{N}_2 \bar{c} \ , \qquad \tilde{N}_2 \bar{b} = M \bar{f} \ . \tag{7.12b}$$

Note that the relation $\tilde{N}_1 \bar{a} = \tilde{N}_2 \bar{c}$ means that the total numbers of females born to males and females are the same, as is obvious. Inserting (7.10), (7.11), and (7.12) into (7.8a) and (7.9), we find

$$p'_i = \sum_{k} \tilde{q}_k b_k \bar{b}^{-1} S_{ki} \ , \tag{7.13a}$$

$$q'_i = \tfrac{1}{2} \sum_{k} (\tilde{p}_k a_k \bar{a}^{-1} R_{ki} + \tilde{q}_k c_k \bar{c}^{-1} S_{ki}) \ . \tag{7.13b}$$

We define

$$x_i = u_i a_i \ , \qquad y_{ij} = v_{ij} b_{ij} \ , \qquad z_{ij} = v_{ij} c_{ij} \ . \tag{7.14}$$

These quantities are the expected numbers of female progeny of a juvenile (i.e., pre-selection) A_i male, male progeny of a juvenile $A_i A_j$ female, and female progeny of a juvenile $A_i A_j$ female; they may be referred to as absolute genotypic fitnesses. Observe that

$$\bar{x} = \sum_{i} x_i p_i = \sum_{i} \bar{u} \tilde{p}_i a_i = \bar{u} \bar{a} \ , \tag{7.15}$$

from (7.14), (7.4a), and (7.12a). The average contributions to males and females of females who carry A_i are given by

$$q_i y_i = \sum_j Q_{ij} y_{ij} , \qquad q_i z_i = \sum_j Q_{ij} z_{ij} . \tag{7.16}$$

Using successively (7.16), (7.14), (7.4b), (7.11a), and (7.4b) again, we derive

$$q_i y_i = \sum_j \overline{v} \tilde{Q}_{ij} b_{ij} = \overline{v} \tilde{q}_i b_i = q_i v_i b_i . \tag{7.17}$$

Hence, and by identical reasoning on z_i, we get

$$y_i = v_i b_i , \qquad z_i = v_i c_i . \tag{7.18}$$

Utilizing (7.4b) and (7.12), with the device employed in (7.15) we obtain

$$\overline{y} = \sum_i q_i y_i = \overline{v}\,\overline{b} , \tag{7.19a}$$

$$\overline{z} = \sum_i q_i z_i = \overline{v}\,\overline{c} . \tag{7.19b}$$

Although the factorization properties (7.15), (7.18), and (7.19) of the mean and allelic fitnesses are expected from intuition concerning survival and reproduction, these relations require proof because they depend on the genotypic definitions (7.14). Since $\tilde{N}_1 = N_1 \overline{u}$ and $\tilde{N}_2 = N_2 \overline{v}$, therefore (7.12b), (7.15), and (7.19) show that the mean absolute fitnesses have the relationship $N_1 \overline{x} = N_2 \overline{z}$, which signifies the identity of the total numbers of female progeny of male and female zygotes.

It remains only to substitute (7.4), (7.14), (7.15), (7.18), and (7.19) into (7.13) to derive our basic equations for the gene frequencies:

$$p_i' = q_i^{\star\star} , \qquad q_i' = \tfrac{1}{2}(p_i^\star + q_i^\star) , \tag{7.20}$$

where

$$p_i^\star = \sum_k (p_k x_k / \overline{x}) R_{ki} , \tag{7.21a}$$

$$q_i^{\star\star} = \sum_k (q_k y_k / \overline{y}) S_{ki} , \tag{7.21b}$$

$$q_i^\star = \sum_k (q_k z_k / \overline{z}) S_{ki} . \tag{7.21c}$$

Note that the system (7.20), (7.21) is not complete: it depends on $X_{i,jk}$ and Q_{ij}. The interpretation of the starred allelic frequencies is immediate: $q_i^{\star\star}$, $q_i^\star$, and $p_i^\star$ are the frequencies of A_i in eggs that produce males, eggs that produce females, and sperm that produce females, respectively. The analysis of random mating below will support this assertion. If, as is generally the case, the sex ratio at birth, λ, is independent of the parental genotypes, $f_{i,jk} = \lambda g_{i,jk}$ for all i, j, and k, then $b_{ij} = \lambda c_{ij}$, so $y_{ij} = \lambda z_{ij}$, whence $q_i^* = q_i^{**}$. In the usual simple situation of constant λ, we have $N_1 = \lambda N_2$, so $N_1\overline{x} = N_2\overline{z}$ implies that $\overline{z} = \lambda\overline{x}$. From (7.12b) we infer also that $\overline{y} = \lambda\overline{z} = \lambda^2\overline{x}$, which leaves us with only one independent mean absolute fitness. If the zygotic sex ratio is unity, all three mean absolute fitnesses are the same.

It is important to recognize that the genotypic fitnesses x_i, y_{ij}, and z_{ij} may be scaled independently without altering (7.21). If they are multiplied by genotype-independent quantities, they become relative genotypic fitnesses, whose constancy may often be reasonably assumed. The three mean relative fitnesses are, of course, independent.

To obtain a complete system of difference equations, let us suppose that *mating is random*:

$$X_{i,jk} = \tilde{p}_i \tilde{Q}_{jk} \ . \tag{7.22}$$

Substituting (7.22) into (7.10) gives

$$\tilde{N}_1 a_i = M \sum_{kl} \tilde{Q}_{kl} g_{i,kl} \ , \tag{7.23a}$$

$$\tilde{N}_2 b_{ij} = M \sum_k \tilde{p}_k f_{k,ij} \ , \qquad \tilde{N}_2 c_{ij} = M \sum_k \tilde{p}_k g_{k,ij} \ . \tag{7.23b}$$

Let us posit that the rates at which various unions produce females may be expressed as *products* of factors that depend on the male and female genotypes:

$$g_{i,jk} = \beta_i \gamma_{jk} \ . \tag{7.24}$$

The corresponding restriction for the rates at which males are born is not required. Substituting (7.24) into (7.23) yields

$$\tilde{N}_1 a_i = M \overline{\gamma} \beta_i \ , \qquad \tilde{N}_2 c_{ij} = M \overline{\beta} \gamma_{ij} \ , \tag{7.25}$$

where

$$\overline{\beta} = \sum_i \beta_i \tilde{p}_i \ , \qquad \overline{\gamma} = \sum_{ij} \gamma_{ij} \tilde{Q}_{ij} \ . \tag{7.26}$$

Inserting (7.22) and (7.24) into (7.7b) produces

$$\overline{g} = \overline{\beta}\overline{\gamma} \ . \tag{7.27}$$

Equations (7.24), (7.25), (7.27), and (7.12b) lead to

$$g_{i,jk} = \overline{g} \, (\overline{a}\,\overline{c})^{-1} a_i c_{jk} \ . \tag{7.28}$$

Substituting (7.22) and (7.28) into (7.8b) and recalling (7.4), (7.11a), (7.14), (7.18), (7.15), (7.19b), and (7.21), we deduce the generalized Hardy-Weinberg Law:

$$Q_{ij}' = \tfrac{1}{2}(p_i^* q_j^* + p_j^* q_i^*) \ . \tag{7.29}$$

The equations are conveniently expressed in terms of gametic frequencies. From (7.21a), (7.20), and (7.15),

$$p_i^{*\prime} = \sum_k (q_k^{**} x_k / \overline{x}') R_{ki} \; , \tag{7.30a}$$

with

$$\overline{x}' = \sum_i x_i q_i^{**} \qquad . \tag{7.30b}$$

Similarly, from (7.21b), (7.21c), (7.16), and (7.29) we derive

$$q_i^{**\prime} = \sum_k (q_k' y_k' / \overline{y}') S_{ki} \; , \tag{7.31a}$$

where

$$q_k' y_k' = \tfrac{1}{2} \sum_l (p_k^* q_l^* + p_l^* q_k^*) y_{kl} \; ; \tag{7.31b}$$

$$q_i^{*\prime} = \sum_k (q_k' z_k' / \overline{z}') S_{ki} \; , \tag{7.32a}$$

where

$$q_k' z_k' = \tfrac{1}{2} \sum_l (p_k^* q_l^* + p_l^* q_k^*) z_{kl} \; ; \tag{7.32b}$$

and, of course,

$$\overline{y}' = \sum_i q_i' y_i' , \qquad \overline{z}' = \sum_i q_i' z_i' \; , \tag{7.33}$$

If the zygotic sex ratio is independent of the parental genotypes, then $q_i^* = q_i^{**}$, and hence (7.30) and (7.31) specify the evolution of the population. A direct argument shows that (7.30) to (7.33) apply also if, instead of mating, each genotype sheds gametes and these gametes fuse at random. In this case, $2a_i$, $2b_{ij}$, and $2c_{ij}$ represent the numbers of gametes shed by an A_i male that produce females, an $A_i A_j$ female that produce males, and an $A_i A_j$ female that produce females, respectively.

Finally, if μ_{ij} and ν_{ij} are the respective mutation rates in males and females, with the convention that $\mu_{ii} = \nu_{ii} = 0$ for all i, we have

$$R_{ij} = \delta_{ij} \left(1 - \sum_k \mu_{ik} \right) + \mu_{ij} \; , \tag{7.34a}$$

$$S_{ij} = \delta_{ij} \left(1 - \sum_k \nu_{ik} \right) + \nu_{ij} \; . \tag{7.34b}$$

Therefore, (7.30a) may be rewritten as

$$p_i^{*\prime} = (q_i^{**} x_i / \overline{x}') \left(1 - \sum_k \mu_{ik} \right) + \sum_k (q_k^{**} x_k / \overline{x}') \mu_{ki} \; , \tag{7.35}$$

with similar equations for q_i^* and q_i^{**}.

In the *absence of selection*, (7.20) and (7.21) reduce to the linear system

$$p_i' = \sum_k q_k S_{ki} \; , \tag{7.36a}$$

$$q_i' = \tfrac{1}{2} \sum_k (p_k R_{ki} + q_k S_{ki}) \; . \tag{7.36b}$$

Recalling (7.34), we can solve at once the diallelic case explicitly for $p_1(t)$ and $q_1(t)$. If $\mu_{ij}, \nu_{ij} \ll 1$ for $i, j = 1, 2$, we find the equilibrium

$$\hat{p}_1 \approx \hat{q}_1 \approx (\mu_{21} + 2\nu_{21})/(\tilde{\mu} + 2\tilde{\nu}) \; , \tag{7.37a}$$

where

$$\tilde{\mu} = \mu_{12} + \mu_{21} \; , \qquad \tilde{\nu} = \nu_{12} + \nu_{21} \; . \tag{7.37b}$$

For weak mutation, one expects the male and female gene frequencies at equilibrium to be equal. The factor 2 appears in (7.37a) because females have two X chromosomes to the males' one. The ultimate rate of approach to the equilibrium is

$$[1 - \tfrac{1}{3}(\tilde{\mu} + 2\tilde{\nu})]^t \approx e^{-(\tilde{\mu}+2\tilde{\nu})t/3} \; , \tag{7.38}$$

which displays the same factor 2. For $\mu_{ij} = \nu_{ij}$, (7.37) and (7.38) reduce to the autosomal case, (2.22) and (2.23).

If there is *no mutation and the juvenile sex ratio is genotype independent*, then (7.30) and (7.31a) simplify to

$$p_i^{\star\prime} = q_i^{\star} x_i/\overline{x}' \, , \qquad q_i^{\star\prime} = q_i' y_i'/\overline{y}_i' \; , \tag{7.39}$$

in which .

$$\overline{x}' = \sum_i x_i q_i^{\star} \; , \tag{7.40}$$

and $q_i' y_i'$ and $\overline{y}'$ are given by (7.31b) and (7.33). Sprott (1957) has discussed the stability of multiallelic polymorphisms of this system, and Cannings (1968a) showed that these polymorphisms, if they exist, may be calculated by matrix inversion, as for the autosomal case.

7.2 Selection with Two Alleles

With two alleles, no mutation, and genotype-independent zygotic sex ratio, from (7.31b), (7.33), (7.39) and (7.40) we obtain the recursion relations for the gene-frequency ratios (Haldane, 1926):

$$\xi = p_1^{\star}/p_2^{\star} \, , \qquad \eta = q_1^{\star}/q_2^{\star} \; , \tag{7.41}$$

$$\xi' = \rho\eta \quad , \qquad \eta' = (2\sigma\xi\eta + \xi + \eta)/(\xi + \eta + 2\tau) \; , \tag{7.42}$$

where the fitnesses appear only in the ratios

$$\rho = x_1/x_2 \, , \qquad \sigma = y_{11}/y_{12} \, , \qquad \tau = y_{22}/y_{12} \; . \tag{7.43}$$

We shall suppose that ρ, σ, and τ are constant.

If one of the alleles is lethal in males, it is easy to solve (7.42) for $\xi(t)$ and $\eta(t)$ (Problem 7.2). Haldane (1924) obtained the special case of this solution for recessives lethal also in females.

The mapping (7.42) of the first quadrant of the $\xi\eta$-plane into itself has the trivial fixed points $(0,0)$ and (∞,∞), corresponding to the fixation of a and A, respectively. If it exists (and a degenerate situation treated below is excluded), the unique polymorphic equilibrium

$$\hat{\xi} = \rho\hat{\eta}, \qquad \hat{\eta} = (\rho+1-2\tau)/(\rho+1-2\rho\sigma) \qquad (7.44)$$

can be located directly from (7.42). Manifestly, $0 < \hat{\eta} < \infty$, whence $0 < \hat{\xi} < \infty$, if and only if either

$$\tau < \tfrac{1}{2}(1+\rho) \qquad \text{and} \qquad \sigma < \tfrac{1}{2}(1+\rho^{-1}) , \qquad (7.45a)$$

or

$$\tau > \tfrac{1}{2}(1+\rho) \qquad \text{and} \qquad \sigma > \tfrac{1}{2}(1+\rho^{-1}) . \qquad (7.45b)$$

Thus, (7.45) is necessary and sufficient for the existence of a nontrivial equilibrium.

Now, $\sigma, \tau \ll 1$ corresponds to extreme heterozygote advantage, which we expect to lead to a stable equilibrium. Therefore, intuition suggests that, as Bennett (1957) has proved, (7.44) is locally asymptotically stable whenever (7.45a) holds. The analogous underdominance argument disposes us to guess that (7.45b) implies instability of (7.44).

Let us show that (7.45a) guarantees a protected polymorphism (Kimura, 1960; see Crow and Kimura, 1970, pp. 278–281). Near the origin, (7.42) has the linearization

$$\mathbf{Y}' = B\mathbf{Y} ,$$

where

$$\mathbf{Y} = \begin{pmatrix} \xi \\ \eta \end{pmatrix} , \qquad B = \begin{pmatrix} 0 & \rho \\ 1/(2\tau) & 1/(2\tau) \end{pmatrix} .$$

From (6.32) we find immediately that the maximal eigenvalue of the irreducible nonnegative matrix B exceeds unity if

$$\tau < \tfrac{1}{2}(1+\rho) . \qquad (7.46)$$

By the argument in Sect. 6.2, (7.46) suffices for protection of the allele A. We can find the condition for maintaining a by the interchange $A \leftrightarrow a$. According to (7.43), this corresponds to $\rho \leftrightarrow \rho^{-1}$ and $\sigma \leftrightarrow \tau$. Applying these transformations to (7.46) produces the second half of (7.45a).

We describe below the global behavior of (7.42). Case 4 and part of Case 5 were first analyzed by Cannings (1967) and the remainder by Palm (1974). The reader may consult their papers for proofs. For convenience, we set

$$\alpha = \sigma , \qquad \beta = \tau\rho^{-1} , \qquad \gamma = \tfrac{1}{2}(1+\rho^{-1}) . \qquad (7.47)$$

1. $\alpha \leq \gamma \leq \beta, \alpha \neq \beta$

Since (7.45) is violated, only the trivial equilibria exist; A is lost.

2. $\alpha \geq \gamma \geq \beta, \alpha \neq \beta$

Again, (7.45) fails, so a polymorphism does not exist, but now A is fixed.

3. $\alpha = \beta = \gamma$

In this degenerate case, we can easily check that all points on the line $(\rho\eta, \eta)$ through the origin are equilibria; there is global convergence to this line.

4. $\alpha, \beta < \gamma$

Now (7.45a) is satisfied and there is global convergence to the polymorphism (7.44).

5. $\alpha, \beta > \gamma$

Owing to (7.45b), the nontrivial equilibrium (7.44) is unstable. There is convergence to (7.44) from its unidimensional stable manifold, S. If the population is initially not on S, one of the alleles will be ultimately lost. (Of course, this will always happen in practice.) The curve S through (7.44) separates the regions of attraction of the trivial equilibria. If $\eta(0), \eta(1) < \hat{\eta}$, then A is lost; if $\eta(0), \eta(1) > \hat{\eta}$, A is fixed.

The following special cases are illuminating.

(a) No Selection in Males

With $\rho = 1$, the male and female gene frequencies are ultimately equal. In the five cases above, we have $\alpha = \sigma$, $\beta = \tau$, and $\gamma = 1$. The fate of the population (though not its evolution) is the same as if the locus were autosomal and all individuals had the female fitnesses.

(b) No Selection in Females

Clearly, $\alpha = 1$, $\beta = \rho^{-1}$, and $\gamma = \frac{1}{2}(1 + \rho^{-1})$. No polymorphism is possible, the eventual fate of the population being that of a haploid one with the male fitnesses.

(c) A Dominant

Using $\sigma = 1$ in (7.45a) yields $\rho < 1$ and $\tau < 1$. Hence (excluding the case without selection), for a stable polymorphism the recessive allele must be advantageous in males and deleterious in females.

(d) No Dominance

From (7.43) we get $\sigma + \tau = 2$, whence the necessary and sufficient condition (7.45a) for a stable polymorphism in the presence of selection becomes

$$\tfrac{1}{2}(3 - \rho) < \sigma < \tfrac{1}{2}(1 + \rho^{-1}) \ .$$

This can be satisfied only if A is favored in males and harmful in females ($\rho > 1$ and $\sigma < 1$) or *vice versa* ($\rho < 1$ and $\sigma > 1$).

(e) No Multiplicative Dominance

Now $\sigma\tau = 1$, so (7.45a) simplifies to

$$2(1 + \rho)^{-1} < \sigma < \tfrac{1}{2}(1 + \rho^{-1}) \ ,$$

from which the conclusion d follows.

7.3 Mutation-Selection Balance

We shall relate at equilibrium the frequencies of mutant alleles in males and females to the mean selection coefficients and mutation rates in the two sexes, as we did for an autosomal locus in Sect. 4.9. Our treatment follows Nagylaki (1977b). We suppose that the juvenile sex ratio is independent of the parental genotypes and choose A_1 as the normal allele. We impose no restrictions on the fertilities or the mating system. From (7.20), (7.21), and (7.34) we obtain at equilibrium

$$p_1 = (q_1 y_1 / \overline{y})(1 - \nu) + \sum_i (q_i y_i / \overline{y}) \nu_{i1} \ , \tag{7.48a}$$

$$q_1 = \tfrac{1}{2}\Big[(p_1 x_1 / \overline{x})(1 - \mu) + (q_1 y_1 / \overline{y})(1 - \nu) \\ + \sum_i (p_i x_i / \overline{x}) \mu_{i1} + \sum_i (q_i y_i / \overline{y}) \nu_{i1} \Big] \ , \tag{7.48b}$$

where

$$\mu = \sum_i \mu_{1i}, \qquad \nu = \sum_i \nu_{1i} \tag{7.49}$$

are the total forward mutation rates for A_1 in males and females.

We consider only deleterious mutants, so we choose the fitnesses $x_i = 1 - r_i$ and $y_{ij} = 1 - s_{ij}$, with $r_i \geq 0$, $s_{ij} \geq 0$, $r_1 = s_{11} = 0$, and the biologically reasonable restriction that r_i and s_{1i} cannot both vanish for any $i > 1$. For recessives, $s_{1i} = 0$, but it would be highly unusual to have $r_i = 0$ for $i > 1$. For weak mutation, which we may define by $\mu_{1i} + 2\nu_{1i} \ll r_i + 2s_{1i}$ for all $i > 1$, the autosomal analysis of Sect. 4.9 leads us to expect that mutant frequencies will be ratios of mutation rates to selection coefficients. Then the reverse-mutation terms, the sums in (7.48), may be neglected to the first order in the mutation rates.

To proceed, we calculate the fitnesses in the leading terms of (7.48). We have

$$\bar{x} = 1 - \sum_i r_i p_i = 1 - (1 - p_1)\bar{r} \ , \tag{7.50a}$$

where

$$\bar{r} = (1 - p_1)^{-1} \sum_i r_i p_i \tag{7.50b}$$

is the mean selection coefficient of the mutants in males. For the females, notice that for $i > 1$

$$q_i = Q_{i1} + \sum_{j>1} Q_{ij} \approx Q_{i1} \ , \tag{7.51}$$

because $Q_{ij} \ll Q_{i1}$ for $j > 1$ unless the inbreeding coefficient is quite substantial. Therefore,

$$q_1 y_1 = \sum_i (1 - s_{\bar{1}i}) Q_{1i} \approx q_1 - \sum_i s_{1i} q_i = q_1 - (1 - q_1)\bar{s} \ , \tag{7.52a}$$

where

$$\bar{s} = (1 - q_1)^{-1} \sum_i q_i s_{1i} \tag{7.52b}$$

is the mean selection coefficient of normal-mutant females. Also,

$$\bar{y} = \sum_{ij} (1 - s_{ij}) Q_{ij} \approx 1 - 2 \sum_i s_{1i} Q_{1i} \approx 1 - 2(1 - q_1)\bar{s} \ . \tag{7.53}$$

Substituting (7.50a), (7.52a), and (7.53) into (7.48) yields

$$p_1 \approx \left[\frac{q_1 - (1 - q_1)\bar{s}}{1 - 2(1 - q_1)\bar{s}} \right] (1 - \nu) \ , \tag{7.54a}$$

$$q_1 \approx \frac{1}{2} \left[\frac{p_1}{1 - (1 - p_1)\bar{r}} \right] (1 - \mu) + \frac{1}{2} \left[\frac{q_1 - (1 - q_1)\bar{s}}{1 - 2(1 - q_1)\bar{s}} \right] (1 - \nu) \ . \tag{7.54b}$$

Recalling that $1 - p_1$ and $1 - q_1$ are of the order of the mutation rates, to first order we obtain from (7.54)

$$p_1 \approx q_1 + (1 - q_1)\bar{s} - \nu \ , \tag{7.55a}$$
$$2q_1 \approx p_1 + q_1 + (1 - p_1)\bar{r} + (1 - q_1)\bar{s} - \mu - \nu \ . \tag{7.55b}$$

Inserting (7.55a) into (7.55b), we find immediately

$$1 - q_1 \approx \frac{\mu + (2 - \bar{r})\nu}{\bar{r} + 2\bar{s} - \bar{r}\,\bar{s}} \ , \tag{7.56a}$$

and substituting (7.56a) into (7.55a) gives

$$1 - p_1 \approx \frac{(1 - \bar{s})\mu + 2\nu}{\bar{r} + 2\bar{s} - \bar{r}\,\bar{s}} \ . \tag{7.56b}$$

Several *special cases* of our general result, (7.56), are of interest.
If the total mutation rate is the same in the two sexes, (7.56) reduces to

$$1 - p_1 \approx \frac{(3 - \bar{s})\mu}{\bar{r} + 2\bar{s} - \overline{r\,s}} , \qquad 1 - q_1 \approx \frac{(3 - \bar{r})\mu}{\bar{r} + 2\bar{s} - \overline{r\,s}} . \qquad (7.57)$$

With equal total mutation rates and mean selection coefficients in the two sexes, (7.57) simplifies to

$$1 - p_1 \approx 1 - q_1 \approx \mu/\bar{r} . \qquad (7.58)$$

If A_1 is dominant, then $\bar{s} = 0$, so (7.56) reads

$$1 - p_1 \approx (\mu + 2\nu)/\bar{r}, \qquad 1 - q_1 \approx [\mu + (2 - \bar{r})\nu]/\bar{r} . \qquad (7.59)$$

Haldane (1935) derived the special case of (7.59) with two alleles and generalized Hardy-Weinberg proportions, (7.29). His formulae are for $p_2^\star$ and $q_2^\star$, with the definitions of μ and ν interchanged. If the mutants are lethal, (7.59) becomes

$$1 - p_1 \approx \mu + 2\nu , \qquad 1 - q_1 \approx \mu + \nu .$$

Finally, for weak selection, (7.56) yields

$$1 - p_1 \approx 1 - q_1 \approx \frac{\mu + 2\nu}{\bar{r} + 2\bar{s}} ,$$

both the approximate equality of p_1 and q_1 and the factors of 2 being expected. See Szucs (1991) for an exact global analysis of the diallelic case.

7.4 Weak Selection

We shall now investigate the evolution of a multiallelic locus under weak selection, following Nagylaki (1979b), who treats also the dioecious autosomal case. We assume that mating is random, the fertilities factor, and the zygotic sex ratio is independent of the parental genotypes. To simplify the notation, we write p_i and q_i for the frequencies of A_i in sperm and eggs, respectively, and redefine $\bar{x}'$ as $\bar{x}$. Then (7.30) and (7.31) have the form

$$p_i' = q_i x_i/\bar{x} , \qquad (7.60a)$$

$$q_i' = (2\bar{y})^{-1} \sum_j (p_i q_j + p_j q_i) y_{ij} , \qquad (7.60b)$$

$$\bar{x} = \sum_i x_i q_i , \qquad \bar{y} = \sum_{ij} y_{ij} p_i q_j . \qquad (7.60c)$$

The male and female relative genotypic fitnesses, x_i and y_{ij}, may be functions of the gametic frequencies, $\mathbf{p}$ and $\mathbf{q}$, and time, t. For weak selection, we may choose all the relative genotypic fitnesses to be close to unity: with selection intensity s,

$$x_i = 1 + s\xi_i, \qquad y_{ij} = 1 + s\eta_{ij}, \qquad\qquad (7.61)$$

in which ξ_i and η_{ij} are bounded as $s \to 0$. In (7.61), s is the order of magnitude of a typical selective difference. The most conservative definition would be

$$s = \sup\left[\, |x_i(\mathbf{p},\mathbf{q},t) - x_j(\mathbf{p},\mathbf{q},t)|, \ |y_{ij}(\mathbf{p},\mathbf{q},t) - y_{kl}(\mathbf{p},\mathbf{q},t)|\,\right],$$

where the supremum is over all i, j, k, l, $\mathbf{p}$, $\mathbf{q}$, and t.

From (7.60) and (7.61) we obtain

$$p_i' = q_i + sq_i(\xi_i - \bar{\xi}) + O(s^2), \qquad\qquad (7.62a)$$

$$q_i' = \tfrac{1}{2}(p_i + q_i) + \tfrac{1}{2}s\left[\sum_j (p_i q_j + p_j q_i)\eta_{ij} - (p_i + q_i)\bar{\eta}\right] + O(s^2), \qquad (7.62b)$$

with

$$\bar{\xi} = \sum_i \xi_i q_i, \qquad \bar{\eta} = \sum_{ij} \eta_{ij} p_i q_j. \qquad\qquad (7.62c)$$

Let P_i and Q_i represent the average frequency of A_i in gametes and one third of the frequency difference between sperm and eggs, respectively:

$$P_i = \tfrac{1}{3}(p_i + 2q_i), \qquad Q_i = \tfrac{1}{3}(p_i - q_i). \qquad\qquad (7.63)$$

Subtracting (7.62b) from (7.62a) gives

$$Q_i' = -\tfrac{1}{2}Q_i + sg_i(\mathbf{P},\mathbf{Q},t), \qquad\qquad (7.64)$$

where g_i is bounded as $s \to 0$. Iteration of (7.64) leads to

$$Q_i(t) = Q_i(0)(-\tfrac{1}{2})^t + s(-\tfrac{1}{2})^t \sum_{\tau=1}^{t}(-\tfrac{1}{2})^{-\tau} g_i[\mathbf{P}(\tau-1), \mathbf{Q}(\tau-1), \tau-1]. \quad (7.65)$$

(The sum is absent for $t = 0$.)

We define t_1 as the shortest time such that

$$|Q_i(0)|(\tfrac{1}{2})^{t_1} \leq s \qquad\qquad (7.66)$$

for all i. If the initial gametic frequencies in males and females are sufficiently close, then $t_1 = 0$. Otherwise, (7.63) and (7.66) yield the conservative estimate $t_1 \approx -\ln s$, which rarely exceeds 5 or 10 generations. Since biology allows us to assume that $|g_i|$ is uniformly bounded, (7.65) implies

$$Q_i(t) = O(s), \qquad t \geq t_1. \qquad\qquad (7.67)$$

Observe that the gene-frequency change during the time t_1 is very small, roughly $st_1 \approx -s\ln s$.

We can now approximate the dynamics of our sex-linked locus by that of an autosomal one in a monoecious population. From (7.61), (7.62), (7.63), and (7.67) we deduce

$$\Delta P_i = P_i(w_i - \tilde{w}) + O(s^2), \qquad t \geq t_1, \qquad\qquad (7.68a)$$

where

$$y_i = \sum_j y_{ij} P_j, \qquad w_i = \tfrac{1}{3}(x_i + 2y_i), \qquad \tilde{w} = \sum_i w_i P_i . \qquad (7.68b)$$

If the genotype $A_i A_j$ has fitness W_{ij} in a monoecious population, then the gene frequencies, π_i, satisfy

$$\Delta \pi_i = \pi_i (\tilde{W}_i - \tilde{W}) + O(s^2), \qquad \tilde{W}_i = \sum_j W_{ij} \pi_j, \qquad \tilde{W} = \sum_i \tilde{W}_i \pi_i .$$
$$(7.69)$$

We choose $\pi_i(t_1) = P_i(t_1)$ and

$$W_{ij} = \tfrac{1}{3}(x_i + x_j - 1) + \tfrac{2}{3} y_{ij} , \qquad (7.70)$$

for every i and j, and compare (7.68) with (7.69). From (7.69) and (7.70) we get

$$\tilde{W}_i - \tilde{W} = \tfrac{1}{3}(x_i - \tilde{x}) + \tfrac{2}{3}(\tilde{y}_i - \tilde{y}) , \qquad (7.71a)$$

where

$$\tilde{x} = \sum_i x_i \pi_i, \qquad \tilde{y}_i = \sum_j y_{ij} \pi_j, \qquad \tilde{y} = \sum_i \tilde{y}_i \pi_i . \qquad (7.71b)$$

Therefore, the leading terms in ΔP_i and $\Delta \pi_i$, which are of $O(s)$, have the same form. It follows that

$$P_i(t) = \pi_i(t) + O(s), \qquad t \geq t_1 . \qquad (7.72)$$

In view of (7.63) and (7.67), it is evident that (7.72) holds also with P_i replaced by p_i or q_i. Thus, with an error of the order of the selection intensity, after a short time a sex-linked locus evolves like an autosomal one in a monoecious population. If $\boldsymbol{\pi}(t)$ does not necessarily converge to some equilibrium point or if $\boldsymbol{\pi}(t_1)$ happens to be on the stable manifold of an unstable equilibrium, then small perturbations may cause large deviations in its ultimate state. In this case, (7.72) is still valid if $t_1 \leq t \leq k/s$ for some constant k.

We have included the x_j term in (7.70) in order to enforce the biological symmetry property $W_{ij} = W_{ji}$; the constant term ensures that $W_{ij} = 1$ for every i and j if $x_i = 1$ and $y_{ij} = 1$ for every i and j. Neither term changes $\Delta \pi_i$ in (7.69). According to (7.69) and (7.71), the evolution of the population is controlled by the ploidy-weighted average of the male and female allelic fitnesses. Recalling (7.61), we can rewrite (7.70) in terms of selection coefficients as

$$W_{ij} = 1 + s\zeta_{ij}, \qquad \zeta_{ij} = \tfrac{1}{3}(\xi_i + \xi_j) + \tfrac{2}{3}\eta_{ij} . \qquad (7.73a)$$

Specializing (7.70) to two alleles is instructive. Suppose A and a males have fitnesses $1 + s_1$ and $1 - s_1$, respectively. Let $1 + s_2$, $1 + hs_2$, and $1 - s_2$ designate the respective fitnesses of AA, Aa, and aa females. In the approximately equivalent monoecious population, these genotypes have fitnesses $1 + s_0$, $1 + h_0 s_0$, and $1 - s_0$, respectively. Then (7.70) yields [cf. Owen (1986)]

$$s_0 = \tfrac{2}{3}(s_1 + s_2), \qquad h_0 = h s_2 / (s_1 + s_2) . \qquad (7.73b)$$

Hence, the degree of dominance in the monoecious population is the same as in females ($h_0 = h$) if and only if there is no dominance in females ($h = 0$) or no selection in males ($s_1 = 0$). Even with no overdominance in females ($|h| \leq 1$), selection in opposite directions in the two sexes ($s_1 s_2 < 0$) can induce overdominance in the monoecious population ($|h_0| > 1$). If selection is in the same direction in the two sexes ($s_1 s_2 > 0$), overdominance in females may not induce overdominance in the monoecious population. Thus, overdominance in females is neither necessary nor sufficient for the existence of a stable polymorphism.

Let us establish next that the gene-frequency differences rapidly become almost constant. From (7.64) we derive for $t \geq t_1$

$$(\Delta Q_i)' = \Delta Q_i' = -\tfrac{1}{2}\Delta Q_i + s\Delta g_i(\mathbf{P}, \mathbf{Q}, t) . \qquad (7.74)$$

But

$$\Delta g_i = \{g_i[\mathbf{P}(t+1), \mathbf{Q}(t+1), t+1] - g_i[\mathbf{P}(t), \mathbf{Q}(t), t+1]\}$$
$$+ \{g_i[\mathbf{P}(t), \mathbf{Q}(t), t+1] - g_i[\mathbf{P}(t), \mathbf{Q}(t), t]\} . \qquad (7.75)$$

From (7.67) and (7.68a) it follows that for $t \geq t_1$

$$\Delta P_i = O(s), \qquad \Delta Q_i = O(s) , \qquad (7.76)$$

so Taylor's theorem informs us that the first brace in (7.75) is of $O(s)$. We assume that the second brace is also of $O(s)$; (7.62) and (7.64) show that this restricts the explicit time dependence of the fitnesses to $O(s^2)$. Then (7.64) gives

$$\Delta Q_i' = -\tfrac{1}{2}\Delta Q_i + O(s^2), \qquad t \geq t_1 ,$$

whence the analysis applied to (7.64) yields

$$\Delta Q_i(t) = O(s^2), \qquad t \geq 2t_1 . \qquad (7.77)$$

Turning to the changes in the *mean fitness*, we note first that the mean male and female fitnesses can decrease simultaneously. Indeed, with two alleles, $x_1 < x_2$ and $y_{22} < y_{12}$, and $p_1 = 1$ and $q_1 = 0$, it follows that $\bar{x}' < \bar{x}$ and $\bar{y}' < \bar{y}$. Therefore, we restrict ourselves to weak selection and seek results for $t \geq t_1$.

Define

$$\tilde{x} = \sum_i P_i x_i, \qquad y_i = \sum_j y_{ij} P_j, \qquad \tilde{y} = \sum_{ij} y_{ij} P_i P_j . \qquad (7.78)$$

Noting that

$$\sum_i \Delta P_i = 0$$

and using (7.68a) and (7.76), we obtain for $t \geq t_1$

$$\Delta \tilde{x} = \widetilde{\Delta x} + \sum_i P_i(x_i - \tilde{x})(w_i - \tilde{w}) + O(s^3) \ , \qquad (7.79a)$$

$$\Delta \tilde{y} = \widetilde{\Delta y} + 2 \sum_i P_i(y_i - \tilde{y})(w_i - \tilde{w}) + O(s^3) \ , \qquad (7.79b)$$

where

$$\widetilde{\Delta x} = \sum_i P_i' \Delta x_i \ , \qquad \widetilde{\Delta y} = \sum_{ij} P_i' P_j' \Delta y_{ij} \ . \qquad (7.80)$$

Hence, in order to obtain a variance, we must weight $\tilde{x}$ and $\tilde{y}$ equally. Recalling (7.68b) and adding (7.79a) to (7.79b), we get

$$\Delta \tilde{z} = V_g + \widetilde{\Delta z} + O(s^3) \qquad (7.81a)$$

for $t \geq t_1$, where

$$\tilde{z} = \tfrac{1}{2}(\tilde{x} + \tilde{y}), \qquad \widetilde{\Delta z} = \tfrac{1}{2}(\widetilde{\Delta x} + \widetilde{\Delta y}) \ ,$$

and the genic variance is reasonably defined by

$$V_g = \tfrac{3}{2} \sum_i P_i(w_i - \tilde{w})^2 \ . \qquad (7.81b)$$

Equations (2.17), (4.57), and (7.81b) display the mean number of genes (1, 2, and $\tfrac{3}{2}$, respectively) per individual at the locus under consideration.

We can gain more insight into the meaning of the genic variance by generalizing Hartl's (1972) least-squares analysis to multiple alleles. To this end only, assume that A_i has frequency P_i in both sexes; according to (7.67), this holds with an error of $O(s)$. Write $X_i = x_i - \tilde{x}$ and $Y_{ij} = y_{ij} - \tilde{y}$ for the deviations of the fitnesses from their means and weight males and females in the total variance in the population in proportion to their sex-linked genetic contribution to the next generation:

$$V = \tfrac{1}{3} \sum_i P_i X_i^2 + \tfrac{2}{3} \sum_{ij} P_i P_j Y_{ij}^2 \ .$$

We define the additive effects α_i with complete compensation (Hartl, 1972):

$$X_i = 2\alpha_i + d_i \ , \qquad Y_{ij} = \alpha_i + \alpha_j + D_{ij} \ .$$

Then the genic and "dominance" (better, "non-dosage-compensation") variances read

$$V_g = \tfrac{1}{3} \sum_i P_i(2\alpha_i)^2 + \tfrac{2}{3} \sum_{ij} P_i P_j(\alpha_i + \alpha_j)^2 \ ,$$

$$V_d = \tfrac{1}{3} \sum_i P_i d_i^2 + \tfrac{2}{3} \sum_{ij} P_i P_j D_{ij}^2 \ .$$

Minimizing V_d with respect to α_i, as in Sect. 4.10, produces $\alpha_i = \frac{3}{4}(w_i - \tilde{w})$ and $V = V_g + V_d$ and demonstrates that V_g is given by (7.81a).

It is also enlightening to examine $\bar{z} = \frac{1}{2}(\bar{x} + \bar{y})$. From (7.63) and (7.67) we find

$$\bar{z} = \tilde{z} + O(s^2), \qquad t \geq t_1 . \tag{7.82a}$$

Using (7.63), (7.67), (7.76), and (7.77), we find for $t \geq 2t_1$

$$\Delta\bar{z} = \Delta\tilde{z} + \frac{1}{2}\sum_{ij} P_i' Q_j' \Delta y_{ij} - \frac{1}{2}\sum_i Q_i' \Delta x_i + O(s^3) . \tag{7.82b}$$

Suppose now that the rate of change of the genotypic fitnesses x_i and y_{ij} is $o(s^2)$. Then for $t \geq t_1$ the mean fitness $\tilde{z}$ will generally increase at a rate close to the additive component of the genetic variance, V_g. The same form of Fisher's Fundamental Theorem of Natural Selection is correct for $\bar{z}$ after the doubled adjustment period $2t_1$.

The mean fitness may decrease in special cases even with constant genotypic fitnesses. Since $\bar{z}$ is not generally stationary at equilibrium, it must decrease along some orbits sufficiently close to an asymptotically stable equilibrium. This happens because the genic variance is particularly small near equilibrium. Then the third-order terms in (7.82b), which can be negative, may dominate. But the mean fitness generally increases while almost all the gene-frequency change occurs, and this is the period of greatest evolutionary interest.

Hartl (1972) presented an intuitive slow-selection analysis of the changes in mean fitness for two alleles. He assumed equal allelic frequencies in males and females and replaced differences by derivatives. Hartl investigated the geometric-mean fitness $z^* = (\tilde{x}\tilde{y})^{1/2}$; with our choice of scale, $z^* = \tilde{z} + O(s)$. His formula for Δz^* agrees with our result (7.81a) if a minor correction is made: his Eq. (4) should have no $\frac{2}{3}$ and $\frac{1}{3}$; in consequence, $\frac{1}{3}$ and $\frac{1}{6}$ in his Eqs. (10) to (12) should both be replaced by $\frac{1}{2}$. This agreement occurs because, under the very weak assumption that the genotypic fitnesses x_i and y_{ij} change at a rate $o(s)$, a Taylor series and (7.79) yield $\Delta z^* = \Delta\tilde{z} + o(s^2)$ for $t \geq t_1$.

Thus, we have established the following properties of evolution at a multiallelic X-linked locus under weak selection.

In time $t_1 \approx -\ln s$, the sperm-egg gene-frequency differences are reduced to $O(s)$; thereafter, evolution at an X-linked locus is close to the dynamics of an autosomal one in a monoecious population with appropriately averaged allelic fitnesses. With constant or slowly varying genotypic fitnesses, the mean fitness generally increases and the change per generation is close to the genic variance. As observed by Hartl (1972), the latter must be evaluated by assuming complete dosage compensation for the additive effects on fitness. Finally, after $2t_1$ generations have elapsed, the gametic frequency differences are very nearly constant and change at the rate of only $O(s^2)$.

The theory of this section applies equally well to species with arrhenotokous parthenogenesis. Consult Hartl (1971) for biological discussion.

Refer to Abugov (1985) and Nagylaki (1987) for generalization of the analysis in this section to arbitrary fertilities.

7.5 Problems

7.1 Verify (7.37) and (7.38).

7.2 Solve (7.42) for $\rho = 0$ (A lethal in males). Show that A is lost for $\tau \geq \frac{1}{2}$, but there is a globally stable polymorphism for $\tau < \frac{1}{2}$.

7.3 Show that if there is no selection in males, at equilibrium (7.39) and (7.40) reduce to the equations for selection at an autosomal locus with the female fitnesses. What is the corresponding result for no selection in females?

7.4 Give a complete formulation of selection in a dioecious population with discrete, nonoverlapping generations. Consider a single autosomal locus with multiple alleles. In terms of mating frequencies, fertilities, and viabilities, write the most general recursion relation for the zygotic genotypic frequencies. Show that with suitably defined fitnesses the zygotic allelic frequencies always satisfy

$$p_i' = \frac{1}{2} \left(\frac{p_i w_i}{\overline{w}} + \frac{q_i y_i}{\overline{y}} \right) , \tag{7.83a}$$

$$q_i' = \frac{1}{2} \left(\frac{p_i x_i}{\overline{x}} + \frac{q_i z_i}{\overline{z}} \right) . \tag{7.83b}$$

Reduce the equations for the zygotic proportions with the additional assumptions of random mating and multiplicative fertilities. Prove that the zygotic frequencies are in generalized Hardy-Weinberg proportions in terms of the frequencies p_i^*, p_i^{**}, q_i^*, and q_i^{**} of A_i in sperm that produce males, sperm that produce females, eggs that produce males, and eggs that produce females, respectively. Demonstrate that if the sex ratio is independent of parental genotypes, then $p_i^{**} = p_i^*$, $q_i^{**} = q_i^*$, and the zygotic frequencies are the same in the two sexes after one generation of panmixia. Show that in this simple, standard case, the model is completely specified by the equations

$$p_i^{*\prime} = B_i/B , \qquad q_i^{*\prime} = C_i/C , \tag{7.84a}$$

$$B_i = \frac{1}{2} \sum_j (p_i^* q_j^* + p_j^* q_i^*) w_{ij} , \qquad B = \sum_i B_i , \tag{7.84b}$$

$$C_i = \frac{1}{2} \sum_j (p_i^* q_j^* + p_j^* q_i^*) y_{ij} , \qquad C = \sum_i C_i . \tag{7.84c}$$

Prove that if there is no selection in females, these equations reduce at equilibrium to those for an autosomal locus with the male fitnesses. For further results, see Owen (1953), Bodmer (1965), Kidwell *et al.* (1977), Karlin (1978), Nagylaki (1979b), Karlin and Lessard (1986, Chap. 7), Selgrade and Ziehe (1987), and Szucs (1991a,b).

7.5 Assume that the sex ratio in Problem 7.4 is independent of genotype and investigate the conditions for the existence of a protected polymorphism in the diallelic case.

(a) Define $\alpha = w_{11}/w_{12}$, $\beta = w_{22}/w_{12}$, $\gamma = y_{11}/y_{12}$, and $\delta = y_{22}/y_{12}$ and prove that A_1 is protected if

$$\tfrac{1}{2}(\beta^{-1} + \delta^{-1}) > 1 \qquad (7.85a)$$

(Parsons, 1961), whereas A_2 is protected if

$$\tfrac{1}{2}(\alpha^{-1} + \gamma^{-1}) > 1 . \qquad (7.85b)$$

Derive (7.85) from both the one-dimensional linearization of (7.83) and the two-dimensional difference equation (7.84).

(b) If A_1 is recessive ($\beta = \delta = 1$), show that it cannot be lost if

$$s + \sigma > 0 , \qquad (7.86a)$$

where $s = \alpha - 1$ and $\sigma = \gamma - 1$ (Nagylaki, 1979a). [Recall the derivation of (6.44).]

(c) Deduce that there is a protected polymorphism if (7.86a) and

$$s^{-1} + \sigma^{-1} > -2 \qquad (7.86b)$$

both hold (Nagylaki, 1979a).

(d) Sketch the region defined by (7.86).

7.6 Formulate a general model with adult migration for a dioecious population that occupies an arbitrary number of niches. Assume that the sex ratio is independent of the genotype at the autosomal locus under consideration, there is random mating in every niche, and generations are discrete and nonoverlapping. Investigate the conditions for the protection of A_1 in the diallelic case (Nagylaki, 1979a).

(a) Let the soft-selection matrix Q in (6.30) refer to males; use $\overline{Q}$ for the corresponding quantity for females; and retain p_i for the frequency of A_1 in zygotes in niche i. Show that Q in (6.30) must be replaced by $\tfrac{1}{2}(Q + \overline{Q})$, as asserted by Christiansen (1974).

(b) Deduce for the dioecious island model with influx of $A_2 A_2$ males and females at rates m and $\overline{m}$, respectively, and $A_2 A_2$ male and female fitness ratios v and $\overline{v}$ relative to $A_1 A_2$ that the condition

$$\tfrac{1}{2}[(1-m)v^{-1} + (1-\overline{m})\overline{v}^{-1}] > 1 \qquad (7.87)$$

suffices to protect A_1.

(c) Prove that the analogous criterion in the dioecious Levene model reads

$$\tfrac{1}{2}\sum_i (c_i v_i^{-1} + \overline{c}_i \overline{v}_i^{-1}) > 1 \ . \qquad (7.88)$$

(d) For hard selection, demonstrate that R in (6.34) becomes $\tfrac{1}{2}(R + \overline{R})$.

(e) Write M and $\overline{M}$ for the male and female backward migration matrices, respectively, and $\mathbf{u}$ or $\overline{\mathbf{u}}$ for the corresponding $A_1 A_1$ to $A_1 A_2$ fitness ratios. Suppose that $\mathbf{U}$ is the unique left eigenvector of $M + \overline{M}$ with eigenvalue two; normalize $\mathbf{U}$ so that

$$\sum_i U_i = 1 \ .$$

Establish that with soft selection the recessive A_1 cannot be lost if

$$\tfrac{1}{2}\mathbf{U}^T(M\mathbf{u} + \overline{M}\,\overline{\mathbf{u}}) > 1 \ . \qquad (7.89)$$

Show that for the Levene model (7.89) simplifies to the arithmetic-mean condition

$$\tfrac{1}{2}\sum_i (c_i u_i + \overline{c}_i \overline{u}_i) > 1 \ . \qquad (7.90)$$

(f) Prove that for hard selection the recessive A_1 is preserved if

$$\tfrac{1}{2}\mathbf{Y}^T(M^\star \mathbf{u} + \overline{M}^\star \overline{\mathbf{u}}) > 1 \ , \qquad (7.91)$$

where $M^\star$ and $\overline{M}^\star$ are defined for each sex as in (6.37), and $\mathbf{Y}$ is the normalized maximal left eigenvector of $M^\star + \overline{M}^\star$.

7.7 Repeat Problem 7.6 for juvenile migration with constant male and female backward migration matrices M and $\overline{M}$ (Nagylaki (1979a)).

(a) Show that T in (6.52) must be replaced by $\tfrac{1}{2}(T + \overline{T})$. Infer that, in contradistinction to the monoecious case, the conditions for protection of an allele are generally not the same for juvenile and adult migration. Prove that a sufficient condition for their identity is symmetry of both backward migration matrices. Establish that the protection criteria are also identical if either the migration or the selection patterns in the two sexes are the same.

(b) Demonstrate that (7.87) holds.

(c) To obtain the condition for protecting A_1 in the Levene model, use the genotypic frequencies in male and female zygotes just after migration as the basic variables. Establish that this condition is

$$b_{11} + b_{22} > \min(2, \ 1 + b_{11}b_{22} - b_{12}b_{21}) \ , \qquad (7.92)$$

where
$$b_{11} = \tfrac{1}{2}\sum_i c_i/v_i, \qquad b_{12} = \tfrac{1}{2}\sum_i c_i/\bar{v}_i,$$
$$b_{21} = \tfrac{1}{2}\sum_i \bar{c}_i/v_i, \qquad b_{22} = \tfrac{1}{2}\sum_i \bar{c}_i/\bar{v}_i,$$

c_i and $\bar{c}_i$ still being measured just after reproduction. Verify that, in agreement with Part a, (7.88) and (7.92) are identical if either $c_i = \bar{c}_i$ or $v_i = \bar{v}_i$ for all i.

(d) Derive the criterion
$$\tfrac{1}{2}\mathbf{U}^T(\mathbf{u}+\bar{\mathbf{u}}) > 1 \tag{7.93}$$

for protecting the recessive A_1. Observe that this generally differs from (7.89), but (7.89) reduces to (7.93) if any one of the three sufficient conditions in Part a is fulfilled.

Demonstrate that for the Levene model, (7.93) simplifies to
$$\tfrac{1}{4}\sum_i (c_i+\bar{c}_i)(u_i+\bar{u}_i) > 1, \tag{7.94}$$

which agrees with (7.90) if either $c_i = \bar{c}_i$ or $u_i = \bar{u}_i$ for all i.

7.8 Prove that with the assumptions of Problem 7.6, the dioecious Levene model with hard selection is equivalent to selection in a panmictic population with the genotypic fitnesses given by averages over the niches for males and females.

8. Two Loci

We are interested in the two-locus system because some characters are determined by two loci and the understanding of this system is essential for the analysis of multilocus traits. The latter, which exhibit continuous or almost continuous variation, are of the greatest evolutionary significance. In view of our second motivation, it will be useful to devote Sect. 8.1 to establishing generalized Hardy-Weinberg proportions and deriving the recursion relations satisfied by the gametic frequencies for an arbitrary number of autosomal loci. We shall specialize our equations to two multiallelic loci and analyze this system in Sect. 8.2. In Sect. 8.3, examples that concern two diallelic loci will further elucidate two-locus behavior. Section 8.4 deals with two multiallelic loci in continuous time.

8.1 General Formulation for Multiple Loci

We consider an arbitrary number of autosomal loci and posit random mating and multiplicative fertilities. In the process of demonstrating that panmixia implies random union of gametes, we shall deduce the difference equations for the gametic frequencies. Our treatment slightly generalizes that of Bodmer (1965).

We designate the frequencies of the gamete G_i in the gametic outputs of males and females of generation t by $p_i(t)$ and $q_i(t)$, respectively. Let the genotype $G_i G_j$ have frequencies $P_{ij}(t)$ and $Q_{ij}(t)$, respectively, in adult males and females in generation t. We assume that the fertility of the union of a male of genotype $G_i G_j$ with a female of genotype $G_k G_l$ has the form $\alpha_{ij}\beta_{kl}$, for some α_{ij} and β_{kl}, for all i, j, k, and l. In those rare species in which parental genotypes influence the sex ratio, we have to distinguish gametic frequencies in sperm from those in eggs and fertilities for producing males from those for producing females. We could carry out this extension easily in the manner of Sect. 7.1 and Problem 7.4. Let $R_{i,jk}$ and $S_{i,jk}$ denote the respective probabilities that a gamete produced at random by males and females of genotype $G_j G_k$ is G_i.

Since

$$\sum_i R_{i,jk} = 1 \quad \text{and} \quad \sum_i S_{i,jk} = 1 \ ,$$

the frequency of the genotype G_iG_j in zygotes in generation $t+1$ reads

$$T_{ij}(t+1) = (\overline{\alpha}\overline{\beta})^{-1} \sum_{klmn} P_{kl}(t)Q_{mn}(t)\alpha_{kl}\beta_{mn}\tfrac{1}{2}(R_{i,kl}S_{j,mn} + R_{j,kl}S_{i,mn}) ,$$

where

$$\overline{\alpha} = \sum_{ij} P_{ij}\alpha_{ij} , \qquad \overline{\beta} = \sum_{ij} Q_{ij}\beta_{ij} .$$

This has the generalized Hardy-Weinberg form

$$T_{ij}(t+1) = \tfrac{1}{2}[p_i(t)q_j(t) + p_j(t)q_i(t)] \tag{8.1}$$

in terms of the gametic frequencies

$$p_i = \overline{\alpha}^{-1} \sum_{jk} P_{jk}\alpha_{jk}R_{i,jk} , \qquad q_i = \overline{\beta}^{-1} \sum_{jk} Q_{jk}\beta_{jk}S_{i,jk} . \tag{8.2}$$

If the male and female viabilities are u_{ij} and v_{ij}, respectively, the adult frequencies in generation $t+1$ will be

$$P_{ij} = u_{ij}T_{ij}/\overline{u} , \qquad Q_{ij} = v_{ij}T_{ij}/\overline{v} , \tag{8.3}$$

with the mean viabilities

$$\overline{u} = \sum_{ij} T_{ij}u_{ij} , \qquad \overline{v} = \sum_{ij} T_{ij}v_{ij} . \tag{8.4}$$

Writing (8.2) at time $t+1$ and substituting (8.3) yields

$$p_i(t+1) = (\overline{\alpha}\,\overline{u})^{-1} \sum_{jk} \alpha_{jk}u_{jk}T_{jk}(t+1)R_{i,jk} , \tag{8.5a}$$

$$q_i(t+1) = (\overline{\beta}\overline{v})^{-1} \sum_{jk} \beta_{jk}v_{jk}T_{jk}(t+1)S_{i,jk} . \tag{8.5b}$$

By virtue of (8.1), the system (8.5) is a set of recursion relations for the gametic frequencies. Since the latter are normalized, from (8.5) we obtain

$$\overline{\alpha}\,\overline{u} = \overline{\alpha u} = \sum_{jk} \alpha_{jk}u_{jk}T_{jk} , \tag{8.6a}$$

$$\overline{\beta}\,\overline{v} = \overline{\beta v} = \sum_{jk} \beta_{jk}v_{jk}T_{jk} . \tag{8.6b}$$

Relations equivalent to (8.6) were proved directly in special cases in Sects. 4.1 and 7.1. The products $\alpha_{ij}u_{ij}$ and $\beta_{ij}v_{ij}$ represent the male and female fitnesses, respectively. If these are proportional, $\alpha_{ij}u_{ij} = (\text{const})\,v_{ij}\beta_{ij}$ for all i and j, and the linkage maps in the two sexes are the same, $R_{i,jk} = S_{i,jk}$ for all i, j, and k, then (8.5) shows that after one generation of panmixia, the gametic frequencies in the two sexes will be the same, and (8.5) reduces after two generations to

$$p_i' = \overline{w}^{-1} \sum_{jk} w_{jk} R_{i,jk} p_j p_k \ , \tag{8.7a}$$

where $w_{ij} = \alpha_{ij} u_{ij}$ and

$$\overline{w} = \sum_{jk} w_{jk} p_j p_k \ . \tag{8.7b}$$

Henceforth, we shall confine ourselves to the monoecious model (8.7). Observe that for a single locus, the Mendelian formula

$$R_{i,jk} = \tfrac{1}{2}(\delta_{ij} + \delta_{ik})$$

immediately simplifies (8.7) to (4.25).

8.2 Analysis for Two Multiallelic Loci

As in Sect. 3.3, we denote the alleles at the two loci by A_i and B_j and designate the frequency of the gamete $A_i B_j$ in the gametic output of generation t by $P_{ij}(t)$. Let $w_{ij,kl}$ $(= w_{kl,ij})$ represent the fitness of the genotype $A_i B_j / A_k B_l$. The fitnesses of $A_i B_j / A_k B_l$ and $A_i B_l / A_k B_j$ are almost invariably the same, i.e., $w_{ij,kl} = w_{il,kj}$. This means that there is no *position effect*: given the composition of the genotype, the arrangement of the alleles is irrelevant. Nevertheless, we shall not impose this restriction on the fitnesses. Let $R_{ij;kl,mn}$ signify the probability that a gamete produced at random by an $A_k B_l / A_m B_n$ individual is $A_i B_j$. In our new notation, (8.7) reads

$$P_{ij}' = \overline{w}^{-1} \sum_{klmn} w_{kl,mn} P_{kl} P_{mn} R_{ij;kl,mn} \ , \tag{8.8a}$$

with

$$\overline{w} = \sum_{klmn} w_{kl,mn} P_{kl} P_{mn} \ . \tag{8.8b}$$

Suppose that c is the recombination frequency between the A and B loci. To calculate R, first consider drawing A_i, and then take into account recombination:

$$R_{ij;kl,mn} = \tfrac{1}{2}\delta_{ik}[(1-c)\delta_{jl} + c\delta_{jn}] + \tfrac{1}{2}\delta_{im}[(1-c)\delta_{jn} + c\delta_{jl}] \ . \tag{8.9}$$

Since the sum in (8.8a) is symmetric under the simultaneous interchanges $k \leftrightarrow m$ and $l \leftrightarrow n$, the two terms in (8.9) contribute equally. Therefore, substituting (8.9) into (8.8a) leads to our fundamental recursion relation:

$$\overline{w} P_{ij}' = P_{ij} w_{ij} - c D_{ij} \ , \tag{8.10}$$

where the linkage disequilibria D_{ij} are defined by

$$D_{ij} = \sum_{kl} (w_{ij,kl} P_{ij} P_{kl} - w_{il,kj} P_{il} P_{kj}) \ , \tag{8.11}$$

and

$$w_{ij} = \sum_{kl} w_{ij,kl} P_{kl} \qquad (8.12)$$

is the fitness of the gamete $A_i B_j$.

Kimura (1956a) deduced the differential equations for the gametic frequencies for two diallelic loci in a continuous-time Hardy-Weinberg model. Lewontin and Kojima (1960) obtained the corresponding difference equations for discrete, nonoverlapping generations. For a more concrete derivation of (8.10), consult Kimura and Ohta (1971).

From (8.10) we find that the change in the frequency of $A_i B_j$ is given by

$$\overline{w} \Delta P_{ij} = P_{ij}(w_{ij} - \overline{w}) - c D_{ij} \ . \qquad (8.13)$$

We infer at once from (8.11) that the linkage disequilibria satisfy

$$\sum_i D_{ij} = 0 , \qquad \sum_j D_{ij} = 0 \ . \qquad (8.14)$$

Note that in each term of the sum in (8.11), the two genotypes are related to each other by crossing over. As in Sect. 3.3, we define linkage equilibrium by the random combination of alleles within gametes:

$$P_{ij} = p_i q_j \qquad (8.15)$$

for all i and j, where

$$p_i = \sum_j P_{ij} \qquad \text{and} \qquad q_j = \sum_i P_{ij} \qquad (8.16)$$

are the frequencies of A_i and B_j in gametes. Substituting (8.15) into (8.11), we see readily that without position effect, in linkage equilibrium $D_{ij} = 0$ for all i and j. In the absence of selection, we may choose all the fitnesses to be unity, in which case (8.11) immediately reduces to (3.28), and (8.13) simplifies to (3.31).

We shall now analyze the dynamics of the two-locus system for *weak selection* (Nagylaki, 1976). The concepts are the same as those below (4.150) for a single locus with overlapping generations. Random combination now refers to linkage equilibrium, rather than Hardy-Weinberg proportions, and our measures of departure are the D_{ij}, rather than the Q_{ij} of (4.149). Our treatment and its interpretation follow Sect. 4.10.

As in (4.58), we may take, for all i, j, k, and l,

$$w_{ij,kl} = 1 + O(s) \qquad (8.17)$$

as the selection intensity $s \to 0$. Our slow-selection analysis will apply if selection is much weaker than recombination, $s \ll c$. Since most pairs of loci are either on different chromosomes ($c = \frac{1}{2}$) or quite far apart on the same chromosome (say, $c \gg 0.01$) and, excluding lethals, selection coefficients rarely exceed several percent, we expect our treatment to possess wide applicability.

At this point, the fitnesses may depend on the gametic frequencies and time in an arbitrary manner.

To show that the linkage disequilibria, D_{ij}, settle down rapidly to approximate constancy, we must first derive a difference equation for them. From (8.10), with the aid of (8.17), we obtain

$$P'_{ij} = P_{ij} - cD_{ij} + O(s) \ . \tag{8.18}$$

Rewriting (8.11) for D'_{ij}, substituting (8.17) and (8.18), and employing (8.14), we find

$$D'_{ij} = (1-c)D_{ij} + O(s) \ .$$

Hence,

$$D'_{ij} = (1-c)D_{ij} + sg_{ij}(\mathbf{P},t) \ , \tag{8.19}$$

where g_{ij} is a complicated function of fitnesses (and therefore possibly of time) and gametic frequencies that is uniformly bounded as $s \to 0$. Iterating (8.19) yields

$$D_{ij}(t) = D_{ij}(0)(1-c)^t + s(1-c)^t \sum_{\tau=1}^{t}(1-c)^{-\tau}g_{ij}[\mathbf{P}(\tau-1),\tau-1] \ . \tag{8.20}$$

(The sum is absent for $t=0$.)

We define t_1 as the shortest time such that

$$|D_{ij}(0)|(1-c)^{t_1} \le s \tag{8.21}$$

for all i and j. If the population starts sufficiently close to linkage equilibrium, then $t_1 = 0$. Otherwise, (8.21) gives the conservative estimate $t_1 \approx \ln s / \ln(1-c)$. For tight linkage ($s \ll c \ll 1$), this reduces to $t_1 \approx -c^{-1}\ln s$. With loose linkage or independent assortment, t_1 will usually be no more than 5 or 10 generations. From (8.20) we conclude that the linkage disequilibria are reduced to the order of the selection intensity in this short time: $D_{ij}(t) = O(s)$ for $t \ge t_1$. Thus, we have $D_{ij}(t) = sD^0_{ij}(t)$, with $D^0_{ij}(t) = O(1)$ for $t \ge t_1$. Summing (8.10) over i and j and using (8.14) produces the usual recursion relations for the gene frequencies, so during the time t_1, the gene-frequency change is very small, roughly $st_1 \approx s \ln s / \ln(1-c)$. For $t \ge t_1$, (8.19) can be reexpressed as

$$\Delta D^0_{ij} = -cD^0_{ij} + g_{ij}(\mathbf{P},t) \ . \tag{8.22}$$

From (8.18) we infer that

$$\Delta P_{ij} = O(s), \qquad t \ge t_1 \ . \tag{8.23}$$

Equation (8.22) informs us that

$$\Delta[D^0_{ij} - c^{-1}g_{ij}(\mathbf{P},t)] = -cD^0_{ij} + g_{ij}(\mathbf{P},t) - c^{-1}\Delta g_{ij}(\mathbf{P},t) \ . \tag{8.24}$$

We decompose the change in g_{ij} into parts due to its dependence on the gametic frequencies and on time:

$$\Delta g_{ij}(\mathbf{P}, t) = \{g_{ij}[\mathbf{P}(t+1), t+1] - g_{ij}[\mathbf{P}(t), t+1]\}$$
$$+ \{g_{ij}[\mathbf{P}(t), t+1] - g_{ij}[\mathbf{P}(t), t]\} . \qquad (8.25)$$

In view of (8.23), the first brace is of $O(s)$. Since the selection term in (8.19) is sg_{ij}, therefore, if we assume that the explicit time dependence of the fitnesses is of $O(s^2)$, the second brace in (8.25) will also be of $O(s)$. Then $\Delta g_{ij} = O(s)$, and (8.24) becomes

$$\Delta[D_{ij}^0 - c^{-1}g_{ij}(\mathbf{P}, t)] = -c[D_{ij}^0 - c^{-1}g_{ij}(\mathbf{P}, t)] + O(s) . \qquad (8.26)$$

Comparing (8.19) with (8.26), we choose t_2 ($\geq t_1$) as the shortest time such that

$$|D_{ij}^0(t_1) - c^{-1}g_{ij}[\mathbf{P}(t_1), t_1]|(1 - c)^{t_2 - t_1} \leq s \qquad (8.27)$$

for all i and j. From (8.27) and an equation like (8.20) for the bracket in (8.26), we obtain

$$D_{ij}^0(t) = c^{-1}g_{ij}(\mathbf{P}, t) + O(s), \qquad t \geq t_2 .$$

Now $\Delta g_{ij} = O(s)$ implies $\Delta D_{ij}^0 = O(s)$, whence

$$\Delta D_{ij}(t) = O(s^2), \qquad t \geq t_2 . \qquad (8.28)$$

Crudely, (8.27) tells us that $t_2 - t_1 \approx t_1$, so $t_2 \approx 2t_1$. Thus, in a short time, generally less than 10 or 20 generations, the linkage disequilibria become nearly constant. This important qualitative observation was first made by Kimura (1965) in a model with two alleles at each locus. Conley (1972) was the first to prove the analogue of (8.28) in a diallelic, continuous-time Hardy-Weinberg model. Since the amount of gene-frequency change during the second period ($t_1 \leq t < t_2$) is approximately the same as during the first, at time t_2 the population is still very far from gene-frequency equilibrium. The allelic frequencies require a time $t_3 \approx s^{-1}$ to approach equilibrium. The change in the linkage disequilibria during the third period, however, is quite small, viz., $s^2(t_3 - t_2) \approx s^2 t_3 \approx s$.

Before we deduce the implications of (8.28) for the change in mean fitness, we shall show that after a time t_1, the population evolves approximately as if it were in linkage equilibrium, the difference between the exact gametic frequencies and those of the much simpler hypothetical system on the linkage-equilibrium surface being of $O(s)$. The exact allelic frequencies, $p_i(t)$ and $q_j(t)$, evolve according to the complicated law derived from (8.10); at time t_1, they have the values $p_i(t_1)$ and $q_j(t_1)$. The allelic frequencies, $\pi_i(t)$ and $\rho_j(t)$, on the linkage-equilibrium surface evolve according to the much simpler law obtained by imposing linkage equilibrium on (8.10); we choose $\pi_i(t_1) = p_i(t_1)$ and $\rho_j(t_1) = q_j(t_1)$ for all i and j. We wish to prove that $P_{ij}(t) = \pi_i(t)\rho_j(t) + O(s)$ for $t \geq t_1$.

Substituting (8.17) into (8.11) gives

$$D_{ij} = P_{ij} - p_i q_j + O(s) . \qquad (8.29)$$

Since $D_{ij}(t) = O(s)$ for $t \geq t_1$, (8.29) informs us that

$$P_{ij}(t) = p_i(t)q_j(t) + O(s), \qquad t \geq t_1 . \tag{8.30}$$

By virtue of (8.17), we can write

$$w_{ij,kl}(\mathbf{P}, t) = 1 + s u_{ij,kl}(\mathbf{P}, t) , \tag{8.31a}$$

whence

$$w_{ij}(\mathbf{P}, t) = 1 + s u_{ij}(\mathbf{P}, t) , \tag{8.31b}$$

$$\overline{w}(\mathbf{P}, t) = 1 + s \overline{u}(\mathbf{P}, t) , \tag{8.31c}$$

with

$$u_{ij}(\mathbf{P}, t) = \sum_{kl} u_{ij,kl}(\mathbf{P}, t) P_{kl} , \tag{8.32a}$$

$$\overline{u}(\mathbf{P}, t) = \sum_{ij} u_{ij}(\mathbf{P}, t) P_{ij} . \tag{8.32b}$$

Summing (8.13) over j and using (8.14) and (8.31), we obtain

$$\overline{w}(\mathbf{P}, t)\Delta p_i = s p_i [u_i(\mathbf{P}, t) - \overline{u}(\mathbf{P}, t)] , \tag{8.33}$$

where

$$p_i u_i(\mathbf{P}, t) = \sum_j u_{ij}(\mathbf{P}, t) P_{ij} . \tag{8.34}$$

For $t \geq t_1$, (8.30) yields

$$w_{ij,kl}(\mathbf{P}, t) = w_{ij,kl}(\mathbf{p} * \mathbf{q}, t) + O(s) , \tag{8.35a}$$

in which $\mathbf{p} * \mathbf{q}$ indicates evaluation at $P_{ij} = p_i q_j$ for all i and j. From (8.35a) it follows that

$$\overline{w}(\mathbf{P}, t) = \overline{w}(\mathbf{p} * \mathbf{q}, t) + O(s) , \tag{8.35b}$$

where

$$\overline{w}(\mathbf{p} * \mathbf{q}, t) = \sum_{ijkl} w_{ij,kl}(\mathbf{p} * \mathbf{q}, t) p_i q_j p_k q_l . \tag{8.36}$$

Defining

$$u_{ij}(\mathbf{p} * \mathbf{q}, t) = \sum_{kl} u_{ij,kl}(\mathbf{p} * \mathbf{q}, t) p_k q_l , \tag{8.37a}$$

$$u_i(\mathbf{p} * \mathbf{q}, t) = \sum_j u_{ij}(\mathbf{p} * \mathbf{q}, t) q_j , \tag{8.37b}$$

$$\overline{u}(\mathbf{p} * \mathbf{q}, t) = \sum_i u_i(\mathbf{p} * \mathbf{q}, t) p_i \tag{8.37c}$$

in analogy with (8.32) and (8.34), we deduce from (8.33), (8.35b), and analogous relations for u_i and $\overline{u}$ that

$$\overline{w}(\mathbf{p} * \mathbf{q}, t)\Delta p_i = s p_i [u_i(\mathbf{p} * \mathbf{q}, t) - \overline{u}(\mathbf{p} * \mathbf{q}, t)] + O(s^2) \tag{8.38}$$

for $t \geq t_1$. From (8.38) we infer that on the linkage-equilibrium surface the frequency of A_i satisfies

$$\overline{w}(\boldsymbol{\pi} * \boldsymbol{\rho}, t)\Delta\pi_i = s\pi_i[u_i(\boldsymbol{\pi} * \boldsymbol{\rho}, t) - \overline{u}(\boldsymbol{\pi} * \boldsymbol{\rho}, t)] ; \qquad (8.39)$$

there is a similar equation for ρ_j.

Comparing (8.38) with (8.39), and the analogous equations for q_j with those for ρ_j, we conclude that for $t \geq t_1$

$$p_i(t) = \pi_i(t) + O(s), \qquad q_j(t) = \rho_j(t) + O(s) . \qquad (8.40)$$

Inserting (8.40) into (8.30) directly yields the desired result,

$$P_{ij}(t) = \pi_i(t)\rho_j(t) + O(s), \qquad t \geq t_1 , \qquad (8.41)$$

for all i and j. If the system $[\boldsymbol{\pi}(t), \boldsymbol{\rho}(t)]$ on the linkage-equilibrium surface does not necessarily converge to some equilibrium point or if $[\boldsymbol{\pi}(t_1), \boldsymbol{\rho}(t_1)]$ happens to be on the stable manifold of an unstable equilibrium, then small perturbations may cause large deviations in its ultimate state. In this case, (8.41) still holds if $t_1 \leq t \leq k/s$ for some constant k.

We shall now examine the *evolution of the mean fitness*. From (8.8b) and (8.12) we have

$$\Delta\overline{w} = \sum_{ijkl}(w'_{ij,kl}P'_{ij}P'_{kl} - w_{ij,kl}P_{ij}P_{kl})$$

$$= \overline{\Delta w} + 2\sum_{ij}w_{ij}\Delta P_{ij} + \sum_{ijkl}w_{ij,kl}\Delta P_{ij}\Delta P_{kl} , \qquad (8.42)$$

where

$$\overline{\Delta w} = \sum_{ijkl}\Delta w_{ij,kl}P'_{ij}P'_{kl} \qquad (8.43)$$

is the mean of the fitness changes over the next generation. Substituting (8.13) into (8.42) and recalling (8.14) leads to

$$\Delta\overline{w} = \overline{\Delta w} + \overline{w}^{-1}(V_{gam} - 2c\overline{X})$$
$$+ \overline{w}^{-2}\sum_{ijkl}P_{ij}P_{kl}(w_{ij,kl} - \overline{w})(w_{ij} - \overline{w})(w_{kl} - \overline{w})$$
$$- 2c\overline{w}^{-2}\sum_{ij}X_{ij}P_{ij}(w_{ij} - \overline{w}) + c^2\overline{w}^{-2}\sum_{ij}X_{ij}D_{ij} , \qquad (8.44)$$

where the *gametic* variance reads

$$V_{gam} = 2\sum_{ij}P_{ij}(w_{ij} - \overline{w})^2 , \qquad (8.45)$$

and

$$X_{ij} = \sum_{kl}w_{ij,kl}D_{kl} = \sum_{kl}(w_{ij,kl} - \overline{w})D_{kl} , \qquad (8.46a)$$

$$\overline{X} = \sum_{ij} P_{ij}X_{ij} = \sum_{kl} w_{kl}D_{kl} = \sum_{kl}(w_{kl} - \overline{w})D_{kl} \ . \tag{8.46b}$$

Note that (8.44) reduces to the single-locus formula (4.56), with alleles replaced by gametes, in the absence of recombination ($c = 0$). Comparing (8.45) with (4.189), we observe that, since the gametes are combined in Hardy-Weinberg proportions, V_{gam} is indeed the additive component of the total genetic variance in the least-squares decomposition [cf. (4.184)]

$$V = V_{gam} + V_{gam,dom} \ , \tag{8.47}$$

where $V_{gam,dom}$ signifies the variance that arises from nonadditivity of the gametic fitnesses. Consequently, if we decompose the gametic variance into its additive (V_g) and epistatic (V_e) components, then V_g will be the additive component of V, and hence may be justly called the genic variance. Kimura (1965) proved this result algebraically for two diallelic loci.

To carry out this analysis of V_{gam}, we set

$$v_{ij} = w_{ij} - \overline{w} = \alpha_i + \beta_j + E_{ij} \ , \tag{8.48}$$

where α_i and β_j denote the average effects of A_i and B_j, respectively, and E_{ij} designates the epistatic deviation. Minimizing the epistatic variance

$$V_e = 2 \sum_{ij} P_{ij}E_{ij}^2 = 2 \sum_{ij} P_{ij}(v_{ij} - \alpha_i - \beta_j)^2 \tag{8.49}$$

with respect to the average effects, we obtain

$$\sum_j P_{ij}(v_{ij} - \alpha_i - \beta_j) = \sum_j P_{ij}E_{ij} = 0 \ , \tag{8.50a}$$

$$\sum_i P_{ij}(v_{ij} - \alpha_i - \beta_j) = \sum_i P_{ij}E_{ij} = 0 \ . \tag{8.50b}$$

The average excesses of A_i and B_j are given by

$$p_i a_i = \sum_j P_{ij}v_{ij} \, , \qquad q_j b_j = \sum_i P_{ij}v_{ij} \ . \tag{8.51}$$

Hence, (8.48) shows that the means of the average excesses vanish:

$$\sum_i p_i a_i = 0 \, , \qquad \sum_j q_j b_j = 0 \ . \tag{8.52}$$

We rewrite (8.50) in the form

$$p_i a_i = p_i \alpha_i + \sum_j P_{ij}\beta_j \ , \tag{8.53a}$$

$$q_j b_j = \sum_i P_{ij}\alpha_i + q_j \beta_j \ , \tag{8.53b}$$

and deduce from (8.52) and (8.53)

$$\sum_i p_i \alpha_i + \sum_j q_j \beta_j = 0 \ . \tag{8.54}$$

Since α_i and β_j occur only in the sum $\alpha_i + \beta_j$, therefore (8.54) implies that, by shifting the appropriate constant between them, we can arrange that the average effects have zero means:

$$\sum_i p_i \alpha_i = 0 , \qquad \sum_j q_j \beta_j = 0 \ . \tag{8.55}$$

From (8.53) we compute the genic variance:

$$V_g = 2 \sum_{ij} P_{ij}(\alpha_i + \beta_j)^2 \tag{8.56}$$

$$= 2 \sum_{ij} P_{ij}(\alpha_i + \beta_j)\alpha_i + 2 \sum_{ij} P_{ij}(\alpha_i + \beta_j)\beta_j$$

$$= 2 \sum_i p_i a_i \alpha_i + 2 \sum_j q_j b_j \beta_j \ , \tag{8.57}$$

which manifestly generalizes (4.188) to two loci. Substituting (8.48) into (8.45) and using (8.49), (8.50), and (8.56) proves the additivity property for the gametic variance:

$$V_{gam} = 2 \sum_{ij} P_{ij}(\alpha_i + \beta_j + E_{ij})^2$$

$$= V_g + V_e + 4 \sum_{ij} P_{ij}(\alpha_i + \beta_j)E_{ij}$$

$$= V_g + V_e \ . \tag{8.58}$$

Since $D_{ij} = O(s)$ for $t \geq t_1$, (8.58) allows us to rewrite (8.44) as

$$\Delta \overline{w} = V_g + \overline{\Delta w} + V_e - 2c\overline{X} + O(s^3), \qquad t \geq t_1 \ . \tag{8.59}$$

To calculate $V_e - 2c\overline{X}$, we evaluate ΔD_{ij} for $t \geq t_1$, explaining the manipulations below:

$$\Delta D_{ij} = \sum_{kl}(P_{kl}\Delta P_{ij} + P_{ij}\Delta P_{kl} - P_{kj}\Delta P_{il} - P_{il}\Delta P_{kj}) + O(s^2)$$

$$= \sum_{kl} \{P_{kl}P_{ij}[(w_{ij} - \overline{w}) + (w_{kl} - \overline{w})] - P_{kj}P_{il}[(w_{il} - \overline{w}) + (w_{kj} - \overline{w})]\}$$

$$\quad - cD_{ij} + O(s^2)$$

$$= \sum_{kl} p_i q_j p_k q_l [(w_{ij} - \overline{w}) + (w_{kl} - \overline{w}) - (w_{il} - \overline{w}) - (w_{kj} - \overline{w})]$$

$$\quad - cD_{ij} + O(s^2)$$

$$= \sum_{kl} p_i q_j p_k q_l (E_{ij} + E_{kl} - E_{il} - E_{kj}) - cD_{ij} + O(s^2)$$

$$= \sum_{kl} [P_{kl}P_{ij}(E_{ij} + E_{kl}) - P_{il}P_{kj}(E_{il} + E_{kj})] - cD_{ij} + O(s^2)$$

$$= P_{ij}E_{ij} - cD_{ij} + O(s^2), \qquad t \geq t_1 , \tag{8.60}$$

where we employed in the successive equations (8.11), (8.17), (8.23), and the assumption that the fitnesses are approximately constant,

$$\Delta w_{ij,kl} = O(s^2) ; \tag{8.61}$$

(8.13), (8.14), and (8.17); (8.17) and (8.30); (8.48); (8.17), (8.30), and (8.48); finally, (8.50). From (8.48) and (8.60) we obtain

$$2 \sum_{ij} E_{ij} \Delta D_{ij} = 2 \sum_{ij} P_{ij} E_{ij}^2 - 2c \sum_{ij} D_{ij}(v_{ij} - \alpha_i - \beta_j) + O(s^3)$$

$$= V_e - 2c\overline{X} + O(s^3), \qquad t \geq t_1 , \tag{8.62}$$

where the second step follows from (8.14), (8.46b), (8.48), and (8.49). Inserting (8.62) into (8.59) yields

$$\Delta \overline{w} = V_g + \overline{\Delta w} + 2 \sum_{ij} E_{ij} \Delta D_{ij} + O(s^3), \qquad t \geq t_1 . \tag{8.63}$$

Now (8.28) informs us that

$$\Delta \overline{w} = V_g + \overline{\Delta w} + O(s^3), \qquad t \geq t_2 , \tag{8.64}$$

as for a single locus [see (4.59)].

Slightly strengthening (8.61) to $\Delta w_{ij,kl} = o(s^2)$ gives the approximate form of the Fundamental Theorem of Natural Selection for two loci: $\Delta \overline{w} \approx V_g$. Therefore, the mean fitness will generally not decrease after the initial adjustment of linkage relations during $t < t_2$. This assertion will be false if the genic variance is particularly small. From (8.13), (8.14), (8.48), and (8.51) we deduce that

$$\overline{w} \Delta p_i = p_i a_i , \qquad \overline{w} \Delta q_j = q_j b_j , \tag{8.65}$$

whence (8.57) becomes

$$V_g = 2\overline{w} \left(\sum_i \alpha_i \Delta p_i + \sum_j \beta_j \Delta q_j \right) . \tag{8.66}$$

Hence, $V_g = 0$ if the allelic frequencies are constant, as is true at equilibrium. Thus, the mean fitness is most likely to decrease in the neighborhood of an equilibrium and in special cases in which symmetry conditions dictate the constancy of the gene frequencies. For further discussion, examples of pathological behavior, and other results, the reader may refer to Nagylaki (1977).

8.3 Two Diallelic Loci

The purpose of this section is to present some examples of the properties of two diallelic loci. Karlin (1975, 1978) gives many other results in his extensive reviews. It will be convenient to simplify the notation of Sect. 8.2. Let x_1, x_2, x_3, and x_4 represent the frequencies of the gametes AB, Ab, aB, and ab, respectively, as in Sect. 3.3. If w_{ij} denotes the fitness of the genotype composed of gametes i and j ($i, j = 1, 2, 3, 4$), the gametic fitnesses and the mean fitness read

$$w_i = \sum_j w_{ij}x_j , \qquad \overline{w} = \sum_{ij} w_{ij}x_ix_j .\tag{8.67}$$

It is easy to show that (8.10) and (8.11) become (Problem 8.4)

$$\overline{w}\Delta x_i = x_i(w_i - \overline{w}) - \varepsilon_i cD ,\tag{8.68}$$

where $\varepsilon_1 = \varepsilon_4 = 1$, $\varepsilon_2 = \varepsilon_3 = -1$, and

$$D_{11} = D_{22} = -D_{12} = -D_{21} \equiv D = w_{14}x_1x_4 - w_{23}x_2x_3 .\tag{8.69}$$

1. The Increase in Frequency of a Rare Allele

Suppose that the allele a is fixed and linked to an overdominant polymorphism at the B-locus. Then (4.29) tells us that the equilibrium frequencies of the gametes aB and ab are

$$\hat{u} = (w_{34} - w_{44})/(2w_{34} - w_{33} - w_{44})$$

and $\hat{v} = 1 - \hat{u}$, respectively. The equilibrium fitness of the genotype $aa--$ is just

$$\hat{\overline{w}} = \hat{u}^2 w_{33} + 2\hat{u}\hat{v}w_{34} + \hat{v}^2 w_{44} ,$$

the mean fitness of the population at equilibrium.

If a rare new mutant, A, appears in the population, its frequency will increase if the two-locus polymorphism is unstable. Since the Bb polymorphism is stable in the absence of A, the equilibria with the gametes aB or ab fixed are evidently unstable. Therefore, if the introduction of A makes the Bb polymorphism unstable, then A will not only increase initially in frequency, but (possibly excluding some very special initial conditions) it will also be permanently established in the population. According to the general result in Sect. 5.5, two-locus instability will occur if at least one eigenvalue of the linearized system exceeds unity in modulus. We reproduce below the results of Bodmer and Felsenstein (1967).

The fitnesses of the gametes AB and Ab at equilibrium (i.e., as their frequencies tend to zero) read

$$\hat{w}_1 = \hat{u}w_{13} + \hat{v}w_{14} , \qquad \hat{w}_2 = \hat{u}w_{23} + \hat{v}w_{24} .$$

The allele A has frequency p and fitness $p^{-1}(x_1w_1 + x_2w_2)$. In linkage equilibrium, $x_1 = pu$ and $x_2 = pv$. Therefore, we may interpret

$$\hat{w}_A = \hat{u}\hat{w}_1 + \hat{v}\hat{w}_2$$

as the stationary fitness of A at linkage equilibrium. We define the critical value

$$c^\star = \left| \frac{(\hat{\bar{w}} - \hat{w}_1)(\hat{\bar{w}} - \hat{w}_2)}{\hat{u}w_{23}(\hat{\bar{w}} - \hat{w}_1) + \hat{v}w_{14}(\hat{\bar{w}} - \hat{w}_2)} \right|$$

of the cross-over fraction. Bodmer and Felsenstein (1967) obtained the following conditions for the increase of A.

(i) $\hat{w}_1 = \hat{w}_2$

Here $\hat{w}_A = \hat{w}_1 = \hat{w}_2$, and the frequency of A increases if $\hat{w}_A > \hat{\bar{w}}$, i.e., roughly, the frequency of the new mutant increases if it is fitter than the original allele, a.

(ii) $\hat{w}_1 > \hat{w}_2$

(a) $\hat{w}_1 > \hat{w}_2 > \hat{\bar{w}}$
Both gametes that carry A are fitter than the population, so A spreads.

(b) $\hat{\bar{w}} > \hat{w}_1 > \hat{w}_2$
Since A decreases the fitness of every gamete, it is lost.

(c) $\hat{w}_1 > \hat{\bar{w}} \geq \hat{w}_2$
(α) $\hat{\bar{w}}(\hat{u}w_{23} + \hat{v}w_{14}) \leq \hat{u}w_{23}\hat{w}_1 + \hat{v}w_{14}\hat{w}_2$: A increases.
(β) $\hat{\bar{w}}(\hat{u}w_{23} + \hat{v}w_{14}) > \hat{u}w_{23}\hat{w}_1 + \hat{v}w_{14}\hat{w}_2$: A increases if $c < c^\star$.

(iii) $\hat{w}_1 < \hat{w}_2$

This is the same as Case ii with the interchange $B \leftrightarrow b$ throughout.

If there is no position effect, then $w_{14} = w_{23}$ and the most interesting case, iic, simplifies to the following two situations.

(α) $\hat{\bar{w}} \leq \hat{w}_A$:   The frequency of A increases.
(β) $\hat{\bar{w}} > \hat{w}_A$:   The allele A will spread if $c < c^\star$. Thus, for sufficiently tight linkage, the frequency of the less fit allele can increase. We saw at the end of Sect. 4.3 that this is not possible if fitness depends only on a single locus. Note that for weak selection $c^\star = O(s)$.

2. The Increase in Frequency of Two Rare Alleles

We assume that the gamete ab is fixed and the alleles A and B are introduced at low frequencies. If the equilibrium $x_4 = 1$ is stable, A and B will be lost. If it is unstable, then, though the frequency of one of the alleles may decrease temporarily (Bodmer and Felsenstein, 1967), both alleles will usually spread after a short time. Whether one of the alleles is ultimately lost cannot be determined by local analysis.

The linearized recursion relation satisfied by the vector

$$\mathbf{x} = \begin{pmatrix} x_1 \\ x_2 \\ x_3 \end{pmatrix}$$

reads (Problem 5.8) $\mathbf{x}' = C\mathbf{x}$, with

$$C = \begin{pmatrix} \lambda_1 & 0 & 0 \\ \alpha & \lambda_2 & 0 \\ \alpha & 0 & \lambda_3 \end{pmatrix} , \qquad (8.70a)$$

where $\alpha = cw_{14}/w_{44}$, and the eigenvalues of C are

$$\lambda_1 = (1-c)w_{14}/w_{44} , \qquad \lambda_2 = w_{24}/w_{44} , \qquad \lambda_3 = w_{34}/w_{44} . \qquad (8.70b)$$

The equilibrium is stable if all the eigenvalues are less than one. If $\lambda_2, \lambda_3 < 1$, we have instability for sufficiently tight linkage, $c < 1 - (w_{44}/w_{14}) \equiv \tilde{c}$. Thus, linkage matters if Ab and aB are less fit at equilibrium than ab ($w_{24}, w_{34} < w_{44}$) and the equilibrium fitness of AB is neither too small nor too large, $w_{44} < w_{14} \le 2w_{44}$ ($0 < \tilde{c} \le \frac{1}{2}$). Observe that for weak selection $\tilde{c} = O(s)$.

3. The Additive and Multiplicative Models

The results given below, as well as many more concerning other models, are presented by Karlin (1975), who gives references to the pertinent original papers. Assuming the absence of a position effect ($w_{14} = w_{23}$), we can use loci instead of gametes to construct a 3×3 fitness matrix instead of our former 4×4 symmetric one:

$$\begin{array}{cccc} & BB & Bb & bb \\ W = (W_{ij}) = \begin{matrix} AA \\ Aa \\ aa \end{matrix} & \begin{pmatrix} w_{11} & w_{12} & w_{22} \\ w_{13} & w_{14} & w_{24} \\ w_{33} & w_{34} & w_{44} \end{pmatrix} \end{array} ,$$

$i, j = 1, 2, 3$. For instance, $W_{22} = w_{14} = w_{23} \ne w_{22}$. The contributions of the individual loci to the genotypic fitnesses are

$$\begin{array}{cccccc} AA & Aa & aa & BB & Bb & bb \\ \alpha_1 & \alpha_2 & \alpha_3 & \beta_1 & \beta_2 & \beta_3 . \end{array}$$

Define

$$\hat{x}_1 = \hat{p}\hat{u} , \qquad \hat{x}_2 = \hat{p}\hat{v} , \qquad \hat{x}_3 = \hat{q}\hat{u} , \qquad \hat{x}_4 = \hat{q}\hat{v} , \qquad (8.71a)$$

where the equilibrium frequencies of A and B read

$$\hat{p} = (\alpha_2 - \alpha_3)/(2\alpha_2 - \alpha_1 - \alpha_3), \qquad \hat{u} = (\beta_2 - \beta_3)/(2\beta_2 - \beta_1 - \beta_3) ; \qquad (8.71b)$$

as above, $\hat{q} = 1 - \hat{p}$ and $\hat{v} = 1 - \hat{u}$. Clearly, the equilibrium (8.71) can exist only if there is either overdominance or underdominance at each locus. Evidently, $D = 0$ at this equilibrium.

(i) Additive Loci

With no additive epistasis, we have $W_{ij} = \alpha_i + \beta_j$ for all i and j, and the following results hold.

(a) An internal equilibrium (i.e., one with $x_i > 0$ for all i) exists only if there is either overdominance or underdominance at each locus, in which case (8.71) is the unique internal equilibrium.

(b) The equilibrium (8.71) is globally asymptotically stable if and only if there is overdominance at each locus, i.e.,

$$\alpha_2 > \alpha_1, \alpha_3 , \qquad \beta_2 > \beta_1, \beta_3 . \tag{8.72}$$

(c) The mean fitness is nondecreasing, and the change in mean fitness is zero only at equilibrium.

(d) If only one of the conditions (8.72) holds, say the first, then the frequency of A converges globally to $\hat{p}$, whereas B or b is lost.

(e) Even with multiple alleles at each locus, there exists at most one internal equilibrium, and it is globally stable if and only if it is locally stable.

(ii) Multiplicative Loci

In the absence of multiplicative epistasis, we have $W_{ij} = \alpha_i \beta_j$ for all i and j, and the system has the following properties (see also Moran, 1968).

(a) With either overdominance or underdominance at each locus, the equilibrium (8.71) exists.

(b) If (8.72) holds, the equilibrium (8.71) is locally stable if and only if the recombination frequency, c, exceeds a critical value, c_0; i.e., with over-dominance at each locus, the expected equilibrium with $D = 0$ is stable if and only if linkage is sufficiently loose. From the explicit expression for c_0 (Bodmer and Felsenstein, 1967), it is easy to see that for weak selection $c_0 = O(s^2)$, so that the condition $c > c_0$ is not very stringent.

(c) If (8.72) applies and $c < c_0$, there exist exactly two locally stable internal equilibria, one with $D > 0$ and one with $D < 0$. Since the quadratic $\overline{w}$ can have at most one internal maximum, the mean fitness can decrease in the neighborhood of at least one of the two stable equilibria. The discussion at the end of Sect. 8.2 is pertinent here.

(d) The equilibrium (8.71) is not a local maximum of $\overline{w}$. Hence, $\overline{w}$ can decrease in the neighborhood of (8.71) when that equilibrium is stable. Refer again to the end of the previous section.

(e) The sign of the linkage disequilibrium D is invariant. Therefore, by continuity, $D = 0$ implies $D' = 0$. Thus, depending on its initial state, a population remains forever in one of the regions $D > 0$ or $D < 0$, or on the surface $D = 0$.

(f) The mean fitness is nondecreasing on the surface $D = 0$. When the equilibrium (8.71) exists, it is a local maximum of $\overline{w}$ on that surface.

(g) There exist at most two stable equilibria (internal and boundary).

(h) If $c > c_0$, (8.71) and an equilibrium with $D \neq 0$ can be simultaneously stable (Karlin and Feldman, 1978; Hastings, 1981a).

4. Cycling

According to Sect. 3.3, convergence to an equilibrium point always occurs for two neutral loci. In Sect. 4.3, we established convergence for selection at a single locus (provided all the equilibria are isolated). Therefore, we might expect the two-locus system with selection to converge to some equilibrium point. In fact, stable cycling is also possible. Akin (1979) has proved the existence of limit cycles in continuous time; consult Akin (1982, 1983) and Hastings (1981) for examples in continuous and discrete models, respectively. These examples are structurally stable under sufficiently small perturbations of the fitnesses and involve neither position effect nor lethality, but they do require strong epistasis. Even with strong selection, the period of the cycles is rather long. In some cases, almost all of the gametic frequency change occurs during a small fraction of the cycle. Thus, cycling may be difficult to observe in laboratory or natural populations.

8.4 Continuous Model with Overlapping Generations

We shall now discuss the behavior of the multiallelic two-locus system in continuous time. We shall assume that either there is no age structure or a stable age distribution has been reached, and therefore we commence with the two-locus generalization of (4.131), in which time is the only independent variable. Moody (1978) has demonstrated for multiple loci that either assumption allows this reduction. For two alleles at each locus, the formulation and analysis to follow have been carried out in Nagylaki and Crow (1974) and Nagylaki (1976).

Our basic variables are the genotypic frequencies. Let $S_{ij,kl}(t)$ represent the frequency of $A_i B_j / A_k B_l$ individuals at time t. From this we compute the frequencies of the gamete $A_i B_j$ and the alleles A_i and B_j:

$$P_{ij} = \sum_{kl} S_{ij,kl}, \qquad p_i = \sum_j P_{ij}, \qquad q_j = \sum_i P_{ij} . \tag{8.73}$$

We denote the number of $A_i B_j / A_k B_l$ individuals by $n_{ij,kl}$ and the total population size by N. Hence,

$$S_{ij,kl} = n_{ij,kl}/N . \tag{8.74}$$

We designate the frequency of matings between $A_\alpha B_\beta / A_\gamma B_\delta$ and $A_\varepsilon B_\zeta / A_\eta B_\theta$ by $X_{\alpha\beta,\gamma\delta;\varepsilon\zeta,\eta\theta}$ and the fertility of such a union, normalized relative to individuals as in (4.130), by $f_{\alpha\beta,\gamma\delta;\varepsilon\zeta,\eta\theta}$. Equation (8.9) gives $R_{ij;\alpha\beta,\gamma\delta}$, the probability that a gamete produced at random by an $A_\alpha B_\beta / A_\gamma B_\delta$ individual is $A_i B_j$.

The differential equation for the genotypic numbers reads

$$\dot{n}_{ij,kl} = N \sum X_{\alpha\beta,\gamma\delta;\varepsilon\zeta,\eta\theta} f_{\alpha\beta,\gamma\delta;\varepsilon\zeta,\eta\theta} R_{ij;\alpha\beta,\gamma\delta} R_{kl;\varepsilon\zeta,\eta\theta} - d_{ij,kl} n_{ij,kl} \ , \quad (8.75)$$

where $d_{ij,kl}$ is the death rate of $A_i B_j / A_k B_l$ individuals. To simplify the notation, in (8.75) we introduced the convention that unspecified summations run over all indices that do not appear on the left-hand side of the equation. We define the average mortality, fertility and fitness by

$$\overline{d} = \sum d_{ij,kl} S_{ij,kl} \ , \quad (8.76a)$$

$$\overline{f} = \sum X_{\alpha\beta,\gamma\delta;\varepsilon\zeta,\eta\theta} f_{\alpha\beta,\gamma\delta;\varepsilon\zeta,\eta\theta} \ , \quad (8.76b)$$

$$\overline{m} = \overline{f} - \overline{d} \ . \quad (8.76c)$$

The recombination functions yield the Mendelian single-locus formulae [see (8.9)]

$$\sum_j R_{ij;kl,mn} = \tfrac{1}{2}(\delta_{ik} + \delta_{im}), \qquad \sum_i R_{ij;kl,mn} = \tfrac{1}{2}(\delta_{jl} + \delta_{jn}) \ , \quad (8.77a)$$

and the normalization

$$\sum_{ij} R_{ij;kl,mn} = 1 \ . \quad (8.77b)$$

From (8.75), (8.76), and (8.77b) we obtain

$$\dot{N} = \overline{m} N \ . \quad (8.78)$$

Differentiating (8.74) and substituting (8.75) and (8.78) gives our fundamental equations:

$$\dot{S}_{ij,kl} = \sum X_{\alpha\beta,\gamma\delta;\varepsilon\zeta,\eta\theta} f_{\alpha\beta,\gamma\delta;\varepsilon\zeta,\eta\theta} R_{ij;\alpha\beta,\gamma\delta} R_{kl;\varepsilon\zeta,\eta\theta} - (d_{ij,kl} + \overline{m}) S_{ij,kl} \ . \quad (8.79)$$

The evolution of the population is determined by (8.79) as soon as the mating frequencies, fecundities, and mortalities are specified as functions of time and the genotypic proportions. The average death rate of an individual who carries the gamete $A_i B_j$ is d_{ij}, where

$$P_{ij} d_{ij} = \sum_{kl} S_{ij,kl} d_{ij,kl} \ . \quad (8.80)$$

From (8.73), (8.77b), (8.79), and (8.80) we deduce

$$\dot{P}_{ij} = \sum X_{\alpha\beta,\gamma\delta;\varepsilon\zeta,\eta\theta} f_{\alpha\beta,\gamma\delta;\varepsilon\zeta,\eta\theta} R_{ij;\alpha\beta,\gamma\delta} - (d_{ij} + \overline{m}) P_{ij} \ . \quad (8.81)$$

We define the rate at which an A_iB_j/A_kB_l individual gives birth, $b_{ij,kl}$, by

$$S_{ij,kl}b_{ij,kl} = \sum X_{ij,kl;\alpha\beta,\gamma\delta} f_{ij,kl;\alpha\beta,\gamma\delta} \ . \tag{8.82a}$$

Then the fertilities of individuals who carry A_iB_j and of those who carry A_i are given by

$$P_{ij}b_{ij} = \sum_{kl} S_{ij,kl}b_{ij,kl} \ , \tag{8.82b}$$

$$p_ib_i = \sum_j P_{ij}b_{ij} \ . \tag{8.82c}$$

Of course, the allelic mortality of A_i satisfies

$$p_id_i = \sum_j P_{ij}d_{ij} \ , \tag{8.83}$$

and

$$m_i = b_i - d_i \tag{8.84}$$

is the Malthusian parameter of A_i. Then (8.73), (8.77a), (8.81), (8.82), (8.83), and (8.84) lead easily to the usual differential equation,

$$\dot{p}_i = p_i(m_i - \overline{m}) \ , \tag{8.85}$$

for the frequency of A_i. Evidently, q_j satisfies a similar equation.

As pointed out in Sects. 4.1 and 4.10, equations for marginal frequencies, such as (8.81) and (8.85), though highly informative, depend on genotypic proportions, and are consequently incomplete. By inserting (8.9), it is trivial to display explicitly in (8.79) and (8.81) the terms that correspond to recombination in 0, 1, or 2 individuals in the mated pair. Since the ensuing analysis will be restricted to weak selection, we shall save writing by decomposing (8.79) and (8.81) only in that case. We suppose henceforth that *mating is random*:

$$X_{\alpha\beta,\gamma\delta;\varepsilon\zeta,\eta\theta} = S_{\alpha\beta,\gamma\delta}S_{\varepsilon\zeta,\eta\theta} \ . \tag{8.86}$$

This reduces (8.82a) to

$$b_{ij,kl} = \sum S_{\alpha\beta,\gamma\delta} f_{ij,kl;\alpha\beta,\gamma\delta} \ .$$

Let us now study *slow selection*:

$$f_{\alpha\beta,\gamma\delta;\varepsilon\zeta,\eta\theta} = b + O(s), \qquad d_{ij,kl} = d + O(s) \ , \tag{8.87}$$

where b is a constant and s is the intensity of selection. If time is measured in generations, b will usually be close to unity (as will d). Inserting (8.9) and (8.87) into (8.79) yields

$$\dot{S}_{ij,kl} = b\Big\{(1-c)^2 P_{ij}P_{kl} + c(1-c)\Big[P_{ij}\sum S_{k\zeta,\eta l} + P_{kl}\sum S_{i\beta,\gamma j}\Big]$$
$$+ c^2 \sum S_{i\beta,\gamma j}S_{k\zeta,\eta l} - S_{ij,kl}\Big\} + O(s) \ . \tag{8.88}$$

Hence,

$$\dot{P}_{ij} = -bcD_{ij} + O(s) \ , \tag{8.89}$$

with the linkage disequilibrium

$$D_{ij} = \sum_{kl}(S_{ij,kl} - S_{il,kj}) = P_{ij} - \sum_{kl} S_{il,kj} \ . \tag{8.90}$$

Differentiating (8.90) and substituting (8.88), (8.89), and then

$$\sum_{kl} S_{il,kj} = P_{ij} - D_{ij} \ , \tag{8.91}$$

we find

$$\dot{D}_{ij} = -b(1+c)D_{ij} + bD_{ij}^r + O(s) \ , \tag{8.92}$$

where

$$D_{ij}^r = P_{ij} - p_i q_j \tag{8.93}$$

would be the linkage disequilibrium if the population were in Hardy-Weinberg proportions. Note that for weak selection (8.11) and (8.93) agree. From (8.85) and the corresponding equation for q_j, we infer

$$\dot{p}_i = O(s) \qquad \dot{q}_j = O(s) \ . \tag{8.94}$$

Using (8.89) and (8.94) in the derivative of (8.93), we obtain

$$\dot{D}_{ij}^r = -bcD_{ij} + O(s) \ . \tag{8.95}$$

We shall also require

$$Q_{ij,kl} = S_{ij,kl} - P_{ij}P_{kl} \tag{8.96}$$

as a measure of the deviation of gametes from Hardy-Weinberg proportions. This is the natural generalization of (4.149). Differentiating (8.96) and inserting (8.88), (8.89), and (8.91) yields

$$\dot{Q}_{ij,kl} = -bQ_{ij,kl} + bc^2 D_{ij}D_{kl} + O(s) \ . \tag{8.97}$$

Concentrating our attention first on the linkage disequilibria, we rewrite (8.92) and (8.95) in matrix form:

$$\mathbf{D}_{ij} = \begin{pmatrix} D_{ij} \\ D_{ij}^r \end{pmatrix}, \qquad C = b\begin{pmatrix} 1+c & -1 \\ c & 0 \end{pmatrix}, \tag{8.98a}$$

$$\dot{\mathbf{D}}_{ij} = -C\mathbf{D}_{ij} + s\mathbf{g}_{ij}(S,t) \ . \tag{8.98b}$$

Here, $\mathbf{g}_{ij}$ is a complicated function of genotypic frequencies and possibly of time. By variation of parameters (see, e.g., Brauer and Nohel, 1969, p. 72),

$$\mathbf{D}_{ij}(t) = e^{-Ct}\mathbf{D}_{ij}(0) + s\int_0^t e^{-(t-\tau)C}\mathbf{g}_{ij}[S(\tau),\tau]d\tau \ . \tag{8.99}$$

From (8.98a) we compute the eigenvalues,

$$\lambda_1 = b, \qquad \lambda_2 = bc , \tag{8.100a}$$

and the corresponding eigenvectors,

$$\mathbf{V}_1 = \begin{pmatrix} 1 \\ c \end{pmatrix}, \qquad \mathbf{V}_2 = \begin{pmatrix} 1 \\ 1 \end{pmatrix} , \tag{8.100b}$$

of C. Then the method employed in Sect. 5.1 produces a more explicit form of (8.99), viz.,

$$D_{ij}(t) = (1 - c)^{-1} \{ [D_{ij}(0) - D_{ij}^r(0)]e^{-bt} + [D_{ij}^r(0) - cD_{ij}(0)]e^{-bct} \}$$
$$+ O(s) , \tag{8.101a}$$

$$D_{ij}^r(t) = (1 - c)^{-1} \{ c[D_{ij}(0) - D_{ij}^r(0)]e^{-bt} + [D_{ij}^r(0) - cD_{ij}(0)]e^{-bct} \}$$
$$+ O(s) . \tag{8.101b}$$

In the same manner, from (8.97) we find

$$Q_{ij,kl}(t) = e^{-bt}Q_{ij,kl}(0) + bc^2 e^{-bt} \int_0^t e^{b\tau} D_{ij}(\tau)D_{kl}(\tau)d\tau + O(s) . \tag{8.102}$$

In the neutral case $(s = 0)$, the gene frequencies are constant, so we can explicitly calculate the gametic frequencies from (8.93) and (8.101b). Substituting (8.101a) into (8.102) and performing the elementary integration, we get $\hat{Q}_{ij,kl}$, with which we deduce the genotypic proportions from (8.96). Observe that the deviations from linkage equilibrium and Hardy-Weinberg proportions approach zero exponentially. For the linkage disequilibria to remain zero as the system evolves, their initial values must be zero, i.e., for $D_{ij}(t) = 0$ or $D_{ij}^r(t) = 0$, we must have $D_{ij}(0) = D_{ij}^r(0) = 0$. If also $D_{kl}(0) = D_{kl}^r(0) = 0$, then the integral in (8.102) is absent, and (8.102) reduces to the single-locus formula. Note that even if the population is initially in Hardy-Weinberg proportions, linkage disequilibrium will generate departures from Hardy-Weinberg proportions; i.e., if $D_{ij}(0) \neq 0$ or $D_{ij}^r(0) \neq 0$, and if $D_{kl}(0) \neq 0$ or $D_{kl}^r(0) \neq 0$, then $Q_{ij,kl}(t) \neq 0$ for $t > 0$.

Returning to slow selection, we infer from (8.101) that

$$\mathbf{D}_{ij}(t) = O(s), \qquad t \geq t_1 , \tag{8.103}$$

for all i and j, where $t_1 \approx -(bc)^{-1} \ln s$. For loose linkage, t_1 is about 5 or 10 generations, as in Sect. 8.2, and for tight linkage $(s \ll c \ll 1)$, it is the same as the corresponding time with discrete, nonoverlapping generations. The fact that

$$Q_{ij,kl}(t) = O(s), \qquad t \geq t_1 , \tag{8.104}$$

for all i, j, k, and l, is established by decomposing the integral in (8.102) into parts from 0 to t_1 and from t_1 to t and using (8.101a) and (8.103). Thus, the rapid approach to nearly random combination that we found for a single locus in continuous time and for two loci with discrete, nonoverlapping generations still applies here.

To demonstrate that the deviations from random combination tend to approximate constancy, we put $\mathbf{D}_{ij} = s\mathbf{D}_{ij}^0$ and $Q_{ij,kl} = sQ_{ij,kl}^0$ and express (8.97) and (8.98b) for $t \geq t_1$ in the form

$$\dot{Q}_{ij,kl}^0 = -bQ_{ij,kl}^0 + h_{ij,kl}(S,t) , \qquad (8.105a)$$

$$\dot{\mathbf{D}}_{ij}^0 = -C\mathbf{D}_{ij}^0 + \mathbf{g}_{ij}(S,t) , \qquad (8.105b)$$

where h_{ij} comes from the $O(s)$ term in (8.97). Equation (8.105a) corresponds precisely to (4.156). We suppose

$$\frac{\partial \mathbf{g}_{ij}}{\partial t} = O(s), \qquad \frac{\partial h_{ij,kl}}{\partial t} = O(s) \qquad (8.106)$$

and rewrite (8.105b) as

$$\frac{d}{dt}[\mathbf{D}_{ij}^0 - C^{-1}\mathbf{g}_{ij}(S,t)] = -C[\mathbf{D}_{ij}^0 - C^{-1}\mathbf{g}_{ij}(S,t)] + O(s) . \qquad (8.107)$$

Therefore, exactly as in (8.99), we conclude

$$\mathbf{D}_{ij}^0(t) = C^{-1}\mathbf{g}_{ij}(S,t) + O(s) \qquad (8.108)$$

for $t \geq t_2 \approx 2t_1$. From (8.93), (8.94), (8.96), (8.103), and (8.104) we get

$$\dot{S}_{ij,kl}(t) = O(s), \qquad t \geq t_1 . \qquad (8.109)$$

With the mild assumption, (8.106), that the explicit time dependence of the fertilities and mortalities is of $O(s^2)$ and the aid of (8.108) and (8.109), we obtain $\dot{\mathbf{D}}_{ij}^0(t) = O(s)$ for $t \geq t_2$. Hence

$$\dot{\mathbf{D}}_{ij}(t) = O(s^2), \qquad \dot{Q}_{ij,kl}(t) = O(s^2), \qquad t \geq t_2 . \qquad (8.110)$$

By virtue of (8.103) and (8.104), for $t \geq t_1$ the exact system that satisfies (8.88) is within $O(s)$ of the much simpler system in Hardy-Weinberg proportions and linkage equilibrium. Since the proof of this statement involves little more than a Taylor expansion, the reader can easily generalize the demonstration in Sect. 4.10 to two loci.

Finally, we wish to consider the behavior of the *mean fitness*. Kimura (1958) calculated the rate of change of the mean fitness by decomposing the genotypic fitnesses first with respect to loci and then with respect to alleles within loci. It will be closer to our approach to do first an orthogonal least-squares decomposition relative to gametes, which yields (8.47), and then to derive (8.58) by an orthogonal least-squares decomposition with respect to alleles within gametes. One finds (Nagylaki, 1989)

$$\dot{\overline{m}} = V_g + \overline{\dot{m}} + 2\overline{\overset{\circ}{\phi}E} + \overline{\overset{\circ}{\theta}\Delta} - \sum_{ijkl}(x_i + y_j)E_{kl}Q_{ij,kl} , \qquad (8.111)$$

where V_g is the genic variance,

$$\overline{m} = \sum_{ijkl} \dot{m}_{ij,kl} S_{ij,kl} \ ,$$

the circle indicates the logarithmic derivative [as below (4.191)],

$$\phi_{ij} = P_{ij}/(p_i q_j), \qquad \theta_{ij,kl} = S_{ij,kl}/(P_{ij} P_{kl}) \ , \qquad (8.112)$$

E_{ij} and $\Delta_{ij,kl}$ denote the epistatic deviations within the least-squares gametic fitnesses and the "dominance" deviations between gametes,

$$\overline{\overset{\circ}{\phi}E} = \sum_{ij} \overset{\circ}{\phi}_{ij} E_{ij} P_{ij} \,, \qquad \overline{\overset{\circ}{\theta}\Delta} = \sum_{ijkl} \overset{\circ}{\theta}_{ij,kl} \Delta_{ij,kl} S_{ij,kl} \ ,$$

and x_i and y_j designate the average excesses for fitness of A_i and B_j. The last term in (8.111) arises because in each decomposition the entire residual variance, defined with respect to two-locus genotypic frequencies, must be minimized. The analysis of variance in Kimura (1958) and Crow and Nagylaki (1976) is not orthogonal. If the unique additive component of the variance is extracted in their scheme, their equations corresponding to (8.111) acquire another term, which couples dominance deviations to linkage disequilibria (Nagylaki, 1989).

Suppose that $\overline{m} = o(s^2)$, i.e., the genotypic fitnesses are almost constant. Since (8.93), (8.96), and (8.112) give

$$\phi_{ij} = 1 + \frac{D^r_{ij}}{p_i q_j}, \qquad \theta_{ij,kl} = 1 + \frac{Q_{ij,kl}}{P_{ij} P_{kl}} \ ,$$

therefore (8.109) and (8.110) imply

$$\overset{\circ}{\phi}_{ij}(t) = O(s^2), \qquad \overset{\circ}{\theta}_{ij,kl}(t) = O(s^2)$$

for $t \geq t_2$. Recalling (8.104), we conclude that the last three terms in (8.111) are of $O(s^3)$, and for $t \geq t_2$ we have the approximate Fundamental Theorem of Natural Selection, $\dot{\overline{m}} \approx V_g$.

8.5 Problems

8.1 For the model (8.10), show that an equilibrium is a stationary point of the mean fitness if and only if the linkage disequilibria vanish there. Moran (1967) proved this for two diallelic loci.

8.2 Assume that fitnesses are multiplicative in (8.10): $w_{ij,kl} = \alpha_{ik}\beta_{jl}$ for all i, j, k, and l, where $\alpha_{ik} \ (= \alpha_{ki})$ and $\beta_{jl} \ (= \beta_{lj})$ are the contributions of $A_i A_k$ and $B_j B_l$, respectively. Notice that with multiplicative fitnesses there is no position effect. Suppose the population is initially in linkage equilibrium: $P_{ij}(0) = p_i(0)q_j(0)$ for all i and j. Prove that

(a) it remains in linkage equilibrium: $P_{ij}(t) = p_i(t)q_j(t)$ for all i, j, and $t \geq 0$;

(b) the gene frequencies evolve independently at the two loci according to the single-locus equations

$$p_i' = p_i\alpha_i/\overline{\alpha}, \qquad q_j' = q_j\beta_j/\overline{\beta} \ , \qquad (8.113a)$$

$$\alpha_i = \sum_k \alpha_{ik}p_k, \qquad \beta_j = \sum_l \beta_{jl}q_l \ , \qquad (8.113b)$$

$$\overline{\alpha} = \sum_i \alpha_i p_i, \qquad \overline{\beta} = \sum_j \beta_j q_j \ ; \qquad (8.113c)$$

(c) the mean fitness is $\overline{w} = \overline{\alpha}\overline{\beta}$, and consequently cannot decrease.

Observe that (a) generalizes the preservation of $D = 0$ in the diallelic model, (b) includes the existence of the overdominant and underdominant equilibria (8.71) in that model, and (c) generalizes the diallelic result that $\Delta\overline{w} \geq 0$ on the $D = 0$ surface.

8.3   Assume that the fitnesses in (8.10) are additive: $w_{ij,kl} = \alpha_{ik} + \beta_{jl}$ for all i, j, k, and l, where α_{ik} $(= \alpha_{ki})$ and β_{jl} $(= \beta_{lj})$ are the contributions of A_iA_k and B_jB_l, respectively. Then there is no position effect.

(a) Show that the gene frequencies satisfy

$$\overline{w}p_i' = p_i\alpha_i + \sum_j \beta_j P_{ij} \ ,$$

$$\overline{w}q_j' = q_j\beta_j + \sum_i \alpha_i P_{ij} \ ,$$

where α_i, β_j, $\overline{\alpha}$, and $\overline{\beta}$ are given by (8.113b) and (8.113c), and $\overline{w} = \overline{\alpha} + \overline{\beta}$. Observe that the mean fitness depends only on the allelic frequencies, and that if the gametic frequencies P_{ij} are specified, then the allelic frequencies in the next generation, p_i' and q_j', are independent of the cross-over fraction c. Therefore, given P_{ij} for all i and j, the allelic frequencies p_i' and q_j' are the same as in the single-locus situation, $c = 0$. But then the same holds for $\overline{w}$, which shows that the mean fitness is nondecreasing (Ewens, 1969a).

(b) Prove that linkage equilibrium is generally not preserved.

(c) Demonstrate that if

$$P_{ij} = p_iq_j, \qquad \alpha_i = \overline{\alpha}, \qquad \beta_j = \overline{\beta}$$

for all i and j, then the population is in equilibrium. This generalizes the existence of (8.71) in the two-allele model.

8.4   Derive (8.68) and (8.69).

8.5   Deduce (8.70).

8.6   Prove that the epistatic parameters

$$E_i = \sum_j \varepsilon_j w_{ij}$$

vanish for the diallelic model (8.68) if the fitnesses are additive between loci.

8.7   Demonstrate that the multiplicative epistatic parameters

$$e_i = \sum_j \varepsilon_j \ln w_{ij}$$

vanish for the diallelic model (8.68) if the fitnesses are multiplicative between loci.

8.8   Derive (8.100) and (8.101).

8.9   Prove (8.111).

8.10 Generalize Problems 8.2 and 8.3 to an arbitrary number of loci.

8.11 Write the recursion relation for mutation at two multiallelic loci with discrete, nonoverlapping generations and no selection. Prove that if mutation is independent at the two loci, then the allelic frequencies satisfy the usual single-locus difference equations and the linkage disequilibria ultimately converge to zero at least as fast as $(1-c)^t$, possibly multiplied by polynomials in t. Show that if all alleles at each locus can be connected by a chain of mutations and at least one allele at each locus has positive probability of not mutating, then the asymptotic rate of convergence is no slower than $(1-c)^t$.

8.12 Investigate mutation-selection balance for two diallelic loci. Assume that selection is sufficiently weak compared with recombination that Sect. 8.2 and Problem 8.11 permit the linkage-equilibrium approximation. Suppose that the alleles A and B mutate to a and b at rates u and v, respectively; neglect reverse mutation.

(a) Let the genotypes $AABB$, $AABb$, and $AaBB$ have respective fitnesses 1, $1-r$, and $1-s$ $(r, s > 0)$. If mutation is much weaker than selection, we may ignore selection in the other genotypes. Demonstrate that the frequencies of a and b converge to the single-locus values u/s and v/r, respectively.

(b) With complete dominance at each locus, assign respective fitnesses $1, 1-r$, and $1-s$ $(r,s>0)$ to $A\text{--}B\text{--}$, $A\text{--}bb$, and $aaB\text{--}$ and assume that double mutant homozygotes are negligibly rare. Establish that the frequencies of a and b converge to $\sqrt{u/s}$ and $\sqrt{v/r}$, respectively, as expected.

(c) Posit that A is completely dominant, but B is not: $A\text{--}BB$, $A\text{--}Bb$, and $aaBb$ have respective fitnesses $1, 1-r$, and $1-s$ $(r,s>0)$. Show that for weak mutation the frequencies of a and b tend to $\sqrt{u/s}$ and v/r, respectively.

See Karlin and McGregor (1971) for some other results on mutation-selection balance.

8.13 Under some conditions, a deleterious allele may be fixed by mutation pressure. The present treatment of duplicate genes follows Christiansen and Frydenberg (1977). Assume that generations are discrete and non-overlapping. The genotype $aabb$ has fitness $1-s$, and all other genotypes have fitness unity. The alleles A and B mutate to a and b at the respective rates u and v; mutation is independent at the two loci. The mutation rates, selection coefficient, and recombination frequency fulfill the constraints $0 \leq u \leq v < s \leq 1$ and $0 < c \leq \frac{1}{2}$.

(a) Deduce from (8.68) that for $i = 1, 2, 3$, the gametic frequencies just before fertilization, x_i, satisfy

$$x_i^\star = (x_i - \varepsilon_i cD)/\overline{w} \ ,$$
$$D = x_1 x_4 - x_2 x_3 , \qquad \overline{w} = 1 - s x_4^2 \ ,$$
$$x_1' = (1-u)(1-v)x_1^\star \ ,$$
$$x_2' = (1-u)(vx_1^\star + x_2^\star) \ ,$$
$$x_3' = (1-v)(ux_1^\star + x_3^\star) \ .$$

(b) Show that if $u < v$ and A is initially present, then B is lost and the frequency of a converges to $\sqrt{u/s}$.

(c) Prove that if $u = v$, the equilibria lie on the hyperbola

$$[c(1 - u) + u](\hat{p}_A + \hat{p}_B + \sqrt{u/s} - 1) = c(1-u)\hat{p}_A\hat{p}_B \ ,$$

$\hat{x}_4 = \sqrt{u/s}$ in (p_A, p_B, x_4) space, unless both A and B are absent at equilibrium, in which case $p_A = p_B = 0$ and $x_4 = 1$. Demonstrate that the corresponding linkage disequilibrium is

$$\hat{D} = -u\hat{x}_1/[c(1 - u)] < 0 \ .$$

To make convergence to the hyperbola plausible, establish that the equilibrium $x_4 = 1$ is unstable.

8.14 Consider selection at an arbitrary number of loci with constant fitnesses and no position effect. Let $\overline{w}$ and $p_j^{(i)}$ denote the mean fitness and the frequency of the allele $A_j^{(i)}$ at locus i. Show that if the population is in linkage equilibrium, then

$$p_j^{(i)'} = p_j^{(i)} \frac{\partial \overline{w}}{\partial p_j^{(i)}} \bigg/ \sum_k p_k^{(i)} \frac{\partial \overline{w}}{\partial p_k^{(i)}} \ . \tag{8.114}$$

Apply the inequality of Baum and Eagon (1967) to conclude that $\overline{w}' \geq \overline{w}$, with equality only at equilibrium. Prove that (4.59) holds for weak selection.

Since linkage equilibrium is not generally preserved, the above results need not apply for more than one generation of selection. Nevertheless, much of the early work on selection at multiple loci posited linkage equilibrium as a weak-selection approximation and employed the maximization of the mean fitness to investigate the dynamics.

8.15 For weak selection, the discrete model (8.10) may be approximated by a continuous Hardy-Weinberg model. Assume that $w_{ij,kl} = 1 + s r_{ij,kl}$, where $r_{ij,kl}$ may depend on the gametic frequencies, but not on time or the selection intensity s $(s > 0)$. Rescale time according to $\tau = st$ and set $\pi_{ij}(\tau) = P_{ij}(t)$ for $t = 0, 1, \ldots$. Define t_1 as in (8.21) and put $D_{ij}(t) = s\Delta_{ij}(\tau)$ for $t \geq t_1$. Show that in the limit $s \to 0$

$$\frac{d\pi_{ij}}{d\tau} = \pi_{ij}(r_{ij} - \overline{r}) - c\Delta_{ij}, \qquad \tau > 0 \ ,$$

where

$$r_{ij} = \sum_{kl} r_{ij,kl}\pi_{kl}, \qquad \overline{r} = \sum_{ij} r_{ij}\pi_{ij} \ .$$

8.16 Prove that if dominance and epistasis are absent in the discrete multi-allelic, multilocus model (8.7), then the single-generation change in the mean fitness is exactly equal to the ratio of the variance in fitness to the mean fitness (Nagylaki, 1989).

8.17 Generalize Problem 4.26 to multiple loci without epistasis (Nagylaki, 1991).

9. Inbreeding and Random Drift

In this chapter, we shall study the intimately related processes of inbreeding and random genetic drift. Inbreeding is of great importance in the empirical investigation of genetic mechanisms and in animal and plant breeding. As discussed in Chap. 1, random drift has an essential rôle in Wright's (1931, 1977) shifting-balance theory of evolution and a predominant one in the neutral theory of Kimura (1968, 1971) and of King and Jukes (1969).

Most of the results on inbreeding derived in this chapter were obtained first by Wright (1921, 1922, 1922a, 1931, 1933a) with his method of path coefficients (Wright, 1921, 1921a, 1934, 1968). Fisher (1922, 1930, 1930a) and Wright (1931) began the investigation of random genetic drift; we owe much of our insight into the stochastic evolution of gene frequencies to them. Our treatment of this subject here is introductory; consult Moran (1962), Ewens (1969, 1979, 1990), Crow and Kimura (1970), Kingman (1980), Nagylaki (1990), and Svirezhev and Passekov (1990) for more advanced and far-reaching analyses.

After introducing the inbreeding coefficient in Sect. 9.1, we shall show in Sect. 9.2 how to calculate it from pedigrees. We shall discuss k-coefficients and present applications of the sort useful for genetic counselling and model fitting in Sect. 9.3. Section 9.4 treats the phenotypic effects of inbreeding. Switching from probabilities of identity by descent to probabilities of allelic identity as our basic variables, we shall study regular systems of inbreeding, random mating in finite populations, and the evolution of heterozygosity under mutation in a finite population in Sects. 9.5, 9.6, and 9.7, respectively. An examination of the effects on inbreeding of departures from the simple models of Sect. 9.6 is presented in Sect. 9.8. The remaining sections deal directly with random variation in gene frequencies. In Sect. 9.9, we shall formulate our basic model for random drift. Sections 9.10 and 9.11 comprise investigations of pure random drift and of the joint action of mutation and random drift. The island model is the subject of Sect. 9.12. Section 9.13 treats the influence on gene frequencies of departures from the scheme of Sect. 9.9.

9.1 The Inbreeding Coefficient

We shall follow the probabilistic approach to inbreeding devised independently by Cotterman (1940) and Malécot (1941, 1942, 1946, 1948). Some of the ideas were foreshadowed by Haldane and Moshinsky (1939), and Borel (1943) presented a brief probabilistic treatment. We posit that the population is infinite, generations are discrete and nonoverlapping, mating is independent of genotype, and all evolutionary forces are absent. Unless stated otherwise, we consider autosomal loci and need not distinguish the sexes.

We can trace the ancestors of any individual or gene back to the initial population ($t = 0$). *Related individuals* have at least one common ancestor, or one individual is descended from the other. *Inbreeding* is the mating of related individuals.

Two homologous genes are *identical in state* if and only if they are allelically the same. Thus, an individual is homozygous at a given locus if and only if the two genes he carries at that locus are identical in state. Two homologous genes are *identical by descent* if and only if they are derived from the same gene or one is derived from the other (Cotterman, 1940; Malécot, 1941, 1942, 1946, 1948). Since there is no mutation, this vital relation is stronger than identity in state: identity by descent implies identity in state, but not *vice versa*. If an individual's two genes at a given locus are identical by descent, we say he is *autozygous* at that locus. Otherwise, he is *allozygous* (Cotterman, 1940). Evidently, autozygosity implies homozygosity, but a homozygous individual may be allozygous.

The *inbreeding coefficient*, F (Wright, 1921, 1922), of an individual is the probability that he is autozygous (Cotterman, 1940; Malécot, 1941, 1942, 1946, 1948). Note that with our assumptions the inbreeding coefficient is a function only of ancestry; it is the same for all (autosomal) loci and independent of gene frequencies.

We focus attention on a locus with alleles A_i that have frequencies p_i. Assume that the initial population was in Hardy-Weinberg proportions. Then all deviations from random combination of alleles are due solely to inbreeding. A randomly chosen individual is autozygous with probability F and allozygous with probability $1 - F$; in both cases, p_i represents the probability that a gene chosen from him at random is A_i. In the first case, his other gene is necessarily also A_i, and hence he has probability p_i of having genotype A_iA_i. In the second, his genes are stochastically independent, and therefore his other gene is A_j with probability p_j, which gives him ordered genotype A_iA_j with the Hardy-Weinberg probability p_ip_j. Thus, the ordered genotypic frequencies read

$$P_{ii} = Fp_i + (1 - F)p_i^2 = p_i^2 + Fp_i(1 - p_i) , \qquad (9.1a)$$

$$P_{ij} = (1 - F)p_ip_j , \qquad i \neq j . \qquad (9.1b)$$

Wright (1921) first obtained this formula for two equally frequent alleles and then generalized it to arbitrary gene frequency (Wright, 1922a). Cotterman (1940) deduced (9.1) for two alleles by a probabilistic argument. We followed Malécot's (1948) elegant multiallelic derivation.

Observe that an inbreeding coefficient of zero yields the Hardy-Weinberg proportions for panmixia, whereas only homozygotes appear in a completely inbred population ($F = 1$), in which $A_i A_i$ has frequency p_i. We can rewrite (9.1) more compactly as

$$P_{ij} = F p_i \delta_{ij} + (1 - F) p_i p_j \ . \tag{9.2}$$

With pure inbreeding, (9.2) shows that the frequency of A_i in uniting gametes, and hence in the next generation, is

$$\sum_j P_{ij} = F p_i + (1 - F) p_i = p_i \ .$$

Hence, the allelic frequencies are constant.

From (9.1) we can calculate immediately the homozygosity, f, and heterozygosity, h ($f + h = 1$). We have

$$f = \sum_i P_{ii} = f_r + F h_r \ , \tag{9.3a}$$

$$h = \sum_{ij, i \neq j} P_{ij} = (1 - F) h_r \ , \tag{9.3b}$$

where the homozygosity and heterozygosity with random mating and allelic frequencies p_i are given by

$$f_r = \sum_i p_i^2 \ , \qquad h_r = \sum_{ij, i \neq j} p_i p_j = \sum_i p_i (1 - p_i) \ . \tag{9.3c}$$

We notice that the heterozygosity is decreased by a factor $1 - F$ and the homozygosity is increased correspondingly. Therefore, the offspring of consanguineous matings are more likely to be homozygous, and hence have a higher probability of exhibiting the effects of recessive alleles. If $p_i \ll 1$, then (9.1a) yields

$$P_{ii} \approx p_i^2 + F p_i = p_i (p_i + F) \ ,$$

so the inbreeding effect for the recessive A_i is significant unless $F \ll p_i$.

Wright (1921, 1922, 1951) defined the inbreeding coefficient as the correlation between uniting gametes. Regardless of the numerical values assigned to the alleles A_i, the interpretation as a correlation is readily deduced from the argument that leads to (9.1). Since the genes of autozygous and allozygous individuals are drawn from the same pool, they have the same mean and variance. This allows us to average the correlations – obviously one and zero – in the two cases. Therefore, the correlation between uniting gametes is $F(1) + (1 - F)(0) = F$. For a computational proof, see Wright (1951) and Crow and Kimura (1970, pp. 66-68).

A measure of the relatedness of two individuals is Malécot's (1941, 1942, 1946, 1948) *coefficient of consanguinity*. The coefficient of consanguinity of individuals I and J, F_{IJ}, is the probability that a randomly chosen gene from I and a homologous randomly chosen gene from J are identical by descent. Hence, if O is the offspring of I and J, then $F_{IJ} = F_O$, the inbreeding coefficient of O.

9.2 Calculation of the Inbreeding Coefficient from Pedigrees

Our aim in this section is to reduce the computation of inbreeding coefficients determined by an arbitrary pedigree to the routine application of a simple rule. For this purpose, it is convenient to extend the notion of a common ancestor of the parents of an individual I. We shall refer to A as a *common ancestor* of I if there are two disjoint (i.e., with no common links), descending paths from A to I. Figure 9.1 exemplifies this for the matings of half-siblings and of parent and offspring. The circles refer to individuals who are either monoecious or whose sex need not be specified. Only individuals relevant to inbreeding are shown.

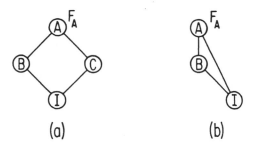

Fig. 9.1. Pedigrees for half-siblings and for parent and offspring.

All inbreeding will be shown in our pedigrees, except that inbreeding of common ancestors who have no ancestors in the pedigree may be indicated, as in Fig. 9.1, just by their inbreeding coefficient. If we trace genes only within the pedigree, the definition of identity by descent implies that two genes are identical by descent if and only if they are derived from the same gene or identical genes, or one is derived from the other.

We signify that genes a and b are identical by descent by writing $a \equiv b$. In Fig. 9.2, $a \equiv b$ and $c \equiv b$ are not mutually exclusive owing to the possibility $a \equiv b \equiv c$. To add probabilities in evaluating inbreeding coefficients, we require a mutually exclusive relation. Let $a \to b$ denote the *direct descent* of gene b in an individual from gene a in one of his parents. This fundamental relation was introduced by Denniston (1967, 1974), who called it *replication*. Obviously, $a \to b$ and $c \to b$ cannot occur simultaneously. Replication is stronger than identity by descent: $a \to b$ implies $a \equiv b$, but $a \equiv b$ does not imply $a \to b$ or $b \to a$.

The essential Mendelian property of replication is that in Fig. 9.2

$$P(a \to b) = P(c \to b) = \tfrac{1}{2} \; . \tag{9.4}$$

By definition,

$$F_A = P(d \equiv g) \; . \tag{9.5}$$

The randomly chosen genes e and f may be identical for two mutually exclusive reasons. With probability $\tfrac{1}{2}$, they are replicates of the same gene (d or g); with probability $\tfrac{1}{2}$, they are replicates of different genes, and these are identical with probability F_A. Therefore,

$$P(e \equiv f) = \tfrac{1}{2}(1 + F_A) \; . \tag{9.6}$$

Viewing Fig. 9.2 as two distinct parts of the same pedigree and using mutual exclusiveness and independence, we have also

$$P(b \equiv e) = P(a \to b)P(a \equiv e) + P(c \to b)P(c \equiv e)$$
$$= \tfrac{1}{2}[P(a \equiv e) + P(c \equiv e)] \tag{9.7a}$$
$$= \tfrac{1}{4}[P(a \equiv d) + P(a \equiv g) + P(c \equiv d) + P(c \equiv g)] \; , \tag{9.7b}$$

where (9.7b) follows from (9.7a). Notice that if we replace b, a, and c by f, d, and g, respectively, and use (9.5) and the self-evident fact $P(a \equiv a) = 1$, we rederive (9.6) from (9.7b). The relations in this paragraph suffice for the analysis of all inbreeding effects at a single autosomal locus.

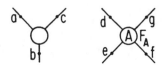

Fig. 9.2. Replication and identity by descent.

We shall infer the rule for calculating inbreeding coefficients by computing them for various types of pedigrees from basic principles.

1. A Simple Pedigree

We evaluate the inbreeding coefficient of I in Fig. 9.3 by appealing to (9.4), (9.5), and (9.6):

$$F_I = P(c \equiv d)$$
$$= P(b \to c)P(a \to b)P(a \equiv e)P(e \to d)$$
$$= (\tfrac{1}{2})^4(1 + F_A) \; .$$

Note that the exponent of $\frac{1}{2}$ is the number of ancestors in the loop. We shall designate each path such as $CBAD$ by a list of ancestors, in which the boldface letter indicates the common ancestor. Each line that does not lead to a common ancestor may be omitted because it cannot form part of a closed loop.

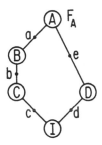

Fig. 9.3. A simple pedigree.

2. Multiple Common Ancestors

We analyze two pedigrees. For Fig. 9.4a, we have

$$F_I = \mathcal{P}(c \equiv d)$$
$$= \mathcal{P}(f \to c)\mathcal{P}(f \equiv g)\mathcal{P}(g \to d) + \mathcal{P}(b \to c)\mathcal{P}(a \to b)\mathcal{P}(a \equiv e)\mathcal{P}(e \to d)$$
$$= (\tfrac{1}{2})^3(1 + F_B) + (\tfrac{1}{2})^4(1 + F_A) \ ;$$

Figure 9.4b yields

$$F_I = \mathcal{P}(c \equiv d)$$
$$= \mathcal{P}(b \to c)\mathcal{P}(b \equiv f)\mathcal{P}(f \to d) + \mathcal{P}(b \to c)\mathcal{P}(a \to b)\mathcal{P}(a \equiv e)\mathcal{P}(e \to d)$$
$$= (\tfrac{1}{2})^3 + (\tfrac{1}{2})^4(1 + F_A) \ .$$

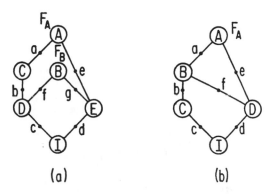

(a)   (b)

Fig. 9.4. Multiple common ancestors.

Evidently, the mutual exclusiveness of replication leads to the addition of the contributions of distinct paths, DBE and $DCAE$ in Fig. 9.4a, and CBD and $CBAD$ in Fig. 9.4b.

3. Inbred Ancestor

In Fig. 9.5,

$$F_I = \mathcal{P}(d \equiv e)$$
$$= \mathcal{P}(c \to d)\mathcal{P}(b \to c)\mathcal{P}(a \to b)\mathcal{P}(a \equiv e)$$
$$\quad + \mathcal{P}(g \to d)\mathcal{P}(f \to g)\mathcal{P}(a \to f)\mathcal{P}(a \equiv e)$$
$$= (\tfrac{1}{2})^4(1 + F_A) + (\tfrac{1}{2})^4(1 + F_A) \ .$$

Thus, the addition of the distinct paths $DCBA$ and $DEBA$ automatically takes into account the inbreeding of D $(F_D = \tfrac{1}{8})$.

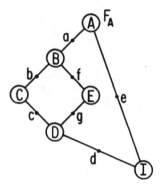

Fig. 9.5. Inbred ancestor.

4. Repeated Inbreeding

For Fig. 9.6, we obtain

$$F_I = \mathcal{P}(d \equiv e)$$
$$= \mathcal{P}(c \to d)\mathcal{P}(c \equiv f)\mathcal{P}(f \to e)$$
$$= (\tfrac{1}{2})^3(1 + F_C) \ ,$$

where

$$F_C = (\tfrac{1}{2})^3(1 + F_A) \ .$$

Our emerging rule applies here if we forbid the path through A by excluding loops that pass through any ancestor more than once.

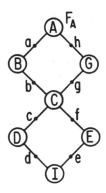

Fig. 9.6. Repeated inbreeding.

5. A Complex Pedigree

Figure 9.7 combines the features of Figs. 9.3 to 9.6. We see that

$$F_I = \mathcal{P}(b \equiv c)$$
$$= \mathcal{P}(b \equiv g)\mathcal{P}(g \to c) + \mathcal{P}(b \equiv d)\mathcal{P}(d \to c)$$
$$= (\tfrac{1}{2})^2(1 + F_B) + \tfrac{1}{2}[\mathcal{P}(a \to b)\mathcal{P}(a \equiv d) + \mathcal{P}(f \to b)\mathcal{P}(f \equiv d)] .$$

But $\mathcal{P}(f \equiv d) = \tfrac{1}{2}$ and

$$\mathcal{P}(a \equiv d) = \mathcal{P}(a \equiv f) = F_B = (\tfrac{1}{2})^2(1 + F_A) .$$

Therefore,
$$F_I = (\tfrac{1}{2})^2(1 + F_B) + (\tfrac{1}{2})^3 + (\tfrac{1}{2})^4(1 + F_A) .$$

The three terms correspond to the paths BC, BDC, and $BADC$, respectively.

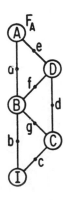

Fig. 9.7. A complex pedigree.

From our examples, we infer

$$F_I = \sum_L (\tfrac{1}{2})^{n_L}(1 + F_{A_L}) \, , \tag{9.8}$$

where the sum is over all distinct closed loops that do not pass through any ancestor of I more than once. When tracing loops, vertical direction must be reversed exactly once. In the loop L, n_L is the number of ancestors, and F_{A_L} represents the inbreeding coefficient of the common ancestor where vertical direction is reversed.

We owe this beautiful result to Wright (1922). Cotterman (1940) and Malécot (1941, 1942, 1948) presented probabilistic derivations independently of each other, and Borel (1943) gave a partial probabilistic analysis. Consult Boucher (1988) for a detailed, rigorous proof.

Recall that F_I is also the coefficient of consanguinity of the parents of I.

In our examples of the application of (9.8), we assume that all inbreeding is shown in the pedigrees. Observe that Fig. 9.1 yields (with $F_A = 0$) inbreeding coefficients of $(\tfrac{1}{2})^3 = \tfrac{1}{8}$ and $(\tfrac{1}{2})^2 = \tfrac{1}{4}$ for the progeny of half-sibling and parent-offspring matings, respectively. For the more complicated pedigree displayed in Fig. 9.8, the paths $EDCJG$, $EDCJHG$, $EDKJG$, $EDKJHG$, $EDCAKJG$, $EDCAKJHG$, $EDCBAKJG$, $EDCBAKJHG$, $EDKACJG$, $EDKACJHG$, $EDKABCJG$, and $EDKABCJHG$ give

$$F_I = (\tfrac{1}{2})^5(1 + F_C) + (\tfrac{1}{2})^6(1 + F_C) + (\tfrac{1}{2})^5 + (\tfrac{1}{2})^6 + (\tfrac{1}{2})^7 + (\tfrac{1}{2})^8$$
$$+ (\tfrac{1}{2})^8 + (\tfrac{1}{2})^9 + (\tfrac{1}{2})^7 + (\tfrac{1}{2})^8 + (\tfrac{1}{2})^8 + (\tfrac{1}{2})^9 \, .$$

Since $F_C = \tfrac{1}{4}$, this simplifies to $F_I = \tfrac{9}{64}$.

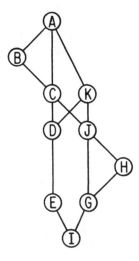

Fig. 9.8. Another complex pedigree.

X-Linkage

For X-linked loci, we must distinguish the sexes. Let squares and circles denote males and females, respectively. The basic replication and identity relations corresponding to Fig. 9.9 are

$$\mathcal{P}(c \to a) = 1, \qquad \mathcal{P}(d \to e) = \mathcal{P}(g \to e) = \tfrac{1}{2} , \tag{9.9a}$$

$$F_A = \mathcal{P}(d \equiv g) , \tag{9.9b}$$

$$\mathcal{P}(a \equiv b) = 1, \qquad \mathcal{P}(e \equiv f) = \tfrac{1}{2}(1 + F_A) , \tag{9.9c}$$

$$\mathcal{P}(a \equiv e) = \tfrac{1}{2}[\mathcal{P}(c \equiv d) + \mathcal{P}(c \equiv g)] . \tag{9.9d}$$

From (9.9a) and (9.9c) we conclude that with the following modifications the rule (9.8) holds for the calculation of the coefficient of consanguinity of two individuals (or equivalently, if their sex is different, for the computation of the inbreeding coefficient of their female offspring). The exponent n_L is now the number of female ancestors of I in the loop L. This interpretation of n_L holds with the *convention* $F_{A_L} = 0$ for male common ancestors of I. Paths with at least two consecutive males do not contribute.

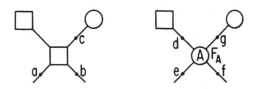

Fig. 9.9. Replication and identity by descent for an X-linked locus.

The above prescription is due to Wright (1933a); it was obtained probabilistically by Crow and Kimura (1970, p. 73) and proved rigorously by Boucher (1988).

Inbreeding at all loci of species with arrhenotokous parthenogenesis follows the rule for X-linked loci. This is a useful fact in apiculture.

Figure 9.1 provides some simple examples. Assume that there is no previous inbreeding. For the half-siblings in Fig. 9.1a, since either B is male and C female, or *vice versa*, therefore $F_I = 0$ if A is male, whereas $F_I = (\tfrac{1}{2})^2 = \tfrac{1}{4}$ if A is female. For the parent-offspring pedigree in Fig. 9.1b, $F_I = \tfrac{1}{2}$ regardless of whether A is male and hence B female, or *vice versa*. For the more complicated pedigree in Fig. 9.10, noting that the path $DCBHGE$ includes the consecutive males H and G and adding the contributions of the paths $DCGE$ and $DCBAGE$, we find

$$F_I = (\tfrac{1}{2})^2 + (\tfrac{1}{2})^4 = \tfrac{5}{16} .$$

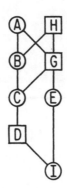

Fig. 9.10. A pedigree for inbreeding at an X-linked locus.

9.3 Identity Relations Between Relatives

At the end of Sect. 9.1, we introduced the coefficient of consanguinity as a measure of the genetic relation between two individuals. Quite different pairs of relatives, however, may have the same coefficient of consanguinity. For example, although parent and offspring share exactly one gene identical by descent, whereas full siblings may have 0, 1, or 2 identical genes, both pairs of relatives have the same coefficient of consanguinity, $\frac{1}{4}$. For single-locus problems of the type that arise in genetic counselling and testing genetic models and for calculating the correlation between relatives, we require a more refined description of the genetic relation between two individuals.

Cotterman (1940) was the first to define probabilities of identity by descent for pairs of relatives in the case of a single autosomal locus. Denniston (1967, 1974) and Cockerham (1971) have presented algorithms for the evaluation of these probabilities. Denniston (1967) has also investigated an X-linked locus. Thompson (1974) has studied probabilities of identity at an autosomal locus for an arbitrary number of relatives.

We shall develop the theory only for pairs of allozygous relatives. In Cotterman's terminology, we confine ourselves to *regular* relatives, i.e., those whose probability of autozygosity is zero. We study only an autosomal locus; consult Li and Sacks (1954) and Campbell and Elston (1971) for results on an X-linked locus, and Campbell and Elston (1971) and Denniston (1975) for two loci.

Consider the regular relatives I and J in Fig. 9.11a. We observe that I and J may have 0, 1, or 2 genes identical by descent and define the *Cotterman (1940) k-coefficients* accordingly:

$$k_0 = \mathcal{P}(I \text{ and } J \text{ have no genes identical by descent})$$
$$= \mathcal{P}(a \not\equiv c,\ a \not\equiv d,\ b \not\equiv c,\ \text{and } b \not\equiv d)\ , \tag{9.10a}$$
$$2k_1 = \mathcal{P}(I \text{ and } J \text{ have exactly one gene identical by descent})$$
$$= \mathcal{P}[(a \equiv c \text{ and } b \not\equiv d) \text{ or } (a \equiv d \text{ and } b \not\equiv c) \text{ or}$$
$$(a \not\equiv c \text{ and } b \equiv d) \text{ or } (a \not\equiv d \text{ and } b \equiv c)]\ , \tag{9.10b}$$
$$k_2 = \mathcal{P}(I \text{ and } J \text{ have both genes identical by descent})$$
$$= \mathcal{P}[(a \equiv c \text{ and } b \equiv d) \text{ or } (a \equiv d) \text{ and } b \equiv c)]\ . \tag{9.10c}$$

Although we assume that I and J are allozygous, their parents, A, B, C, and D, may be autozygous. Since I and J are not inbred, the three events that define k_0, $2k_1$, and k_2 are mutually exclusive and exhaust all possibilities. Hence,

$$k_0 + 2k_1 + k_2 = 1\ . \tag{9.11a}$$

If I or J were inbred, relations such as $a \equiv b \equiv c \not\equiv d$ and $a \equiv b \equiv c \equiv d$ would be possible. Note also that k_1 is the probability that a randomly chosen gene from I is identical to one of the genes of J and that the other gene of I is not identical to the other gene of J.

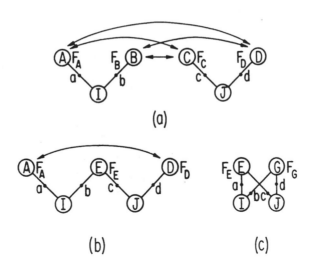

Fig. 9.11. Identity relations (indicated by double-headed arrows) between regular relatives.

To abbreviate probabilities of identity, we write $F_{ac} = \mathcal{P}(a \equiv c)$. A path connecting A and C cannot include an ancestor of B or D, for this would cause inbreeding of I or J. Therefore, every path connecting A and C must be disjoint from every path connecting B and D. Similarly, paths connecting A and D are disjoint from those connecting B and C. Finally, the four events in parentheses in (9.10b) are mutually exclusive, as are the two parenthetical events in (9.10c). We conclude (Cotterman, 1960)

$$k_2 = F_{ac}F_{bd} + F_{ad}F_{bc} \ , \tag{9.11b}$$
$$2k_1 = F_{ac}(1 - F_{bd}) + F_{ad}(1 - F_{bc}) + (1 - F_{ac})F_{bd} + (1 - F_{ad})F_{bc} \ .$$

Inserting (9.7b) and (9.11b) into the last equation, we find

$$k_1 = 2F_{IJ} - k_2 \ . \tag{9.11c}$$

As a check, we rearrange (9.11c) in the form

$$F_{IJ} = \tfrac{1}{2}\tfrac{1}{2}(2k_1) + \tfrac{1}{2}k_2 \ .$$

This can be justified directly. Two randomly chosen genes, one from I and one from J, may be identical by descent for two mutually exclusive reasons, viz., the allozygous individuals I and J may have either one or two genes identical by descent. These events have respective probabilities $2k_1$ and k_2. In the first case, the probability of choosing the identical genes from both I and J is evidently $\tfrac{1}{2}\tfrac{1}{2}$; in the second, whichever gene is drawn from I, the probability of choosing the identical gene from J is $\tfrac{1}{2}$.

From (9.11a) and (9.11c) we see that

$$2F_{IJ} = k_1 + k_2 \le 1 \ ;$$

the maximum value, $\tfrac{1}{2}$, of the coefficient of consanguinity of two regular relatives is attained for monozygotic twins, for whom $k_2 = 1$.

Following Thompson (1976), we now derive an inequality that restricts the possible values of the k-coefficients. Successive application of (9.11b) and (9.7b) produces

$$\begin{aligned}
k_2 &\le \tfrac{1}{4}[(F_{ac} + F_{bd})^2 + (F_{ad} + F_{bc})^2] \\
&= \tfrac{1}{4}[(F_{ac} + F_{bd})^2 + (4F_{IJ} - F_{ac} - F_{bd})^2] \\
&= 4F_{IJ}^2 - \tfrac{1}{2}(F_{ac} + F_{bd})(F_{ad} + F_{bc}) \\
&\le 4F_{IJ}^2 = (k_1 + k_2)^2 \ .
\end{aligned}$$

Hence, (9.11a) yields the desired inequality:

$$k_1^2 \ge k_0 k_2 \ . \tag{9.12}$$

This result is sharp, for our proof shows that equality obtains in (9.12) if and only if $F_{ac} = F_{bd}$, $F_{ad} = F_{bc}$, and either $F_{ac} + F_{bd} = 0$ or $F_{ad} + F_{bc} = 0$. Therefore $k_1^2 = k_0 k_2$ if and only if either

$$F_{ac} = F_{bd} \qquad \text{and} \qquad F_{ad} = F_{bc} = 0$$

or

$$F_{ad} = F_{bc} \qquad \text{and} \qquad F_{ac} = F_{bd} = 0 \ ,$$

i.e., if and only if two probabilities of identity are equal and the other two are zero (Thompson, 1976).

With the definition $K_i = (1 + \delta_{i1})k_i$, (9.12) becomes $K_1^2 \geq 4K_0K_2$. Identifying (K_0, K_1, K_2) with (D, H, R) in the de Finetti (1926) diagram of Fig. 3.1 in Problem 3.1, we see that no relationships fall below the parabola and that those satisfying the restriction displayed in the previous paragraph fall on it. Consult Thompson (1976) for further discussion.

It is not difficult to see that $k_2 > 0$ if and only if I and J are connected by at least two disjoint paths. Hence, Cotterman calls regular relatives with $k_2 > 0$ *bilineal* and those with $k_2 = 0$ *unilineal*. For unilineal relatives, (9.11a) and (9.11c) give

$$k_2 = 0, \qquad k_1 = 2F_{IJ}, \qquad k_0 = 1 - 4F_{IJ} . \tag{9.13}$$

Although we have shown four distinct parents of I and J in Fig. 9.11a, we have done so only for descriptive ease; (9.10) and (9.11) refer only to the genes of I and J and always hold. By separating different parental situations, we can convince ourselves of the generality of the reasoning and reduce the calculation of the k-coefficients to that of coefficients of consanguinity. It will suffice to reexpress (9.11b), for then (9.11a) and (9.11c) complete the computation.

With four distinct parents, Fig. 9.11a allows us to rewrite (9.11b) as (Cotterman, 1960)

$$k_2 = F_{AC}F_{BD} + F_{AD}F_{BC} . \tag{9.14a}$$

If there are three parents, Fig. 9.11b, (9.6), and (9.11b) yield

$$k_2 = \tfrac{1}{2}F_{AD}(1 + F_E) . \tag{9.14b}$$

Similarly, for two parents, we infer from Fig. 9.11c, (9.6), and (9.11b) that

$$k_2 = \tfrac{1}{4}(1 + F_E)(1 + F_G) . \tag{9.14c}$$

Finally, if I is an ancestor of J and $k_2 > 0$, then at least two disjoint paths must connect I and J, so J must be inbred. We conclude with van Aarde (1975) that two regular relatives must be unilineal if one of them is an ancestor of the other. Consequently, the simple result (9.13) applies.

When using (9.11a), (9.11c), and (9.13) or (9.14) to calculate the k-coefficients from a pedigree, one must keep in mind that Wright's rule, (9.8), assumes that all inbreeding, except possibly that of common ancestors with no ancestors in the pedigree, is shown in the pedigree.

Table 9.1 provides easy exercises for computing k-coefficients. In the table, all individuals and all their ancestors are allozygous. For instance, for half-siblings and for parent and offspring, in Sect. 9.2 we found coefficients of consanguinity of $\tfrac{1}{8}$ and $\tfrac{1}{4}$, respectively. The k-coefficients follow at once from (9.13). For full siblings, Fig. 9.11c and (9.14c) yield $k_2 = \tfrac{1}{4}$. This is obvious: draw a and b; for two identical genes, we must independently choose the same genes a second time, so the probability is $(\tfrac{1}{2})^2$. From Fig. 9.11c and the rule (9.8), $F_{IJ} = \tfrac{1}{4}$. Now k_1 follows from (9.11c) and k_0 from (9.11a).

Table 9.1. Identity relations between relatives

Relationship	F_{IJ}	k_2	$2k_1$	k_0
Monozygotic twins	$\frac{1}{2}$	1	0	0
Parent-offspring	$\frac{1}{4}$	0	1	0
Full siblings	$\frac{1}{4}$	$\frac{1}{4}$	$\frac{1}{2}$	$\frac{1}{4}$
Half siblings Uncle-niece Grandparent-grandchild	$\frac{1}{8}$	0	$\frac{1}{2}$	$\frac{1}{2}$
Double first cousins	$\frac{1}{8}$	$\frac{1}{16}$	$\frac{3}{8}$	$\frac{9}{16}$
First cousins	$\frac{1}{16}$	0	$\frac{1}{4}$	$\frac{3}{4}$

As a more complicated example, consider in Fig. 9.12 the double first cousins I and J with some inbred ancestors. Noting that $F_E = \frac{1}{8}$ and $F_{AD} = F_{BC} = 0$, we have

$$F_{IJ} = \left(\tfrac{1}{2}\right)^5 \left(1 + \tfrac{1}{8}\right) + \left(\tfrac{1}{2}\right)^5 + 2\left(\tfrac{1}{2}\right)^5 + 2\left(\tfrac{1}{2}\right)^7 = \tfrac{37}{256} \ ,$$
$$F_{AC} = \left(\tfrac{1}{2}\right)^3 + \left(\tfrac{1}{2}\right)^3 \left(1 + \tfrac{1}{8}\right) = \tfrac{17}{64} \ ,$$
$$F_{BD} = 2\left(\tfrac{1}{2}\right)^3 + 2\left(\tfrac{1}{2}\right)^5 = \tfrac{5}{16} \ ,$$

so (9.14a), (9.11c), and (9.11a) lead to

$$k_2 = \tfrac{85}{1024}, \qquad 2k_1 = \tfrac{211}{512}, \qquad k_0 = \tfrac{517}{1024} \ .$$

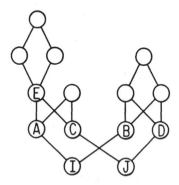

Fig. 9.12. A pedigree for the computation of k-coefficients.

Applications

We present some applications that demonstrate the usefulness of the k-coefficients in genetic counselling and testing genetic hypotheses. The methods we discuss here require less algebra than the direct approach of Chap. 3 and apply at once to all pairs of regular relatives. Consult Murphy and Chase (1975) for an extensive treatment of genetic counselling.

For simplicity, we restrict ourselves to two alleles throughout. As usual, A and a have frequencies p and q; let the genotypes AA, Aa (unordered), and aa be numbered 1, 2, and 3, respectively.

1. Genotype and Phenotype Probabilities

Suppose that individual I has genotype i. What is the probability,

$$Q_{ji} = \mathcal{P}(J = j \,|\, I = i) , \qquad (9.15)$$

that his relative J has genotype j? Observe that

$$Q_{ji} = \sum_{\ell=0}^{2} (1 + \delta_{\ell 1}) k_\ell \mathcal{P}(J = j \,|\, I = i, \text{ and } I \text{ and } J \text{ have exactly } \ell \text{ genes identical by descent}) .$$

Clearly, the coefficients of k_2 and k_0 are δ_{ij} and the Hardy-Weinberg frequency of genotype j, respectively; we can easily calculate that of $2k_1$ in each case. For instance, if $i = j = 2$, the last coefficient is $\frac{q}{2} + \frac{p}{2} = \frac{1}{2}$. To see this, note that the identical gene in the heterozygote I has probability $\frac{1}{2}$ of being A and $\frac{1}{2}$ of being a, and the nonidentical gene in J has probability p of being A and q of being a; the result follows by requiring that J be heterozygous. We may conveniently express (9.15) as the matrix

$$Q = \begin{pmatrix} k_2 + 2pk_1 + p^2 k_0 & p(k_1 + pk_0) & p^2 k_0 \\ 2q(k_1 + pk_0) & k_2 + k_1 + 2pqk_0 & 2p(k_1 + qk_0) \\ q^2 k_0 & q(k_1 + qk_0) & k_2 + 2qk_1 + q^2 k_0 \end{pmatrix} \qquad (9.16)$$

(Cotterman, 1940, 1960; Li and Sacks, 1954).

As a check, notice that (9.11a) and (9.16) yield

$$\sum_j Q_{ji} = 1 , \qquad (9.17)$$

as required by (9.15).

As $q \to 0$, (9.16) informs us that $Q_{31} \to 0$, $Q_{32} \to 0$, and $Q_{33} \to k_2$. We conclude that if a is a rare recessive, a bilineal relative is much more likely to have the recessive phenotype than is a unilineal one.

See Denniston (1967, 1974) for the generalization of (9.16) to inbred relatives.

Joint frequencies of relatives follow immediately from (9.16):

$$S_{ij} = \mathcal{P}(I = i \text{ and } J = j) = \mathcal{P}(I = i)Q_{ji} . \qquad (9.18)$$

Therefore,

$$S_{1j} = p^2 Q_{j1} , \qquad S_{2j} = 2pq Q_{j2} , \qquad S_{3j} = q^2 Q_{j3} . \qquad (9.19)$$

We can easily verify that (9.19) satisfies the obvious relations

$$S_{ij} = S_{ji} , \qquad \sum_j S_{ij} = \mathcal{P}(I = i) . \qquad (9.20)$$

Refer to Harris (1964) and Cockerham (1971) for the extension of (9.19) to multiple alleles and inbred relatives.

It is often useful to have conditional and absolute probabilities with complete dominance. If a is recessive, we denote the dominant and recessive phenotypes by D and R, respectively, and obtain the joint frequencies

$$S_{DD} = S_{11} + 2S_{12} + S_{22} , \qquad (9.21a)$$
$$S_{DR} = S_{RD} = S_{13} + S_{23} , \qquad (9.21b)$$
$$S_{RR} = S_{33} . \qquad (9.21c)$$

From (9.21) we deduce the conditional probabilities such as

$$Q_{DR} = \mathcal{P}(J = D \,|\, I = R) = S_{DR}/\mathcal{P}(I = R) .$$

We find

$$Q_{DD} = S_{DD}/(1 - q^2), \qquad Q_{DR} = S_{DR}/q^2 , \qquad (9.22a)$$
$$Q_{RD} = S_{RD}/(1 - q^2), \qquad Q_{RR} = S_{RR}/q^2 . \qquad (9.22b)$$

2. Genetic Counselling

Suppose that a is recessive and the individuals I and J are dominant. Assume that I has a recessive offspring who is related to J only through paths (or a path) passing through I. What is the probability, b, that J is heterozygous (Denniston, 1974)? We derive the answer as follows (Denniston, 1974):

$$
\begin{aligned}
b(q) &= \mathcal{P}(J = 2 \,|\, I = 2 \text{ and } J = D) \\
&= \frac{\mathcal{P}(I = 2,\, J = 2, \text{ and } J = D)}{\mathcal{P}(I = 2 \text{ and } J = D)} \\
&= \frac{\mathcal{P}(I = 2 \text{ and } J = 2)}{\mathcal{P}(I = 2 \text{ and } J = 1) + \mathcal{P}(I = 2 \text{ and } J = 2)} \\
&= \frac{S_{22}}{S_{21} + S_{22}} \\
&= \frac{Q_{22}}{Q_{12} + Q_{22}} \\
&= \frac{k_2 + k_1 + 2pq k_0}{1 - q(k_1 + q k_0)} .
\end{aligned}
\qquad (9.23)
$$

For recessive diseases, almost invariably $q \ll 1$. If, in addition, the relatives are sufficiently close for $qk_0 \ll F_{IJ}$ to hold, then (9.23) simplifies to

$$b(q) \approx b(0) = k_2 + k_1 = 2F_{IJ} \ . \tag{9.24}$$

We can derive (9.24) directly, because in the limit $q \to 0$ the fact that J is dominant provides no information. Consequently, (9.16) gives

$$b(0) = Q_{22}(0) = k_2 + k_1 \ .$$

We remark that (9.23) is exact in the absence of other information about I and J. Information about other individuals related to I or J would require modification of (9.23). Knowledge that certain relatives of I or J are dominant, however, would have no effect in the limit $q \to 0$, so (9.24) would still apply.

3. Incomplete Penetrance

Since many characters are environmentally influenced, it is desirable to generalize the above applications to genes with partial penetrance. We assume that there are just two phenotypic classes and designate these "normal" (N) and "affected" (T); affected individuals are said to display the trait (e.g., a genetic disease) under consideration.

Let

$$x_i = \mathcal{P}(I = T \,|\, I = i) \tag{9.25}$$

represent the probability that genotype i displays the trait. The frequency of affected individuals in the population (the *incidence* of the trait) is

$$s = \mathcal{P}(I = T) = \sum_i \mathcal{P}(I = T \,|\, I = i)\mathcal{P}(I = i)$$

$$= p^2 x_1 + 2pq x_2 + q^2 x_3 \ . \tag{9.26}$$

Denoting the probability that an affected individual has genotype i by y_i, we have

$$y_i = \mathcal{P}(I = i \,|\, I = T) = \mathcal{P}(I = i \text{ and } I = T)/\mathcal{P}(I = T)$$
$$= \mathcal{P}(I = T \,|\, I = i)\mathcal{P}(I = i)/\mathcal{P}(I = T)$$
$$= x_i \mathcal{P}(I = i)/s \ ,$$

whence

$$y_1 = p^2 x_1/s, \qquad y_2 = 2pq x_2/s, \qquad y_3 = q^2 x_3/s \ . \tag{9.27}$$

Note that (9.26) and (9.27) yield

$$\sum_i y_i = 1 \ , \tag{9.28}$$

as is necessary.

In terms of the vectors $\mathbf{x}$ and $\mathbf{y}$, we deduce the probability that the relative of an affected individual is affected (Elston and Campbell, 1970; Campbell and Elston, 1971):

$$c(q) = \mathcal{P}(J = T \mid I = T)$$
$$= \sum_{ij} \mathcal{P}(J = T \mid J = j)\mathcal{P}(J = j \mid I = i)\mathcal{P}(I = i \mid I = T)$$
$$= \sum_{ij} x_j Q_{ji} y_i = \mathbf{x}^T Q \mathbf{y} \ , \tag{9.29}$$

in which the superscript T signifies transposition. This and the incidence are usually the primary data in clinical investigations. The simple model studied here has four independent parameters, p, x_1, x_2, and x_3.

If A is the normal allele, AA is usually unaffected. With $x_1 = 0$ and $q \ll 1$ we find

$$c(q) \approx c(0) = 2F_{IJ}x_2 = b(0)x_2 \ , \tag{9.30a}$$

provided

$$qk_0 \ll F_{IJ}, \qquad 2qx_3 \ll x_2, \qquad qx_3^2 k_2 \ll 4x_2^2 F_{IJ} \ . \tag{9.30b}$$

Thus, (9.30a) is a good approximation for close relatives as long as the heterozygote has a fairly high probability of being affected. In this approximation, essentially all affected individuals are heterozygous, so

$$c(q) \approx \mathcal{P}(J = T \mid J = 2)\mathcal{P}(J = 2 \mid I = 2) = x_2 Q_{22} \ ,$$

which leads more intuitively to (9.30a).

For a major-gene model of schizophrenia, using (9.29) to fit the data for various relationships gives approximately (Elston and Campbell, 1970; Campbell and Elston, 1971)

$$x_1 \approx 0, \qquad x_2 \approx 0.06, \qquad x_3 \approx 1, \qquad q \approx 0.07,$$
$$y_1 \approx 0, \qquad y_2 \approx 0.61, \qquad y_3 \approx 0.39, \qquad s \approx 0.013 \ .$$

The incidence of schizophrenia is about 1.3%; the gene frequency is about 0.07. All AA individuals are normal; all aa genotypes and about 6% of heterozygotes are schizophrenic. A schizophrenic is Aa with probability 0.61 and aa with probability 0.39. Consult O'Rourke *et al.* (1982) and Risch and Baron (1984) for more recent work on genetic models for schizophrenia.

Finally, let us generalize to incomplete penetrance our genetic counselling example. If I is affected and J is normal, what is the probability, d_j, that J has genotype j? We obtain

$$d_j(q) = \mathcal{P}(J = j \mid I = T \text{ and } J = N)$$
$$= \frac{\mathcal{P}(I = T, J = j, \text{ and } J = N)}{\mathcal{P}(I = T \text{ and } J = N)}$$
$$= \frac{\mathcal{P}(J = N \mid J = j)\mathcal{P}(I = T \text{ and } J = j)}{\mathcal{P}(J = N \mid I = T)\mathcal{P}(I = T)}$$
$$= \frac{(1 - x_j)\mathcal{P}(J = j \mid I = T)\mathcal{P}(I = T)}{[1 - c(q)]\mathcal{P}(I = T)}$$

$$= \frac{(1 - x_j) \sum_i \mathcal{P}(J = j \,|\, I = i) \mathcal{P}(I = i \,|\, I = T)}{1 - c(q)}$$

$$= \frac{(1 - x_j) \sum_i Q_{ji} y_i}{1 - c(q)} = \frac{(1 - x_j)(Q\mathbf{y})_j}{1 - c(q)} \ . \tag{9.31}$$

We leave it as an exercise to prove that (9.17), (9.28), and (9.29) imply

$$\sum_j d_j(q) = 1 \ . \tag{9.32}$$

Assuming that $x_1 = 0$, $q \ll 1$, and (9.30b) holds, we derive the approximation

$$\mathbf{d}^T(q) \approx \left(\frac{k_1 + k_0}{1 - 2F_{IJ}x_2} \quad \frac{2F_{IJ}(1 - x_2)}{1 - 2F_{IJ}x_2} \quad \frac{q(1 - x_3)(2x_2 k_1 + x_3 k_2)}{2x_2(1 - 2F_{IJ}x_2)} \right) \ . \tag{9.33}$$

9.4 Phenotypic Effects of Inbreeding

We have already marked below (9.3) that inbreeding increases the expression of recessives, which are usually harmful, by raising homozygosity. The two related topics covered in this section illustrate in more depth the importance and applicability of this observation.

1. Genetic Disease and Mortality

It is of considerable interest to estimate the total mutational damage in a population. A suitable measure of such damage should weight genes by their detrimental effectiveness. A *lethal equivalent* is a group of genes that, dispersed in different individuals, would cause an average of one death. For example, n mutants, each with a selection coefficient of $-1/n$, constitute one lethal equivalent. Similarly, a dispersed *detrimental equivalent* would cause the appearance of one genetic disease.

Muller (1948) was the first to use data on pre-reproductive mortality of the progeny of consanguineous marriages to estimate the number of lethal equivalents in man. We follow Morton, Crow, and Muller (1956). For much more detailed discussion of such rough estimates, refer to their paper, Schull and Neel (1965), and Cavalli-Sforza and Bodmer (1971, pp. 353-377).

Let $p_i^{(k)}$ and $w_{ij}^{(k)}$ denote the frequency of the allele $A_i^{(k)}$ and the fitness of the genotype $A_i^{(k)} A_j^{(k)}$ at locus k. Put $w_{ij}^{(k)} = 1 - s_{ij}^{(k)}$ and suppose that the selection coefficient of the normal homozygote, $A_1^{(k)} A_1^{(k)}$, at each locus is zero: $s_{11}^{(k)} = 0$. Recalling (9.2), we obtain

$$\overline{w}^{(k)} = 1 - \sum_{ij} s_{ij}^{(k)} [F p_i^{(k)} \delta_{ij} + (1 - F) p_i^{(k)} p_j^{(k)}]$$

for the mean fitness at locus k. We posit that genetic and environmental causes of mortality are independent and that there is no multiplicative epistasis in fitness. We assume also that the genotypic frequencies are products of single-locus genotypic frequencies; Problem 9.17 shows that with inbreeding this can hold only approximately. Then the probability that an individual with inbreeding coefficient F survives to reproductive age is

$$S = \lambda \overline{w} \approx \lambda \prod_k \overline{w}^{(k)} = \lambda \exp\left(\sum_k \ln \overline{w}^{(k)}\right)$$

$$\approx \lambda \exp\left(-\sum_{ijk} s_{ij}^{(k)}\left[F p_i^{(k)} \delta_{ij} + (1-F) p_i^{(k)} p_j^{(k)}\right]\right) ,$$

where λ signifies the probability of surviving the environmental causes of death. The last approximation is accurate if all mutants are very rare or very slightly deleterious. We conclude

$$-\ln S \approx A + BF , \tag{9.34a}$$

where

$$A = -\ln \lambda + C , \qquad B = D - C , \tag{9.34b}$$

$$C = \sum_{ijk} s_{ij}^{(k)} p_i^{(k)} p_j^{(k)} , \qquad D = \sum_{ik} s_{ii}^{(k)} p_i^{(k)} . \tag{9.34c}$$

The approximate linearity of (9.34a) agrees well with observation (Cavalli-Sforza and Bodmer, 1971, pp. 362–364). We define D as the number of lethal equivalents per gamete, for D would be the decrement in fitness if the chromosomes of this gamete were doubled. Since $C > 0$ and $\lambda < 1$, therefore, even though D cannot be calculated from observations of S, from (9.34b) we conclude that $B < D < A + B$. It is usually found that $B \gg A$, which almost determines D; from 18 studies, the median number of lethal equivalents per gamete in man is 1.1, but the scatter is considerable (Cavalli-Sforza and Bodmer, 1971, pp. 363–364). The number of lethal equivalents per zygote is $2D \approx 2.2$.

The estimation of the number of detrimental equivalents requires only a reinterpretation. Now $w_{ij}^{(k)}$ represents the probability that $A_i^{(k)} A_j^{(k)}$ does not display a genetic defect, and hence S becomes the probability that an individual is free of disease. Human data again generally satisfy the condition $B \gg A$; depending on the country and the class of diseases considered, the number of detrimental equivalents per gamete ranges from 0.35 to 2.3 (Cavalli-Sforza and Bodmer, 1971, pp. 368–369).

2. The Effect of Inbreeding on the Mean and Variance

We consider a character determined by a single multiallelic locus. If genotype $A_i A_j$ has value Z_{ij}, from (9.2) we find that the trait has mean

$$\overline{Z} = \sum_{ij} Z_{ij} P_{ij} = \overline{Z}_r + FY , \tag{9.35a}$$

where

$$\overline{Z}_r = \sum_{ij} Z_{ij} p_i p_j \quad \text{and} \quad \overline{Z}_1 = \sum_i Z_{ii} p_i \qquad (9.35b)$$

are the means with random mating and complete inbreeding, respectively, and

$$Y = \overline{Z}_1 - \overline{Z}_r \ . \qquad (9.35c)$$

Thus, the change in the mean under inbreeding is proportional to the inbreeding coefficient. We owe (9.35) to Wright (1951).

To elucidate the biological significance of (9.35), we set

$$Z_{ij} = \tfrac{1}{2}(Z_{ii} + Z_{jj}) + X_{ij} \ . \qquad (9.36)$$

This implies that $X_{ii} = 0$ for every i. Substituting (9.36) into (9.35c) yields

$$Y = -\sum_{ij} X_{ij} p_i p_j \ , \qquad (9.37)$$

from which we conclude that if $X_{ij} \gtreqqless 0$ for all i and j, then $Y \lesseqqgtr 0$. Hence, inbreeding decreases (increases) the mean if the genotypic value of each heterozygote is greater (less) than the average value of the corresponding homozygotes. Consequently, both overdominance and complete or partial dominance of favorable genes lead to the common phenomenon of *inbreeding depression*. If there is no dominance, then $X_{ij} = 0$ for all i and j, so $Y = 0$, and therefore inbreeding does not affect the mean (Wright, 1951). Crow and Kimura (1970, pp. 77–78) established all the results in this paragraph for two alleles.

Equations (9.2) and (9.35) inform us that the variance of the character is given by

$$V = \sum_{ij} Z_{ij}^2 P_{ij} - \overline{Z}^2$$
$$= (1 - F)(\overline{Z}_r^2 + V_r) + F(\overline{Z}_1^2 + V_1) - [(1-F)\overline{Z}_r + F\overline{Z}_1]^2$$
$$= (1 - F)V_r + FV_1 + F(1 - F)Y^2 \ , \qquad (9.38a)$$

where

$$V_r = \sum_{ij} Z_{ij}^2 p_i p_j - \overline{Z}_r^2 \quad \text{and} \quad V_1 = \sum_i Z_{ii}^2 p_i - \overline{Z}_1^2 \qquad (9.38b)$$

are the variances with panmixia and complete inbreeding, respectively. This result (Wright, 1951) shows that the variance depends quadratically on the inbreeding coefficient.

In the absence of dominance, $X_{ij} = 0$ for all i and j; using (9.35b), (9.35c), and (9.36) in (9.38b) leads to

$$V_r = \sum_{ij} \tfrac{1}{4}(Z_{ii} + Z_{jj})^2 p_i p_j - \overline{Z}_1^2 = \tfrac{1}{2}V_1 \ .$$

Substituting this into (9.38a) and recalling that $Y = 0$ yields

$$V = (1 + F)V_r \qquad (9.39)$$

(Malécot, 1941, 1948; Wright, 1951). By increasing homozygosity, inbreeding raises the genetic variance, and the increment is proportional to the inbreeding coefficient.

One might be tempted by (9.39) to conjecture that with heterozygote intermediacy inbreeding always increases the variance. To see that this is not so, suppose that there are two alleles, and assume first that A is completely dominant to a. Choosing the mean and scale of the character so that $Z_{11} = Z_{12} = 0$ and $Z_{22} = 1$, we get

$$V_r = q^2 - (q^2)^2 = pq^2(1 + q), \qquad V_1 = q - q^2 = pq ,$$

whence

$$V_1 - V_r = pq(1 - q - q^2) ,$$

which is negative if $\frac{1}{2}(\sqrt{5} - 1) < q < 1$. By continuity, $V_1 < V_r$ if $Z_{12} > 0$ but sufficiently small, i.e., if dominance is not quite complete.

9.5 Regular Systems of Inbreeding

Wright applied his path coefficients (Wright, 1921, 1921a, 1934, 1968) to the methodical analysis of a large number of regular systems of inbreeding (Wright, 1921, 1922a, 1933, 1933a, 1951). Although his initial investigation (Wright, 1921) was for two alleles with equal frequencies, he generalized to two alleles with arbitrary gene frequency almost immediately (Wright, 1922a) and subsequently to multiple alleles (Wright, 1951). Since the calculation of the inbreeding coefficient does not involve gene frequencies, this generalization is quite straightforward. Results equivalent to Wright's can be derived by probabilistic arguments that involve identity by descent; consult Malécot (1948), Crow and Kimura (1970, pp. 85–94), and Jacquard (1974, pp. 228–248). Arzberger (1985, 1988) and Boucher and Nagylaki (1988) studied large classes of inbreeding systems.

We shall follow the approach of Kimura (1963) and employ probabilities of identity in state. Although formally very similar to the use of identity by descent, Kimura's formulation is more general and deals directly with measurable quantities. This is a significant advantage for practical applications to plant and animal breeding. Furthermore, we are hoping to gain some insight into the random drift of gene frequencies, and this process is related more directly to the evolution of probabilities of identity in state than to that of probabilities of identity by descent.

We indicate that genes a and b are identical in state by writing $a = b$. In Fig. 9.2, the probability that A is homozygous is

$$f_A = \mathcal{P}(d = g) . \qquad (9.40a)$$

The reasoning that leads to (9.6) and (9.7) holds for identity in state, so we have the reinterpreted form of these relations:

$$\mathcal{P}(e = f) = \tfrac{1}{2}(1 + f_A) \; , \tag{9.40b}$$

$$\mathcal{P}(b = e) = \tfrac{1}{2}[\mathcal{P}(a = e) + \mathcal{P}(c = e)] \tag{9.40c}$$

$$= \tfrac{1}{4}[\mathcal{P}(a = d) + \mathcal{P}(a = g) + \mathcal{P}(c = d) + \mathcal{P}(c = g)] \; . \tag{9.40d}$$

Equation (9.40) suffices for the analysis of any mating system in terms of identity in state.

After investigating a variety of regular systems of inbreeding, we shall make some general remarks.

We shall always designate by f_t the probability that a randomly chosen individual in generation t is homozygous. The complementary probability, $h_t = 1 - f_t$, represents the heterozygosity in generation t and is an index of genetic variability in the population.

1. Selfing

Selfing is of great importance in plant breeding and evolution; it occurs also in many animal species. We assume that the population is infinitely large.

Pure self-fertilization is extremely simple. From (9.40b) we infer

$$f_{t+1} = \tfrac{1}{2}(1 + f_t) \; ,$$

whence

$$h_{t+1} = \tfrac{1}{2}h_t \; .$$

Therefore,

$$h_t = h_0(\tfrac{1}{2})^t \; , \tag{9.41}$$

which shows that the heterozygosity converges rapidly to zero. From the more detailed treatment in Sect. 5.2 (with $r = 1$), we see that the equilibrium homozygote frequencies are just those of the constituent alleles. These results go back to Mendel.

In the more general case of partial self-fertilization, we suppose that in each generation a fraction r of the population is produced by selfing. To formulate the recursion relations, we require not only f_t, but also the probability, g_t, that two genes chosen at random from different individuals in generation t are the same allele. Since the probabilities that an individual in generation $t + 1$ was produced by selfing and random cross-fertilization are r and $1 - r$, respectively, we have

$$f_{t+1} = r\tfrac{1}{2}(1 + f_t) + (1 - r)g_t \; . \tag{9.42a}$$

In a infinite population, genes in different individuals in generation $t + 1$ are derived from different individuals in generation t. Therefore,

$$g_{t+1} = g_t \; , \tag{9.42b}$$

whence $g_t = g_0$ is constant, and the solution of (9.42a) reads

$$f_t = \hat{f} + (f_0 - \hat{f})(\tfrac{r}{2})^t \;,$$

(9.43a)

the homozygosity at equilibrium being

$$\hat{f} = \frac{r + 2(1 - r)g_0}{2 - r} \;.$$

(9.43b)

Since the equilibrium heterozygosity,

$$\hat{h} = \frac{2(1 - r)(1 - g_0)}{2 - r} \;,$$

(9.43c)

is positive unless either selfing is complete ($r = 1$) or there is essentially no initial genetic variability ($g_0 = 1$, whereas f_0 is arbitrary), therefore partial selfing does not lead to genetic homogeneity. Observe that (9.43) agrees with the analysis of genotypic frequencies in Sect. 5.2. Wright (1921, 1922a) obtained (9.43c) for initial panmixia ($h_0 = 1 - g_0$). The complete solution (9.43) for two alleles is due to Haldane (1924a). Kimura (1963) derived (9.43c) in general.

2. Sib Mating

Sib mating is widely used to produce inbred lines of laboratory animals.

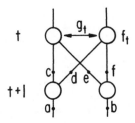

Fig. 9.13. Sib mating.

In Fig. 9.13, let g_t represent the probability that two randomly chosen genes, one from each sibling, are the same allele. Clearly,

$$f_{t+1} = g_t \;.$$

(9.44a)

From (9.40b) and (9.40c) we deduce

$$\begin{aligned}
g_{t+1} &= \mathcal{P}(a = b) \\
&= \tfrac{1}{4}[\mathcal{P}(c = e) + \mathcal{P}(c = f) + \mathcal{P}(d = e) + \mathcal{P}(d = f)] \\
&= \tfrac{1}{2}[\tfrac{1}{2}(1 + f_t) + g_t] \;.
\end{aligned}$$

(9.44b)

Setting $k_t = 1 - g_t$ in (9.44) leads to the homogeneous system

$$h_{t+1} = k_t \; , \tag{9.45a}$$
$$k_{t+1} = \tfrac{1}{4}h_t + \tfrac{1}{2}k_t \; . \tag{9.45b}$$

The general solution (5.10) yields

$$h_t = \tfrac{2}{\sqrt{5}}[(k_0 - \lambda_- h_0)\lambda_+^t + (\lambda_+ h_0 - k_0)\lambda_-^t] \tag{9.46}$$
$$\sim \tfrac{2}{\sqrt{5}}(k_0 - \lambda_- h_0)\lambda_+^t \qquad \text{as } t \to \infty \; , \tag{9.47}$$

where $\lambda_\pm = \tfrac{1}{4}(1 \pm \sqrt{5})$ and $k_t = h_{t+1}$.

Since $\lambda_+ \approx 0.809$ and $\lambda_- \approx -0.309$, (9.47) is a good approximation after 5 to 10 generations. The rate of loss of genetic variability is slower than for pure selfing. If the sib pair is drawn at random from an infinite panmictic population, then $h_0 = k_0$, and hence $h_t \approx 1.17 h_0 (0.809)^t$ for $t \gg 1$. Our conclusion that $h_t \to 0$ as $t \to \infty$ implies that if we consider many sib pairs, then only one allele ultimately survives in each pair, but different alleles may be fixed in different sibships. Thus, each subpopulation, or line, tends to homogeneity, but genetic diversity is preserved in the population.

Pearl (1914), Fish (1914), and Jennings (1914, 1916) studied sib mating numerically. We owe the analytic solution to Robbins (1917).

3. Half-Sib Mating

The mating system displayed in Fig. 9.14, due to Wright (1921), is useful in breeding livestock. The number of generations of inbreeding must be less than the initial number of individuals. Note that an individual in generation t has $\tau + 1$ ancestors in generation $t - \tau$ ($0 \le \tau \le t$). For panmixia in an infinite population, the corresponding number of ancestors, 2^τ, is much larger for $\tau \gg 1$.

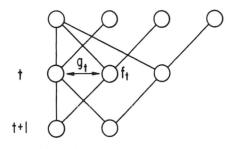

Fig. 9.14. Half-sib mating.

Let g_t denote the probability that two genes chosen at random from two distinct individuals are allelically identical. Arguing as for sib mating, we find

$$f_{t+1} = g_t \; , \tag{9.48a}$$
$$g_{t+1} = \tfrac{1}{4}[\tfrac{1}{2}(1 + f_t) + 3g_t] \; . \tag{9.48b}$$

The homogeneous system satisfied by the probabilities of nonidentity reveals that the heterozygosity approaches zero at the asymptotic rate λ^t, with $\lambda = \tfrac{1}{8}(3 + \sqrt{17}) \approx 0.890$ (Wright, 1969, p. 190). This rate is slower than for the mating of full siblings.

4. Half-First-Cousin Mating

All the inbreeding systems above that involve a finite number of individuals in each generation tend to complete fixation. Roughly, the convergence slows down as the mated relatives become more distant. Wright's (1921; 1969, p. 194) system of half-first-cousin mating demonstrates that with sufficiently distant mates some genetic diversity may be preserved.

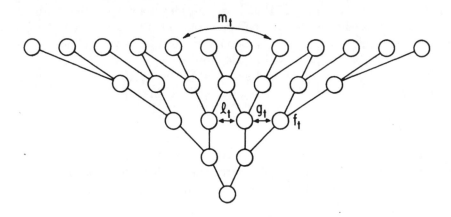

Fig. 9.15. Half-first-cousin mating.

Define the probabilities of identity l_t and g_t for half-sibs and half-first cousins, respectively, as indicated in Fig. 9.15. We designate by m_t the probability that two genes chosen at random from unrelated individuals in generation t are identical in state. Now (9.40) yields

$$f_{t+1} = g_t \; , \tag{9.49a}$$
$$g_{t+1} = \tfrac{1}{4}(l_t + 3m_t) \; , \tag{9.49b}$$
$$l_{t+1} = \tfrac{1}{4}[\tfrac{1}{2}(1 + f_t) + 2g_t + m_t] \; , \tag{9.49c}$$
$$m_{t+1} = m_t \; . \tag{9.49d}$$

Therefore, $m_t = m_0$ and the equilibrium probabilities of identity read

$$\hat{f} = \hat{g} = \tfrac{1}{27}(1 + 26m_0), \qquad \hat{l} = \tfrac{1}{27}(4 + 23m_0) \; . \tag{9.50}$$

The special case with $m_0 = 0$, correct for identity by descent, was derived by Wright (1921; 1969, p. 194). The deviations of the probabilities of allelic identity from their equilibrium values satisfy a homogeneous system of difference equations; the corresponding characteristic equation is

$$32\lambda^3 - 4\lambda - 1 = 0 \ . \tag{9.51}$$

The dominant eigenvalue, $\lambda_0 \approx 0.442$, yields rapid convergence to (9.50).

In contrast to this example, fairly close inbreeding can lead to ultimate loss of genetic variability even with an infinite number of individuals in every generation. This occurs for half-sib and for first-cousin mating by infinitely many individuals (Wright, 1921; 1969, pp. 190–192, in which the two pertinent diagrams are interchanged; Nagylaki, 1980). These systems converge at an asymptotic rate proportional to $t^{-1/2}$ (Nagylaki, 1980). See Arzberger (1985) for other results.

5. Maximum Avoidance of Consanguinity

It is of some conceptual importance to investigate systems of inbreeding in which, given the number of individuals N in each generation, mates are as distantly related as possible. The population number may have the values $N = 2^n$, $n = 1, 2, 3, \ldots$, corresponding to the mating of full siblings, double first cousins, quadruple second cousins, $\ldots$, respectively. Wright (1921; 1933a; 1969, pp. 199–200) derived the recursion relations for $N = 4, 8$, and 16 and wrote them down for arbitrary $N = 2^n$ by analogy. Consult also the discussion of Cockerham (1970). We deduce the difference equations for any $N = 2^n$. In Fig. 9.16 we exhibit the mating system for $N = 8$. The reader may find it helpful to refer to the treatment of double-first-cousin mating by Crow and Kimura (1970, pp. 89–91).

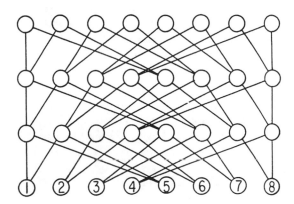

Fig. 9.16. Mating of quadruple second cousins.

Number the individuals in the pedigree in each generation from 1 to N. If we measure all positions and separations modulo N, then the parents of individual i are individuals $2i-1$ and $2i$ (mod N). Let $I_\tau^{(i)}$ designate the set of ancestors (mod N) of individual i ($= 1, 2, \ldots, N$) at a time τ ($= 0, 1, \ldots, n$) generations in the past. By induction, we see that

$$I_\tau^{(i)} = \{2^\tau i - k \ (\mathrm{mod}\, N), \quad k = 0, 1, \ldots, 2^\tau - 1\} \ . \tag{9.52}$$

Since an individual has 2^τ ancestors τ generations in the past, $I_n^{(i)}$ includes the entire population.

We define the degree of relationship, l ($\le n$), of individuals i and j as the smallest value of τ for which

$$I_\tau^{(i)} = I_\tau^{(j)} \ . \tag{9.53}$$

Thus, l is the smallest number of generations in the past for which all ancestors of i and j coincide. By virtue of (9.52), we note that (9.53) is equivalent to

$$2^\tau i = 2^\tau j \ (\mathrm{mod}\, N) \ ,$$

which is the same as

$$2^\tau (i - j) = kN \, , \qquad \text{for some } k = 0, \pm 1, \ldots \ . \tag{9.54}$$

The smallest value of τ in (9.54) must occur for $k = 0$ or k odd, for otherwise $\tau - 1$ also satisfies (9.54). Excluding the trivial case $i = j$ ($l = 0$) and observing that the separation $m = |i - j| < N$, from (9.54) we find that for each l ($1 \le l \le n$), the possible separations are

$$m = (2k + 1)2^{n-l}, \qquad k = 0, 1, \ldots, 2^{l-1} - 1 \ . \tag{9.55}$$

For instance, sibs ($l = 1$) have separation 2^{n-1}, whereas for double first cousins ($l = 2$) $m = 2^{n-2}$ or $m = 3(2^{n-2})$. Since the powers of two in (9.55) differ for different values of l, the sets of separations for different degrees of relationship are disjoint. Hence, m determines l uniquely.

As a check, notice that the total number of possible separations, including zero, is

$$1 + \sum_{l=1}^{n} 2^{l-1} = N \ .$$

With the above formulation, we can deduce recursion relations for maximum avoidance of consanguinity. Let $x_{ij}(t)$ signify the probability that in generation t a gene drawn at random from individual i is identical in state to one chosen at random from j (who may be the same as i). Since the parents of i are $2i - 1$ and $2i$ (mod N), we have

$$x'_{ii} = \tfrac{1}{2}(1 + x_{2i-1, 2i}) \, , \tag{9.56a}$$

$$x'_{ij} = \tfrac{1}{4}(x_{2i-1,2j-1} + x_{2i-1,2j} + x_{2i,2j-1} + x_{2i,2j}), \qquad i \ne j \ , \tag{9.56b}$$

in which all subscripts are mod N, and primes denote generation $t + 1$. From (9.56) it is apparent that if $x_{ij}(t)$ depends only on separation, $x_{ij}(t) = y_m(t)$, then $x_{ij}(t+1) = y_m(t+1)$. Consequently, the assumption of initial dependence only on separation, $x_{ij}(0) = y_m(0)$, which we adopt, implies by induction that $x_{ij}(t) = y_m(t)$ for all t. Therefore, (9.56) simplifies to

$$y_0' = \tfrac{1}{2}(1 + y_1) , \tag{9.57a}$$

$$y_m' = \tfrac{1}{4}(y_{2m-1} + 2y_{2m} + y_{2m+1}), \qquad m \neq 0 , \tag{9.57b}$$

the subscripts being mod N.

Assume that the initial probabilities of identity depend only on the degree of relationship: $y_m(0) = g_l(0)$. Then the type of induction argument applied to (9.56) shows that $y_m(t) = g_l(t)$ for all t, thereby reducing the number of independent variables from N to $n+1$. Thus, $g_l(t)$ designates the probability that two genes drawn at random in generation t, one from each of two individuals with degree of relationship l, are identical in state. For two genes chosen from the same individual,

$$g_0 = y_0 = \tfrac{1}{2}(1 + f) ,$$

where $f(t)$ is the homozygosity in generation t. If m is odd, (9.55) immediately informs us that $l = n$. Finally, (9.55) can be rewritten in the form

$$2m = (2k + 1)2^{n-(l-1)} .$$

Hence, we obtain from (9.57) the difference equations

$$g_0' = \tfrac{1}{2}(1 + g_n) , \tag{9.58a}$$

$$g_l' = \tfrac{1}{2}(g_{l-1} + g_n), \qquad 1 \leq l \leq n . \tag{9.58b}$$

To calculate the rate of convergence to homogeneity, set

$$g_l(t) = 1 - z_l \lambda^t , \qquad \mu = 2\lambda \tag{9.59}$$

and substitute into (9.58):

$$\mu z_0 = z_n , \tag{9.60a}$$

$$\mu z_l = z_{l-1} + z_n , \qquad 1 \leq l \leq n . \tag{9.60b}$$

Inserting (9.60a) into (9.60b) produces the simple recurrence relation

$$\mu z_l = z_{l-1} + \mu z_0 , \qquad 1 \leq l \leq n . \tag{9.61}$$

Before deriving the characteristic equation, we show that $\lambda = 0$ and $\lambda = \tfrac{1}{2}$ ($\mu = 0, 1$) are *not* eigenvalues. If $\mu = 0$, (9.61) implies that $z_l = 0$ for $0 \leq l \leq n - 1$, whence (9.60a) gives $z_n = 0$. If $\mu = 1$, the solution of (9.61) is $z_l = (l+1)z_0$, which satisfies (9.60a) if and only if $z_0 = 0$. But then $z_l = 0$ for $0 \leq l \leq n$.

Since $\mu \neq 0, 1$, the solution of (9.61) reads

$$z_l = \frac{z_0}{1 - \mu}(\mu^{-l} - \mu) . \tag{9.62}$$

By (9.62), $z_0 \neq 0$ for otherwise $z_l = 0$ for $0 \leq l \leq n$. Imposing (9.60a) on (9.62) yields the characteristic equation

$$\mu^2 - 2\mu + \mu^{-n} = 0, \qquad n \geq 1 . \qquad (9.63)$$

The relations

$$N = 2^n , \qquad \mu = 2\lambda \qquad (9.64)$$

enable us to rewrite (9.63) as (Wright, 1921; 1933a; 1969, pp. 199–201)

$$\lambda^{n+2} - \lambda^{n+1} + (4N)^{-1} = 0, \qquad n \geq 1 . \qquad (9.65)$$

It must be kept in mind that $\lambda = \frac{1}{2}$, even though it satisfies (9.65), is not an eigenvalue.

For $n = 1$ (full siblings), 2 (double first cousins), 3 (quadruple second cousins), the dominant eigenvalues λ_m are approximately 0.809, 0.920, and 0.964, respectively (Wright, 1933a; 1969, p. 200).

By rearranging (9.65) in the form

$$1 - \lambda = (4N\lambda^{n+1})^{-1} \qquad (9.66)$$

and taking into account that, owing to (6.31), the real, positive dominant eigenvalue of the irreducible nonnegative matrix in (9.58) satisfies $\lambda_m < 1$, we establish the inequality

$$\lambda_m < 1 - (4N)^{-1} . \qquad (9.67)$$

Cockerham (1970) deduced upper and lower bounds on λ_m with a different method.

We are most interested in approximating the maximal eigenvalue λ_m for large N. Since (9.66) confirms our intuition that $\lambda_m \rightarrow 1$ as $N \rightarrow \infty$, we substitute $\lambda_m = 1 - \varepsilon$ into (9.66) and expand:

$$\varepsilon = (4N)^{-1}[1 + (n+1)\varepsilon + O(n^2\varepsilon^2)] \qquad (9.68)$$

as $N \rightarrow \infty$. Solving for ε produces

$$1 - \lambda_m = \frac{1 + O(n^2/N^2)}{4N - n - 1} \qquad (9.69)$$

as $N \rightarrow \infty$. This is a fair approximation even for $N = 4$. The leading term in (9.69), $1/(4N)$, was derived by Wright (1933a; 1969, p. 201); the correction in the denominator is due to Robertson (1964).

We shall discuss (9.69) after investigating circular mating.

6. Circular Mating

We now examine circular mating by N (≥ 3) individuals, an interesting inbreeding system proposed by Kimura and Crow (1963). Our approach is based on that of Boucher and Nagylaki (1988), whose analysis is more general and complete: they allow the probabilities of identity to depend on the locations of the two individuals sampled (not only on their separation), find all the eigenvalues, and construct explicitly the complete time-dependent solution. Figure 9.17 (best visualized as wrapped around a cylinder) shows that all matings are between half-siblings. Let $g_j(t)$ represent the probability that two genes drawn at random in generation t, one from each of two individuals j $(= 0, 1, 2, \ldots, N-1)$ steps apart, are identical in state. For g_0, we sample with replacement: $g_0 = \frac{1}{2}(1 + f)$. Owing to the circular arrangement, we may assume the symmetry condition $g_j = g_{N-j}$ for every j. Since our main interest is in dioecious organisms, we suppose that N is even, $N = 2n$ $(n = 2, 3, \ldots)$.

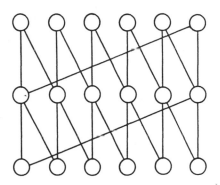

Fig. 9.17. Circular mating of six individuals.

Figure 9.17 informs us immediately that (cf. Kimura and Crow, 1963)

$$g_0' = \tfrac{1}{2}(1 + g_1) \,, \tag{9.70a}$$

$$g_j' = \tfrac{1}{4}(g_{j-1} + 2g_j + g_{j+1}), \qquad 1 \le j \le n - 1 \,, \tag{9.70b}$$

$$g_n' = \tfrac{1}{2}(g_{n-1} + g_n) \,. \tag{9.70c}$$

Note that (9.70c) is obtained by setting $j = n$ in (9.70b) and using the fact that $g_{n+1} = g_{n-1}$. [Only (9.70c) would be different for $N = 2n + 1$, in which case it would be replaced by $g_n' = \tfrac{1}{4}(g_{n-1} + 3g_n)$.] To obtain the rate of convergence to homogeneity, we put

$$g_j(t) = 1 - x_j \lambda^t \,, \tag{9.71}$$

$$\lambda = \tfrac{1}{2}(1 + \cos\theta) \,. \tag{9.72}$$

We may restrict θ to $0 \le \mathrm{Re}\,\theta \le \pi$ and still find all the eigenvalues λ. Substituting (9.71) and (9.72) into (9.70) leads to

$$(1 + \cos\theta)x_0 = x_1 \,, \tag{9.73a}$$

$$x_{j+1} - 2\cos\theta\, x_j + x_{j-1} = 0, \qquad 1 \le j \le n - 1 \,, \tag{9.73b}$$

$$\cos\theta\, x_n = x_{n-1} \,. \tag{9.73c}$$

If $\theta \ne l\pi$ for $l = 0, 1$, then the linear, homogeneous, second-order difference equation (9.73b) has the solutions $e^{\pm ij\theta}$, where $i = \sqrt{-1}$. If $\theta = l\pi$, the two linearly independent solutions of (9.73b) are $(-1)^{lj}$ and $j(-1)^{lj}$. Imposing (9.73a), we find that the components of the eigenvector $\mathbf{x}$ read

$$x_j = \frac{\sin j\theta}{\sin \theta} + \cos j\theta , \qquad \theta \neq l\pi , \qquad (9.74a)$$

$$x_j = (-1)^{lj}[(-1)^l j + 1], \qquad \theta = l\pi , \qquad (9.74b)$$

in which we have suppressed a normalization constant that may depend on θ but not on j. As a check, note that (9.74b) can be deduced from (9.74a) by letting $\theta \to l\pi$. It is easy to see that (9.74b) cannot satisfy (9.73c). Inserting (9.74a) into (9.73c) yields the characteristic equation (Kimura and Crow, 1963)

$$\psi_n(\theta) = \sin \theta \sin n\theta - \cos n\theta = 0 , \qquad (9.75)$$

which holds even for sib mating ($n = 1$).

The maximal eigenvalues, λ_c, for $N = 4$ and $N = 8$ are approximately 0.927 and 0.975, respectively (Kimura and Crow, 1963). Contrary to naive expectation, these values exceed the corresponding ones for maximum avoidance of consanguinity. To see that the ultimate rate of loss of genetic variability for circular mating is always slower than for maximum avoidance with the same number of individuals, note that

$$\psi_n(0) = -1, \qquad \psi_n \left(\frac{\pi}{2n} \right) = \sin \left(\frac{\pi}{2n} \right) > 0 . \qquad (9.76)$$

Consequently, the characteristic equation $\psi_n(\theta) = 0$ has at least one root with $0 < \theta < \pi/N$. Since λ is a monotone decreasing function of θ for $0 < \theta < \pi/2$, we conclude

$$\lambda_c > \tfrac{1}{2} \left(1 + \cos \frac{\pi}{N} \right) > \tfrac{1}{2} \left(1 + 1 - \frac{\pi^2}{2N^2} \right) = 1 - \frac{\pi^2}{4N^2} . \qquad (9.77)$$

Comparing (9.67) with (9.77), we infer that $\lambda_c > \lambda_m$, provided $N > \pi^2 \approx 9.87$. Since $\lambda_c > \lambda_m$ for $N = 4$ and $N = 8$, this establishes the desired result.

To approximate λ_c for large N, observe from (9.77) that $\lambda_c \to 1$ as $N \to \infty$. Therefore, $\theta_c \to 0$ in the characteristic equation

$$\cot n\theta = \sin \theta , \qquad (9.78)$$

which shows that $n\theta_c \to \pi/2$. Expanding (9.78) yields

$$\frac{\pi}{2} - n\theta_c + O \left[\left(\frac{\pi}{2} - n\theta_c \right)^3 \right] = \theta_c + O(\theta_c^3)$$

as $N \to \infty$. Solving for θ_c gives

$$\theta_c = \frac{\pi[1 + O(N^{-3})]}{N + 2} .$$

Then (9.72) informs us that

$$\lambda_c = 1 - \tfrac{1}{4}\theta_c^2 + O(\theta_c^4) = 1 - \frac{\pi^2}{4(N+2)^2} + O(N^{-4}) \tag{9.79}$$

as $N \to \infty$ (Kimura and Crow, 1963).

In Sect. 9.6 we shall see that for large N the characteristic time for convergence to homogeneity with panmixia is $2N$ generations. According to (9.69), the corresponding time for maximum avoidance of consanguinity is about twice as long, viz., $4N$ generations. For circular mating, however, the typical time, $4N^2/\pi^2 \approx 0.405N^2$ generations, is much longer. The fact that it is proportional to N^2, rather than N, is especially remarkable in view of the results of Sects. 9.6 and 9.8.

At first sight, one might imagine that the asymptotic rate of loss of genetic diversity for maximum avoidance of consanguinity is about half that for panmixia because mates are as distantly related as possible. We shall prove in Sect. 9.8, however, that this rate is also near $1/(4N)$ for random mating with exactly two offspring per individual. Thus, the approximate doubling of the convergence time is primarily due to the absence of variation in offspring number.

For circular mating, the close inbreeding of proximate individuals causes a much more rapid initial decrease of heterozygosity than for maximum avoidance (Kimura and Crow, 1963). However, different alleles tend to be fixed in the small, partially isolated mating units (or sublines) of closely related individuals, and this brings about a much slower ultimate rate of convergence to genetic homogeneity than in the case of maximum avoidance. Consult Kimura and Crow (1963), Robertson (1964), and Wright (1965) for further discussion.

As noted below (9.51), a rather complete analysis of circular mating by an infinite number of individuals can be carried out, but the techniques required are beyond the scope of this section.

For a given number of individuals, N, which regular system of inbreeding has the slowest asymptotic rate of loss of genetic variability? This interesting and nontrivial question is meaningful only if we exclude systems that do not converge to genetic homogeneity. For example, two sib lines ($N = 4$) ultimately converge to homozygosity, but not to homogeneity; if they are crossed in each generation to produce a fifth individual who has no offspring ($N = 5$), even some heterozygosity is preserved. We assume that the mating pattern is the same in every generation. Problem 9.18 shows that without this restriction, arbitrarily slow convergence rates can occur. For simplicity, we exclude selfing.

Intuition has been a false guide in this problem. Between Wright's (1921) proposal of maximum avoidance of consanguinity and Kimura and Crow's (1963) formulation of circular mating, the former system was widely thought to be asymptotically the slowest. It was then supplanted in this special rôle by the latter. Much later, however, Boucher and Cotterman (1990) enumerated all inequivalent mating systems for 2, 3, 4, and 5 individuals and calculated their ultimate rates of convergence. Circular mating is, indeed, the slowest for

$N = 2$ (trivially), 3, and 4, but two systems are slower for $N = 5$; partial results yield many such counterexamples for $N = 6$ and 7.

In circular mating, every individual has two offspring. Other things being equal, one might expect variation in progeny number to accelerate convergence. Therefore, it is perhaps surprising that all of Boucher and Cotterman's (1990) counterexamples for $N = 5$ and 6 have unequal progeny numbers. Finally, we note also that both of their counterexamples for $N = 5$ and all but one of them for $N = 6$ are feasible for dioecious organisms.

7. General Remarks

The homozygosity $f(t)$ and heterozygosity $h(t)$ can be interpreted as the expected proportions of homozygotes and heterozygotes in the population, where the expectation is over the possible evolutionary paths of many populations, all of which start with the same initial conditions. Essentially the same argument as that leading to (3.39) can be used to justify this assertion.

Assume now that the initial generation is sampled at random from an infinitely large population in Hardy-Weinberg proportions with allelic frequencies $\overline{p}_i$. With pure inbreeding, the expected frequency of A_i is constant and equal to $\overline{p}_i$, and the reasoning that establishes (9.1) now yields the probability that an individual chosen at random has genotype $A_i A_j$. It follows that the expected frequency of the ordered genotype $A_i A_j$ in generation t is given by (9.2) as

$$\overline{P}_{ij}(t) = F(t)\overline{p}_i\delta_{ij} + [1 - F(t)]\overline{p}_i\overline{p}_j , \tag{9.80}$$

in which $F(t)$ represents the inbreeding coefficient in generation t. Hence, as in (9.3), inbreeding reduces the expected heterozygosity,

$$h(t) = h(0)H(t) , \tag{9.81a}$$

relative to its initial value by Wright's (1951) panmictic index,

$$H(t) = 1 - F(t) . \tag{9.81b}$$

Since the fundamental relation (9.81) encapsulates the significance of the inbreeding coefficient, we shall offer a more explicit proof. Scrutiny of the examples in this section and some reflection reveal that with the same mating pattern in each generation and mating only within generations, k probabilities of homozygosity, $x_i(t) = f_i(t)$ for $1 \leq i \leq k$, and $n - k$ probabilities of allelic identity between individuals, $x_i(t)$ for $k + 1 \leq i \leq n$, suffice for the analysis of any system of inbreeding by finitely many individuals. The positive integers k and n depend on the inbreeding system: e.g., if all individuals have the same probability of homozygosity, as in the above examples, then $k = 1$. If the vector $\mathbf{y}(t)$ is given by

$$y_i(t) = \begin{cases} \frac{1}{2}[1 + x_i(t)], & 1 \leq i \leq k , \\ \\ x_i(t), & k + 1 \leq i \leq n , \end{cases} \tag{9.82}$$

then there exists an $n \times n$ matrix C such that

$$\mathbf{x}' = C\mathbf{y}, \qquad \sum_j c_{ij} = 1 \ . \tag{9.83}$$

From (9.82) and (9.83) we easily establish that the probabilities of allelic nonidentity, $z_i(t) = 1 - x_i(t)$, satisfy

$$\mathbf{z}' = B\mathbf{z} \ , \tag{9.84a}$$

where

$$b_{ij} = \begin{cases} \frac{1}{2}c_{ij}, & 1 \leq j \leq k \ , \\ \\ c_{ij}, & k+1 \leq j \leq n \ . \end{cases} \tag{9.84b}$$

Therefore,

$$\mathbf{z}(t) = B^t\mathbf{z}(0) \ . \tag{9.85}$$

The probabilities of nonidentity by descent, $Z_i(t)$, also satisfy (9.84) and hence (9.85). By definition, initially no genes are identical by descent: $Z_i(0) = 1$, which yields

$$\mathbf{Z}(t) = B^t\mathbf{u} \ , \tag{9.86}$$

where $\mathbf{u}$ designates the n-dimensional vector of ones. For an infinite, panmictic initial population, $\mathbf{z}(0) = h(0)\mathbf{u}$. Substituting this into (9.85) and using (9.86), we obtain

$$\mathbf{z}(t) = h(0)\mathbf{Z}(t) \ . \tag{9.87}$$

Thus, the probabilities of allelic nonidentity are proportional to the probabilities of nonidentity by descent.

If a fraction α_i $(1 \leq i \leq k)$ of the population has probabilities of heterozygosity and allozygosity $h_i(t)$ and $H_i(t)$, respectively, then the corresponding probabilities for a randomly chosen individual read

$$h(t) = \sum_{i=1}^{k} \alpha_i h_i(t), \qquad H(t) = \sum_{i=1}^{k} \alpha_i H_i(t) \ .$$

In view of (9.87), these satisfy (9.81).

For the simple model in Sect. 2.5, Thompson (1976a) presents a detailed and lucid discussion of some of the matters considered above.

Finally, let $L(t)$ represent the (random) number of heterozygous loci of an individual chosen at random in generation t. Suppose that the random variable $L^{(i)}(t)$ is equal to one or zero according as this individual is heterozygous or homozygous at the ith locus. Then

$$L(t) = \sum_i L^{(i)}(t) \ , \tag{9.88}$$

and the expected heterozygosity at the ith locus reads

$$\mathcal{E}[L^{(i)}(t)] = h^{(i)}(t) \ . \tag{9.89}$$

Taking the expectation of (9.88), substituting (9.89), and noting that in (9.81) the same panmictic index applies to all loci, we find

$$\overline{L}(t) = \overline{L}(0)H(t) \tag{9.90}$$

for initial panmixia. Thus, the expected number (or proportion) of heterozygous loci in a randomly chosen member of the population is reduced by a factor $H(t)$. For sib mating, from (9.46) we calculate that after 5, 10, and 25 generations $H(t)$ equals approximately 0.406, 0.141, and 0.00585, respectively.

9.6 Panmixia in Finite Populations

Now we shall apply the methods of the previous section to the study of the decay of genetic variability in finite panmictic populations. Fisher (1922, 1930, 1930a) investigated this problem in the diffusion approximation, whereas Wright (1931) employed path coefficients. Malécot utilized the theory of Markov chains (Malécot, 1944) and the concept of identity by descent (Malécot, 1946, 1948).

We suppose that mating is random and offspring are obtained by independent sampling from the parents with replacement. The latter assumption was discussed at the beginning of Sect. 2.5.

We denote by f_t the probability that an individual chosen at random in generation t is homozygous. All the remarks at the end of Sect. 9.5 apply to f_t. Let g_t represent the probability that two genes chosen at random from distinct individuals in generation t are identical in state.

1. Monoecious Populations

We posit that in each generation a fraction r of the population of N monoecious individuals is produced by selfing and the rest come from random cross-fertilization. Sampling with replacement implies that genes in distinct offspring are derived from the same parent with probability $1/N$. Therefore (Wright, 1951),

$$f_{t+1} = r\tfrac{1}{2}(1 + f_t) + (1 - r)g_t \ , \tag{9.91a}$$

$$g_{t+1} = N^{-1}\tfrac{1}{2}(1 + f_t) + (1 - N^{-1})g_t \ . \tag{9.91b}$$

From (9.91) we see that the probabilities of nonidentity, $h_t = 1 - f_t$ and $k_t = 1 - g_t$, satisfy

$$h_{t+1} = \tfrac{r}{2}h_t + (1 - r)k_t \ , \tag{9.92a}$$

$$k_{t+1} = (2N)^{-1}h_t + (1 - N^{-1})k_t \ . \tag{9.92b}$$

The eigenvalues corresponding to (9.92) read (Wright, 1969, pp. 194–196)

$$\lambda_\pm = \tfrac{1}{2}\left\{1 - \frac{1}{N} + \frac{r}{2} \pm \left[\left(1 - \frac{r}{2}\right)^2 + \frac{1}{N}\left(\frac{1}{N} - r\right)\right]^{1/2}\right\} . \tag{9.93}$$

The heterozygosity converges to zero at the ultimate rate λ_+^t. For $N \gg 1$, we have the approximation

$$\lambda_+ = 1 - [N(2 - r) + 2(1 - r)(2 - r)^{-1}]^{-1} + O(N^{-3}) . \tag{9.94}$$

If self-fertilization has the same probability as any other particular union, then $r = 1/N$ and (9.91) shows that $f_{t+1} = y_{t+1}$. Hence, (9.91a) and (9.92a) yield

$$f_{t+1} = \frac{1}{2N} + \left(1 - \frac{1}{2N}\right) f_t , \tag{9.95a}$$

$$h_{t+1} = \left(1 - \frac{1}{2N}\right) h_t \tag{9.95b}$$

for $t \geq 1$. If we assume that at least one generation of random mating preceded the initial generation, we obtain from (9.95b) for $t \geq 0$ (Wright, 1931; Malécot, 1946, 1948; Kimura, 1963)

$$h_t = k_t = h_0 \left(1 - \frac{1}{2N}\right)^t . \tag{9.96}$$

Thus, the characteristic time for the decay of the expected proportion of heterozygotes is exactly $2N$ generations. Equation (9.96) and Sect. 2.5 (to which the reader should refer for more discussion) inform us that the evolution of the homozygosity of N monoecious diploids with a "random" proportion (i.e., $1/N$) of selfing is the same as the evolution of the probability of identity for $2N$ asexual haploids. Evidently, it is the number of genes that matters, not the number of individuals.

The simplicity of the last model dictates its use as a reference despite its implausibility for diploids. We shall now demonstrate that for large populations it approximates more realistic situations accurately.

If selfing is excluded ($r = 0$), (9.91) reduces to

$$f_{t+1} = g_t , \tag{9.97a}$$
$$g_{t+1} = (2N)^{-1}(1 + f_t) + (1 - N^{-1})g_t . \tag{9.97b}$$

From (9.93) we obtain the eigenvalues

$$\lambda_\pm = \tfrac{1}{2}[1 - N^{-1} \pm (1 + N^{-2})^{1/2}] , \tag{9.98}$$

which for $N \gg 1$ have the asymptotic behavior

$$\lambda_+ = 1 - \frac{1}{2N+1} + O(N^{-3}) , \tag{9.99a}$$

$$\lambda_- = -\frac{1}{2N-1} + O(N^{-3}) . \tag{9.99b}$$

The ultimate rate of convergence, λ_+^t, is extremely close to that for random mating with selfing, (9.96), provided N is replaced by $N + \frac{1}{2}$ in the rate $[1 - (2N)^{-1}]^t$. In large populations, this correction is negligible. We owe all these results to Wright (1931).

The heterozygosity has the explicit form (cf. Malécot, 1948)

$$h_t = (1 + N^{-2})^{-1/2}[(k_0 - \lambda_- h_0)\lambda_+^t + (\lambda_+ h_0 - k_0)\lambda_-^t] \tag{9.100a}$$

$$\sim k_0 \lambda_+^t \tag{9.100b}$$

as $N \to \infty$, provided $t \geq 1$ and $k_0 > 0$. The approximation (9.100b) agrees with (9.96).

2. Dioecious Populations

If N_1 males and N_2 females mate at random, then after one generation the probabilities of identity are sex independent, so (9.97a) still holds. Two genes drawn at random from distinct individuals in generation $t + 1$ are both paternal, both maternal, and one paternal and one maternal with respective probabilities $\frac{1}{4}$, $\frac{1}{4}$, and $\frac{1}{2}$. Multiplying these factors with the corresponding conditional probabilities of allelic identity of the two genes gives

$$g_{t+1} = \frac{1}{4}[N_1^{-1}\frac{1}{2}(1 + f_t) + (1 - N_1^{-1})g_t]$$

$$+ \frac{1}{4}[N_2^{-1}\frac{1}{2}(1 + f_t) + (1 - N_2^{-1})g_t] + \frac{1}{2}g_t .$$

This reduces to (9.97b) with the actual population number N replaced by the effective population number

$$N_e = 4N_1 N_2/(N_1 + N_2) . \tag{9.101}$$

We conclude that the probabilities of identity evolve as in a monoecious population of size N_e with selfing forbidden (Wright, 1931; Malécot, 1946, 1948; Kimura, 1963). (Of course, N_e need not be an integer.)

The effective population number (9.101) cannot exceed the actual population number $N = N_1 + N_2$: $N_e \leq N$, with equality if and only if $N_1 = N_2$. If $N_2 \gg N_1$, as is sometimes the case, then $N_e \approx 4N_1$, which is much less than the actual population size $N = N_1 + N_2 \approx N_2$. As is intuitive, this leads to a more rapid decay of heterozygosity than occurs in a population of the same actual size with a sex ratio close to unity.

3. Deterministically Varying Population Number

Let us generalize the simple monoecious model with a random proportion of selfing to variable population size. The argument that leads to (9.95) now yields

$$h_{t+1} = [1 - (2N_t)^{-1}]h_t \; , \tag{9.102}$$

where $N_t \, (\geq 1)$ is a known function of time. Iteration of (9.102) tells us

$$h_t = h_0 \prod_{\tau=0}^{t-1} [1 - (2N_\tau)^{-1}] \; . \tag{9.103}$$

If the population sizes are *cyclic* with a period of T generations, as will often happen with seasonal variation, from (9.103) we deduce

$$h_{nT} = h_0 \left\{ \prod_{\tau=0}^{T-1} [1 - (2N_\tau)^{-1}] \right\}^n = h_0 [1 - (2N_e)^{-1}]^{nT} \tag{9.104}$$

for $n - 1, 2, \ldots$, in which N_e denotes the effective population number. Hence (Wright, 1938),

$$[1 - (2N_e)^{-1}]^T = \prod_{\tau=0}^{T-1} [1 - (2N_\tau)^{-1}] \; . \tag{9.105}$$

Thus, every T generations the heterozygosity is the same as for a "standard" population of constant size N_e. As for dioecy, and as we shall see in other situations, introduction of an effective population number enables us to incorporate additional biological complexity into our simple models.

Taking $n = 1$ in (9.104) shows that (9.105) holds for T generations without periodicity.

If the population numbers are large, $N_t \gg 1$ for $0 \leq t \leq T - 1$, and the period (in the aperiodic case, the total time T) short, $T \ll N_e$, then from (9.105) we derive

$$N_e \approx \tilde{N} = T \left(\sum_{\tau=0}^{T-1} N_\tau^{-1} \right)^{-1} \; . \tag{9.106}$$

So, the effective population size is close to the harmonic mean of the actual population sizes (Wright, 1938).

Recall that the harmonic mean cannot exceed the arithmetic mean: $\tilde{N} \leq \overline{N}$, with equality if and only if the population number is constant (see Problem 6.8). As repeatedly stressed by Wright, since the harmonic mean is very sensitive to small population numbers, bottlenecks can greatly amplify the evolutionary rôle of random drift. To illustrate, suppose $N_t = N_1$ for $1 \leq t \leq T - 1$ and $N_1 \gg TN_0$. Then $\tilde{N} \approx TN_0 \ll (1 - T^{-1})N_1 \approx \overline{N}$.

For *long times*, we wish to determine whether sufficiently rapid expansion of the population can preserve genetic heterogeneity. According to (9.102), the positive sequence $\{h_t\}$ is monotone decreasing; hence, it has a limit

$$\hat{h} = h_0 \prod_{\tau=0}^{\infty} [1 - (2N_\tau)^{-1}] \tag{9.107}$$

as $t \to \infty$. We seek a necessary and sufficient condition for $\hat{h} > 0$.
From (9.103) we have

$$\ln(h_t/h_0) = \sum_{\tau=0}^{t-1} \ln[1 - (2N_\tau)^{-1}] \; ; \tag{9.108}$$

$\hat{h} > 0$ if and only if the sum in (9.108) converges as $t \to \infty$. Since convergence requires the tth term of the sum to tend to zero as $t \to \infty$, it implies that $N_t \to \infty$ as $t \to \infty$, as is intuitively expected. But then

$$\ln[1 - (2N_t)^{-1}] \sim -(2N_t)^{-1}$$

as $t \to \infty$, whence the Limit Comparison Test for series of positive terms shows that the sum in (9.108) converges if and only if

$$\sum_{\tau=0}^{\infty} N_\tau^{-1} < \infty \; . \tag{9.109}$$

We conclude that convergence of the sum of the reciprocal population sizes is necessary and sufficient for the maintenance of genetic diversity. Malécot (1946, 1948) proved this beautiful result in the more difficult case without selfing.

If the population number does not grow faster than linearly with time, (9.109) yields $\hat{h} = 0$; if N_t increases at least as fast as $\tau^{1+\varepsilon}$ for some $\varepsilon > 0$, then $\hat{h} > 0$. Since resource limitation prevents indefinite population growth, the practical significance of Malécot's result is that rapid increase of population number can greatly decelerate the loss of genetic variability. Two examples will illuminate the dependence of $\hat{h}$ on the rate of population growth.

(a) Asymptotically Geometric Population Growth

Suppose

$$\ln[1 - (2N_t)^{-1}] = -br^{-t} \; , \tag{9.110}$$

with $r > 1$. Then

$$b = -\ln[1 - (2N_0)^{-1}] \sim (2N_0)^{-1} \tag{9.111a}$$

as $N_0 \to \infty$, and

$$N_t = \tfrac{1}{2}[1 - \exp(-br^{-t}]^{-1} \sim (2b)^{-1} r^t \tag{9.111b}$$

as $t \to \infty$. If $N_0 \gg 1$, the asymptotic form in (9.111b) may be further approximated as $N_t \approx N_0 r^t$. Inserting (9.110) and (9.111a) into (9.108) leads to

$$h_t = h_0[1 - (2N_0)^{-1}]^{(1-r^{-t})/(1-r^{-1})} \qquad (9.112a)$$

$$\rightarrow h_0[1 - (2N_0)^{-1}]^{r/(r-1)} = \hat{h} \qquad (9.112b)$$

as $t \rightarrow \infty$. In accord with our intuition, $\hat{h}$ increases as the rate of population growth increases. Note that

$$h_t \geq h_0(\tfrac{1}{2})^{r/(r-1)} \;, \qquad (9.113a)$$

with equality if and only if $N_0 = 1$, and for $t \geq 2$

$$h_t < h_0[1 - (2N_0)^{-1}] \;. \qquad (9.113b)$$

Thus, the loss of variability is no greater than for a single founding individual, and it exceeds that due to the initial generation of panmixia.

(b) Asymptotically Algebraic Population Growth

If

$$\ln[1 - (2N_t)^{-1}] = -b(t+1)^{-s} \;, \qquad (9.114a)$$

with $s > 1$, then b is again given by (9.111a), and

$$N_t = \tfrac{1}{2}\{1 - \exp[-b(t+1)^{-s}]\}^{-1} \sim (2b)^{-1}t^s \qquad (9.114b)$$

as $t \rightarrow \infty$. Inserting (9.111a) and (9.114a) into (9.108) leads to

$$\hat{h} = h_0[1 - (2N_0)^{-1}]^{\zeta(s)} \;, \qquad (9.115a)$$

where

$$\zeta(s) = \sum_{n=1}^{\infty} n^{-s} \qquad (9.115b)$$

is the Riemann zeta function (Haynsworth and Goldberg, 1964, p. 807). Clearly, $\zeta(s) > 1$ and $\zeta(s)$ is a decreasing function of s. Consequently, (9.115a) has the same qualitative behavior as (9.112b). The values $\zeta(2) = \pi^2/6 \approx 1.64$ and $\zeta(4) = \pi^4/90 \approx 1.08$ (Haynsworth and Goldberg, 1964, p. 807) show that (in this model) for population growth at the asymptotic rate t^2, less variability is lost than in two generations of panmixia with the initial population number, and that with the asymptotic rate t^4, essentially all loss of variability occurs in the first generation.

4. Stochastically Varying Population Number

Random variation of the population number greatly increases the difficulty of the analysis. Seneta (1974) and Heyde and Seneta (1975) studied the diallelic case; minor modifications establish all their results for multiple alleles. Assume that the gene frequencies and the population number are jointly Markovian. If the sum in (9.109) diverges with probability one, then some allele is ultimately fixed with probability one (Heyde and Seneta, 1975). Heyde and Seneta present also a partial converse to this theorem. If the population number is stochastically independent of the history of the population, then the almost sure occurrence of (9.109) is necessary and sufficient for the ultimate survival of at least two alleles with positive probability (Heyde and Seneta, 1975). See also Donnelly (1986), Klebaner (1988), and references therein.

9.7 Heterozygosity under Mutation and Random Drift

So far in this chapter we have ignored mutation. Since mutation rates are usually very low, this approximation is generally accurate unless we are concerned with the long-term evolution of at least moderately large populations.

To include mutation in the model of Sect. 9.6, assume that every allele mutates at rate u per generation and each mutant is of a novel allelic type. If the sequence of nucleotides were determined for every allele, the number of possible alleles would be astronomical (greater than 10^{60} even for 100 nucleotides). In this case, granted complete symmetry of the mutation process and absence of selection, our hypothesis would be a reasonable idealization.

This model of infinitely many alleles was proposed by Malécot (1946, 1948, 1951) for identity by descent and by Wright (1949) and Kimura and Crow (1964) for identity in state.

In Sect. 9.6 we showed that in a panmictic population of finite size N, without mutation, fixation eventually occurs. Therefore, with recurrent mutation to new alleles, every allele present in any given generation will be lost in a finite time with probability one. As we shall see, however, the population converges to a stationary state in which the loss of old alleles by random drift and the gain of new alleles by mutation are in equilibrium.

First, we posit *selfing at the rate* $1/N$. Since two genes can be identical only if they have descended without mutation from the same gene or identical genes in the previous generation, (9.95a) generalizes to

$$f_{t+1} = (1-u)^2 \left[\frac{1}{2N} + \left(1 - \frac{1}{2N}\right) f_t \right] . \tag{9.116}$$

Therefore, the expected homozygosity converges at the rate (Malécot, 1948)

$$(1-u)^{2t} \left(1 - \frac{1}{2N}\right)^t \approx \exp\left[-\left(2u + \frac{1}{2N}\right)t \right] \tag{9.117a}$$

to

$$\hat{f} = \frac{(1-u)^2}{2N - (1-u)^2(2N-1)} \approx \frac{1}{1 + 4Nu} \tag{9.117b}$$

(Malécot, 1946, 1948; Kimura and Crow, 1964), where the approximation in (9.117a) holds if $u \ll 1$ and $N \gg 1$, whereas that in (9.117b) presupposes only that $u \ll 1$. In fact, the second approximation is a very slight overestimate: some algebra demonstrates

$$\frac{1}{1 + 4Nu} \geq \hat{f} \geq \frac{1 - \min\left(6u^2 N, 3u\right)}{1 + 4Nu} . \tag{9.118}$$

Essentially, only $\mu = 4Nu$, twice the number of new mutants per generation, matters. The population is very heterogeneous if $\mu \gg 1$ and quite homogeneous if $\mu \ll 1$. Thus, mutation dominates random drift in the first case and the contrary holds in the second.

Next we *exclude selfing*. With mutation, (9.97) becomes

$$f_{t+1} = (1-u)^2 g_t \ , \tag{9.119a}$$

$$g_{t+1} = (1-u)^2 [(2N)^{-1}(1+f_t) + (1-N^{-1})g_t] \ . \tag{9.119b}$$

If the mutation rates in males and females are equal and we replace N by the effective population number (9.101), then after one generation of panmixia, we can apply (9.119) to a dioecious population. Convergence now occurs at the asymptotic rate $(1-u)^{2t}\lambda_+^t$, where λ_+ is given by (9.98). On account of (9.99a), for $N \gg 1$ this rate agrees with (9.117a). At equilibrium (Malécot, 1946, 1948),

$$\hat{f} = \frac{(1-u)^4}{2N - 2(N-1)(1-u)^2 - (1-u)^4} \approx \frac{1}{1+4Nu} \ , \tag{9.120}$$

as with selfing. Noting that $N \geq 2$, we obtain the bounds

$$\frac{1}{1+4Nu} \geq \hat{f} \geq \frac{1-11u}{1+4Nu} \tag{9.121}$$

on $\hat{f}$, and these allow us to conclude that the approximation in (9.120) is again very accurate. Observe that the approximations in (9.117b) and (9.120) require only weak mutation; they do not depend on the magnitude of the population number.

9.8 The Inbreeding Effective Population Number

In Sect. 2.5 we saw that sampling with replacement corresponds to the production of an extremely large gametic pool from which a much smaller number of successful gametes survive to form the next generation. This process clearly implies that the number of successful gametes contributed by a given individual to the next generation has a binomial distribution. As pointed out in Sect. 2.5, however, sundry random factors will usually raise or lower the contribution of some individuals to the large gametic pool, thereby increasing the variance of the number of successful gametes contributed by a particular individual. Indeed, the gametic variance observed in natural and laboratory populations exceeds the binomial value (Crow and Morton, 1955). We shall show in this section that by introducing an inbreeding effective population number, $N_e^{(i)}$, we can incorporate an arbitrary progeny-size distribution into our models.

We permit the number of monoecious individuals, N_t, in generation t to vary deterministically. Let P_t designate the probability that genes chosen at random from distinct individuals in generation t are derived from the same parent in generation $t - 1$. Denote by Q_t the proportion of the population in generation t produced by selfing. Then (9.91) generalizes to (Pollak, 1977)

$$f_{t+1} = Q_{t+1}\tfrac{1}{2}(1 + f_t) + (1 - Q_{t+1})g_t \ , \tag{9.122a}$$

$$g_{t+1} = P_{t+1}\tfrac{1}{2}(1 + f_t) + (1 - P_{t+1})g_t \ . \tag{9.122b}$$

If the population number and reproductive pattern are constant, then $N_t = N$, $P_t = P$, and $Q_t = Q$. From (9.122) we find that the rate of decay of heterozygosity is λ_0^t, with

$$\lambda_0 = 1 - \tfrac{1}{2}P + O(P^2 + Q^2)$$

as $P \to 0$ and $Q \to 0$. In large populations ($N \gg 1$), we have $P \ll 1$, and if selfing occurs at a rate of $O(1/N)$, also $Q \ll 1$. We shall see that (9.124) below will always define the inbreeding effective population number. We conclude

$$\lambda_0 \approx 1 - (2N_e^{(i)})^{-1} \ .$$

1. Monoecious Populations With Selfing

If mating is wholly random, including selfing, then the probability that two genes are derived from the same parent is the same for genes in the same individual as for genes in distinct individuals. Therefore, $Q_t = P_t$, and hence $f_{t+1} = g_{t+1}$. After one generation of panmixia, (9.122) simplifies to

$$f_{t+1} = \tfrac{1}{2}P_{t+1} + (1 - \tfrac{1}{2}P_{t+1})f_t \ . \tag{9.123}$$

Comparing (9.123) with (9.95a), we define the inbreeding effective population number for generation t by

$$N_e^{(i)} = P_t^{-1} \ . \tag{9.124}$$

Let K_i represent the (random) number of successful gametes in generation t from individual i in generation $t - 1$. Given that $\mathbf{K} = \mathbf{k}$ (a fixed vector), the probability that two gametes chosen at random in generation t are derived from the same parent is the ratio of the total number of choices of two gametes from the same parent to the total number of ways of choosing two gametes, viz.,

$$\sum_{i=1}^{N_{t-1}} \binom{k_i}{2} \bigg/ \binom{2N_t}{2} = [2N_t(2N_t - 1)]^{-1} \sum_i k_i(k_i - 1) \ .$$

Then the required unconditional probability reads

$$P_t = \sum_{\mathbf{k}} \mathcal{P}(\mathbf{K} = \mathbf{k})[2N_t(2N_t - 1)]^{-1} \sum_i k_i(k_i - 1)$$

$$= \mathcal{E}\left\{ [2N_t(2N_t - 1)]^{-1} \sum_i K_i(K_i - 1) \right\}$$

$$= [2N_t(2N_t - 1)]^{-1} N_{t-1}(\sigma^2 + \mu^2 - \mu) \ , \tag{9.125}$$

where μ and σ^2 represent the mean and variance of the number of successful gametes (or offspring) produced by an individual:

$$\mu = \mathcal{E}(K_i), \qquad \sigma^2 = \mathrm{Var}\,(K_i) \ . \tag{9.126}$$

The expectation of the obvious relation

$$\sum_{i=1}^{N_{t-1}} K_i = 2N_t \tag{9.127}$$

yields

$$\mu = 2N_t/N_{t-1} \ . \tag{9.128}$$

From (9.124), (9.125), and (9.128) we obtain

$$N_e^{(i)} = \frac{2N_t - 1}{\sigma^2 \mu^{-1} + \mu - 1} \ . \tag{9.129}$$

Wright (unpublished) was the first to derive this formula (see Crow, 1954, p. 550). Consult also Kimura and Crow (1963a) and Pollak (1977). Chia and Pollak (1974) posit that $\{N_t\}$ is a finite Markov chain.

The inbreeding effective population number is a decreasing function of the gametic variance. Hence, the empirical result $\sigma^2 > \mu$ (Crow and Morton, 1955) implies that $N_e^{(i)} < N_{t-1} - \mu^{-1} < N_{t-1}$. The maximum value of σ^2 occurs when only one parent reproduces, in which case

$$\sigma^2 = N_{t-1}^{-1}(2N_t)^2 - \mu^2 = \mu^2(N_{t-1} - 1) = \mu(2N_t - \mu) \ .$$

Inserting this into (9.129) shows that $N_e^{(i)} \geq 1$.

If $N_{t-1} = 1$, so that $\sigma^2 = 0$, then (9.128) and (9.129) give $N_e^{(i)} = 1$.

For sampling with replacement, K_i has a binominal distribution with index $2N_t$ and parameter N_{t-1}^{-1} for each i:

$$\mathcal{P}(K_i = k_i) = \binom{2N_t}{k_i} N_{t-1}^{-k_i}(1 - N_{t-1}^{-1})^{2N_t - k_i} \ . \tag{9.130}$$

The mean, of course, agrees with (9.128); the variance is

$$\sigma^2 = 2N_t N_{t-1}^{-1}(1 - N_{t-1}^{-1}) = \mu(1 - N_{t-1}^{-1}) \ . \tag{9.131}$$

Substituting (9.131) into (9.129), we deduce that $N_e^{(i)} = N_{t-1}$ (Kimura and Crow, 1963a).

If the actual population number is constant, $N_t = N$, then $\mu = 2$ and (9.129) becomes (Wright, 1938, 1939)

$$N_e^{(i)} = \frac{2(2N - 1)}{\sigma^2 + 2} \ . \tag{9.132}$$

With exactly two gametes per individual, $\sigma^2 = 0$, so (9.132) reduces to $N_e^{(i)} = 2N - 1 \sim 2N$ as $N \to \infty$. As discussed below (9.79), this doubling of the effective size compared with an "ideal" population provides an interpretation for the decay rate $1/(4N)$ per generation in the case of maximum avoidance of consanguinity by $N \gg 1$ individuals.

2. Monoecious Populations Without Selfing

In the absence of selfing, (9.122) simplifies to

$$f_{t+1} = g_t , \tag{9.133a}$$
$$g_{t+1} = \tfrac{1}{2}P_{t+1}(1 + f_t) + (1 - P_{t+1})g_t ; \tag{9.133b}$$

comparing this with (9.97) again yields (9.124), but P_t must be recalculated. Since two gametes derived from the same individual in generation $t - 1$ are now automatically in different individuals in generation t, therefore, given that $\mathbf{K} = \mathbf{k}$, the number of possible choices from distinct offspring of gametes derived from the same parent is still

$$\sum_i \binom{k_i}{2} .$$

The total number of ways of choosing two gametes from distinct progeny equals

$$\binom{2N_t}{2} - N_t ,$$

because N_t is the number of pairs in the same individual. Equations (9.126) to (9.128) still hold, but (9.129) is now slightly modified to (Kimura and Crow, 1963a; Crow and Kimura, 1970, pp. 349–352)

$$N_e^{(i)} = \frac{2(N_t - 1)}{\sigma^2 \mu^{-1} + \mu - 1} . \tag{9.134}$$

As a check, notice that if $N_{t-1} = 2$, then $\sigma^2 = 0$, and hence (9.134) gives $N_e^{(i)} = 2$. For large population numbers, (9.129) and (9.134) are quite close. With constant population size, $N_t = N$ and $\mu = 2$, so (9.134) reduces to

$$N_e^{(i)} = \frac{4(N - 1)}{\sigma^2 + 2} . \tag{9.135}$$

3. Dioecious Populations

If the population is dioecious, after one generation of panmixia (9.133) is still valid, but we must recompute the expectation that leads to (9.134). Using subscripts 1 and 2 for males and females, we have $N_{1,t} + N_{2,t} = N_t$ for the total population number. Denote by $K_{j,i}$ the number of successful gametes contributed by the ith individual with sex j $(= 1, 2)$, and define the means and variances

$$\mu_j = \mathcal{E}(K_{j,i}), \qquad \sigma_j^2 = \text{Var}(K_{j,i}) . \tag{9.136}$$

From

$$\sum_{i=1}^{N_{j,t-1}} K_{j,i} = N_t \tag{9.137}$$

we deduce

$$\mu_j = N_t/N_{j,t-1} \ . \tag{9.138}$$

Two gametes from the same parent in distinct individuals can now be chosen conditionally in

$$\sum_{j=1}^{2} \sum_{i=1}^{N_{j,t-1}} \binom{k_{j,i}}{2}$$

ways; the total number of choices is the same as above (9.134). Proceeding as in the derivation of (9.125), we find (Pollak, 1977)

$$N_e^{(i)} = \frac{4N_t(N_t - 1)}{N_{1,t-1}(\sigma_1^2 + \mu_1^2 - \mu_1) + N_{2,t-1}(\sigma_2^2 + \mu_2^2 - \mu_2)} \ . \tag{9.139}$$

The special case of this formula with constant, large population numbers was obtained by Malécot (1951) and Moran and Watterson (1959). Chia and Pollak (1974) derived the inbreeding effective number on the assumption that $\{N_{1,t}, N_{2,t}\}$ is a finite Markov chain.

If $K_{j,i}$ is binomially distributed with index N_t and parameter $N_{j,t-1}^{-1}$, the progeny variances read

$$\sigma_j^2 = \mu_j(1 - N_{j,t-1}^{-1}) \ . \tag{9.140}$$

Inserting (9.140) into (9.139) produces (Crow and Kimura, 1970, p. 350)

$$N_e^{(i)} = \frac{4N_{1,t-1}N_{2,t-1}}{N_{1,t-1} + N_{2,t-1}} \ , \tag{9.141}$$

thereby generalizing (9.101) to time-dependent population numbers.

Finally, let us prove that (9.134) and (9.139) are equivalent (Pollak, 1977). Since an individual in generation $t - 1$ has sex j with probability $N_{j,t-1}N_{t-1}^{-1}$, we obtain from (9.138)

$$\mu = \mathcal{E}(K_i) = N_{1,t-1}N_{t-1}^{-1}\mu_1 + N_{2,t-1}N_{t-1}^{-1}\mu_2 = 2N_t/N_{t-1} \ , \tag{9.142}$$

which confirms (9.128). Similarly,

$$\mathcal{E}(K_i^2) = N_{1,t-1}N_{t-1}^{-1}(\sigma_1^2 + \mu_1^2) + N_{2,t-1}N_{t-1}^{-1}(\sigma_2^2 + \mu_2^2) \ . \tag{9.143}$$

Substituting

$$\sigma^2 = \mathrm{Var}\,(K_i) = \mathcal{E}(K_i^2) - \mu^2 \ , \tag{9.144}$$

(9.142), and (9.143) into (9.139) leads to (9.134). Thus, $N_e^{(i)}$ is the same for a dioecious population as for a monoecious one without selfing.

9.9 The Model for Random Drift of Gene Frequencies

In the previous sections, we were able to investigate rather directly the dynamics of the probabilities of identity in a finite population. A detailed description of the evolution of the population, however, requires the use of genotypic (or, at least, gene) frequencies as the fundamental dependent variables. We shall formulate the basic model of this section quite generally, making the biological assumptions explicit. Our treatment is based on Ethier and Nagylaki (1980) and Nagylaki (1986, 1990).

The life cycle starts with N monoecious adults. We focus attention on a single locus with n alleles and denote the frequency of the unordered genotype $A_i A_j$, where $i \leq j$, just before reproduction by P_{ij}. It is essential to use *unordered* frequencies as the basic variables because random variation would destroy the symmetry of the ordered frequencies. At this stage, the frequency of A_i is

$$p_i = P_{ii} + \tfrac{1}{2} \sum_{j:j>i} P_{ij} + \tfrac{1}{2} \sum_{j:j<i} P_{ji} \ . \tag{9.145}$$

Each genotypic frequency P_{ij} has the possible values k/N, where $k = 0, 1, \ldots, N$, with the constraint

$$\sum_{i \leq j} P_{ij} = 1 \ .$$

Each allelic frequency p_i may be $l/(2N)$, where $l = 0, 1, \ldots, 2N$, subject to the constraint

$$\sum_i p_i = 1 \ .$$

Reproduction is panmictic, including selfing, and without fertility differences. The adults produce an infinite number of gametes, which fuse at random to form zygotes in Hardy-Weinberg proportions with unordered genotypic frequencies $(2 - \delta_{ij})p_i p_j$, in which p_i is still given by (9.145).

Selection acts through viability differences. If w_{ij} represents the viability of $A_i A_j$ individuals, after selection the genotypic frequencies read

$$P_{ij}^\star = (2 - \delta_{ij}) w_{ij} p_i p_j \Big/ \sum_{k \leq l} (2 - \delta_{kl}) w_{kl} p_k p_l \ . \tag{9.146}$$

The population size remains infinite.

We introduce mutation as in Sect. 4.9. Let u_{ij} designate the probability that A_i mutates to A_j; by convention, $u_{ii} = 0$. Then

$$R_{ij} = \left(1 - \sum_k u_{ik} \right) \delta_{ij} + u_{ij} \tag{9.147}$$

is the probability that an A_i allele in a zygote appears as A_j in a gamete. Assume that the two genes carried by an individual mutate independently. Consequently, after mutation the germ-line genotypic frequencies are

$$P_{ij}^{\star\star} = \tfrac{1}{2}(2 - \delta_{ij}) \sum_{k \leq l} (R_{ki}R_{lj} + R_{kj}R_{li})P_{kl}^{\star} \ . \qquad (9.148)$$

The population number is, of course, unaltered.

Random drift operates through population regulation, which reduces the population to N adults with unordered genotypic frequencies P_{ij}', thereby completing the life cycle. Therefore, given the genotypic frequencies $\mathbf{P}$, the distribution of the genotypic numbers NP_{ij}' is multinomial with index N and parameters $P_{ij}^{\star\star}$, i.e., the probability of transition from $\mathbf{P}$ to $\mathbf{P}'$ is

$$N! \prod_{i \leq j} \left[(P_{ij}^{\star\star})^{NP_{ij}'} \Big/ (NP_{ij}')! \right] \ . \qquad (9.149)$$

We summarize the above information in the following formal scheme.

Adult $\xrightarrow{\text{reproduction}}$ Zygote $\xrightarrow{\text{selection}}$ Adult $\xrightarrow{\text{mutation}}$ Adult $\xrightarrow{\text{regulation}}$ Adult

$N, P_{ij}, p_i \qquad \infty, (2 - \delta_{ij})p_i p_j, p_i \qquad \infty, P_{ij}^{\star}, p_i^{\star} \qquad \infty, P_{ij}^{\star\star}, p_i^{\star\star} \qquad N, P_{ij}', p_i'$

Consult Ethier and Nagylaki (1980) for similar models of evolution at multiallelic autosomal and X-linked loci in a dioecious population. These encompass independent multinomial sampling and different viabilities, mutation rates, and numbers of adults in the two sexes.

A model rather similar to (9.149) was formulated by Moran (1958; 1962, pp. 144–152) for a diallelic autosomal locus in a dioecious population. Moran introduced selection by assigning different constant fertilities to the three genotypes. As we saw in Sect. 4.1, however, if matings occur, even in an infinite population this formulation requires multiplicative fertilities; the necessary detailed biological assumptions in a finite population are unknown. Hence, we permitted differential viability rather than fecundity.

Moran (1958a–d, 1962) has also proposed various models in which individuals die and are replaced one by one. Although such birth-and-death schemes with overlapping generations are probably generally less realistic than multinomial models, their greater mathematical tractability ensures their utility. The exact evaluation of many quantities in these schemes allows us to test various approximate procedures analytically.

To some extent, the order of the evolutionary forces in our life cycle is arbitrary. We assume that while selection acts on the phenotype, which develops from the zygotic genotype, the germ cells mutate with no phenotypic effect. Consequently, in any formal scheme, selection must always precede mutation. We are left with two possible sequences in addition to ours: reproduction, selection, regulation, mutation; and reproduction, regulation, selection, mutation. Whereas in our model selection and mutation occur in an infinite population and hence can be treated deterministically, the first of the above alternatives would entail a much more complicated probabilistic formulation of mutation, and the second would necessitate this for both mutation and selection. If all the evolutionary forces were weak, it is plausible, but unproved, that the dynamics of the three models would be quite close.

One important result follows directly from our model. In the absence of mutation, fixation of each allele ($p_i = 1$) is an absorbing state of our finite Markov chain; if we exclude lethality, there are no other absorbing states. We conclude that some allele will be fixed in finite time with probability one. Thus, all genetic diversity is ultimately lost.

It is convenient to express the transition probabilities (9.149) of the Markov chain of genotypic frequencies in terms of probability-generating functions:

$$\mathcal{E}\left(\prod_{i\leq j}\xi_{ij}^{NP'_{ij}}\Bigg|\mathbf{P}\right) = \left(\sum_{i\leq j}P_{ij}^{\star\star}\xi_{ij}\right)^N . \tag{9.150}$$

The conditional means and covariances read

$$\mathcal{E}(P'_{ij}|\mathbf{P}) = P_{ij}^{\star\star} , \tag{9.151}$$

$$\text{Cov}\,(P'_{ik}, P'_{jl}|\mathbf{P}) = N^{-1}P_{ik}^{\star\star}(\delta_{ij}\delta_{kl} - P_{jl}^{\star\star}) . \tag{9.152}$$

Most of our interest centers on the evolution of the gene frequencies. Since the transition probabilities (9.149) depend on the initial genotypic frequencies $\mathbf{P}$ only through the allelic frequencies $\mathbf{p}$, the vector of gene frequencies $\mathbf{p}(t)$ is Markovian. Hence, we can study the Markov chain $\{\mathbf{p}(t)\}$ without analyzing $\{\mathbf{P}(t)\}$; the consequent decrease in dimensionality greatly simplifies many calculations.

Computations *within generations* are facilitated by employing ordered (or symmetrized) genotypic frequencies

$$\tilde{P}_{ij} = \begin{cases} \frac{1}{2}P_{ij}, & i < j , \\ P_{ij}, & i = j , \\ \frac{1}{2}P_{ji}, & i > j . \end{cases} \tag{9.153}$$

Then (9.145) reduces to our customary form,

$$p_i = \sum_j \tilde{P}_{ij} ; \tag{9.154}$$

with the definition $w_{ji} = w_{ij}$ for $i < j$, (9.146) becomes

$$\tilde{P}_{ij}^{\star} = w_{ij}p_ip_j/\overline{w}, \qquad \overline{w} = \sum_{ij} w_{ij}p_ip_j ; \tag{9.155}$$

and (9.148) reads

$$\tilde{P}_{ij}^{\star\star} = \sum_{kl} R_{ki}R_{lj}\tilde{P}_{kl}^{\star} . \tag{9.156}$$

Summing (9.156) over j gives

$$p_i^{\star\star} = \sum_k R_{ki} p_k^\star \ , \tag{9.157}$$

where

$$p_i^\star = p_i w_i / \overline{w} \ , \qquad w_i = \sum_j w_{ij} p_j \ . \tag{9.158}$$

To derive the transition probabilities for the Markov chain of gene frequencies, take $\xi_{ij} = \zeta_i \zeta_j$ in (9.150) for all i and j such that $i \le j$. We find

$$\prod_{i \le j} (\zeta_i \zeta_j)^{N P'_{ij}} = \prod_{ij} (\zeta_i \zeta_j)^{N \tilde{P}'_{ij}}$$

$$= \left(\prod_{ij} \zeta_i^{N \tilde{P}'_{ij}} \right) \left(\prod_{ij} \zeta_i^{N \tilde{P}'_{ji}} \right)$$

$$= \prod_{ij} \zeta_i^{2 N \tilde{P}'_{ij}}$$

$$= \prod_i \zeta_i^{2 N p'_i} \ . \tag{9.159}$$

Substituting (9.159) into (9.150) yields

$$\mathcal{E}\left(\prod_i \zeta_i^{2 N p'_i} \Big| \mathbf{P} \right) = \left(\sum_{ij} \tilde{P}_{ij}^{\star\star} \zeta_i \zeta_j \right)^N \ . \tag{9.160}$$

Since the right-hand side depends only on $\mathbf{p}$, we obtain the generating function of the transition probabilities:

$$\phi(\boldsymbol{\zeta}) = \mathcal{E}\left(\prod_i \zeta_i^{2 N p'_i} \Big| \mathbf{p} \right) = \left(\sum_{ij} \tilde{P}_{ij}^{\star\star} \zeta_i \zeta_j \right)^N \ . \tag{9.161}$$

From (9.161) we can easily deduce the conditional means and covariances of the allelic frequencies. If the random variables N_i have joint probability-generating function

$$\psi(\boldsymbol{\zeta}) = \mathcal{E}\left(\prod_i \zeta_i^{N_i} \right) \ , \tag{9.162}$$

then

$$\mathcal{E}(N_i) = \frac{\partial \psi}{\partial \zeta_i}(\mathbf{u}) \ , \tag{9.163}$$

$$\mathrm{Cov}\,(N_i, N_j) = \frac{\partial^2 \psi}{\partial \zeta_i \partial \zeta_j}(\mathbf{u}) - \frac{\partial \psi}{\partial \zeta_i}(\mathbf{u}) \frac{\partial \psi}{\partial \zeta_j}(\mathbf{u}) + \delta_{ij} \frac{\partial \psi}{\partial \zeta_i}(\mathbf{u}) \ , \tag{9.164}$$

where $\mathbf{u}$ denotes the vector of ones. Applying (9.163) and (9.164) to (9.161) leads to

$$\mathcal{E}(p_i'|\mathbf{p}) = p_i^{\star\star} \; , \tag{9.165a}$$

$$\text{Cov}\,(p_i', p_j'|\mathbf{p}) = (2N)^{-1}(\tilde{P}_{ij}^{\star\star} + \delta_{ij}p_i^{\star\star} - 2p_i^{\star\star}p_j^{\star\star})$$

$$= (2N)^{-1}[p_i^{\star\star}(\delta_{ij} - p_j^{\star\star}) + (\tilde{P}_{ij}^{\star\star} - p_i^{\star\star}p_j^{\star\star})] \; . \tag{9.165b}$$

In Ethier and Nagylaki (1980), (9.165) is derived from (9.151) and (9.152).

We are now in a position to show that the transition probabilities specified by (9.161) have the usual multinomial form if and only if

$$\tilde{P}_{ij}^{\star\star} = p_i^{\star\star}p_j^{\star\star} \tag{9.166}$$

for all i and j, i.e., the adults just before population regulation are in Hardy-Weinberg proportions. On the one hand, (9.166) immediately simplifies (9.161) to

$$\phi(\zeta) = \left(\sum_i p_i^{\star\star}\zeta_i\right)^{2N} \; , \tag{9.167}$$

which reveals that the probability of transition from $\mathbf{p}$ to $\mathbf{p}'$ is

$$(2N)! \prod_i \left[(p_i^{\star\star})^{2Np_i'}\Big/(2Np_i')!\right] \; . \tag{9.168}$$

Thus, given the gene frequencies $\mathbf{p}$, the distribution of the allelic numbers $2Np_i'$ is multinomial with index $2N$ and parameters $p_i^{\star\star}$. On the other hand, (9.168) implies

$$\text{Cov}\,(p_i', p_j'|\mathbf{p}) = (2N)^{-1}p_i^{\star\star}(\delta_{ij} - p_j^{\star\star}) \; ; \tag{9.169}$$

comparing (9.169) with the general formula (9.165b), we infer (9.166).

In Problem 4.1, we established that the genotypic frequencies after viability selection are in Hardy-Weinberg proportions (i.e., $\tilde{P}_{ij}^{\star} = p_i^{\star}p_j^{\star}$ for all i and j) if and only if the viabilities are multiplicative ($w_{ij} = c_i c_j$, for some c_i, for all i and j). According to the calculation that leads to (4.99), mutation preserves Hardy-Weinberg proportions, but generally does not produce them. Hence, (9.166) holds if viabilities are multiplicative, but at most in very special cases otherwise. Since there is generally no reason to posit multiplicative viabilities, we conclude that (9.168) usually applies as an exact model only in the absence of selection.

Since this argument presupposes (9.149), it does not preclude the derivation of (9.168) from a different set of explicit biological postulates, but for diploids it seems unlikely that this can be done. With a formulation in the spirit of (9.149), the reduction to multinomial transition probabilities for the allelic frequencies is never possible for models of random drift in a dioecious population (Ethier and Nagylaki, 1980).

In spite of this negative result, (9.168) has considerable biological value even in the presence of selection. It can, of course, describe the evolution of $2N$ haploid individuals. More important, for slow selection (9.166) is a good approximation, and therefore so is (9.168). This can be proved rigorously if all three evolutionary forces are weak, for in that case (9.149) and (9.168) have the same diffusion limit (Ethier and Nagylaki, 1980).

Fisher (1922, 1930, 1930a) implicitly employed the diallelic form of (9.168) in diffusion analyses. This binomial model was first written down and studied explicitly by Wright (1931). Kimura (1955, 1955a, 1956b) tacitly relied on the multinomial (9.168) for his multiallelic diffusion work. See also Karlin and McGregor (1964, 1965). We shall call (9.168) the Wright-Fisher model.

9.10 Random Drift of Gene Frequencies

In this section, we shall investigate some aspects of pure random genetic drift. Refer to Ewens (1969, 1979, 1990), Crow and Kimura (1970), Kingman (1980), and Nagylaki (1990) for many other results.

Although no exact formula has been obtained for the distribution of gene frequencies even in the diallelic Wright-Fisher model, we can evaluate easily the multiallelic moments of low order. From (9.165a) we have

$$\mathcal{E}(p_i'|\mathbf{p}) = p_i \ , \tag{9.170}$$

whence

$$\mathcal{E}(p_i') = \mathcal{E}[\mathcal{E}(p_i'|\mathbf{p})] = \mathcal{E}(p_i) \ . \tag{9.171}$$

Therefore, the mean gene frequencies are constant, and hence equal to their initial values, which we denote by $\bar{p}_i$:

$$\mathcal{E}[p_i(t)] = \bar{p}_i \ . \tag{9.172}$$

If the initial gene frequencies are fixed, then $\bar{p}_i$ represents the frequency of A_i in the adults of generation zero just before reproduction.

Before calculating the covariances, we establish a simple, useful identity. For any three random variables or random vectors X, Y, and Z, we deduce

$$\begin{aligned}
\mathrm{Cov}\,(X,Y) &= \mathcal{E}(XY) - \mathcal{E}(X)\mathcal{E}(Y) \\
&= \mathcal{E}[\mathcal{E}(XY|Z)] - \mathcal{E}[\mathcal{E}(X|Z)]\mathcal{E}[\mathcal{E}(Y|Z)] \\
&= \mathcal{E}[\mathcal{E}(XY|Z) - \mathcal{E}(X|Z)\mathcal{E}(Y|Z)] + \mathcal{E}[\mathcal{E}(X|Z)\mathcal{E}(Y|Z)] \\
&\quad - \mathcal{E}[\mathcal{E}(X|Z)]\mathcal{E}[\mathcal{E}(Y|Z)] \\
&= \mathcal{E}[\mathrm{Cov}\,(X,Y|Z)] + \mathrm{Cov}\,[\mathcal{E}(X|Z), \mathcal{E}(Y|Z)] \ . \tag{9.173}
\end{aligned}$$

The first term corresponds to the expected covariation with Z fixed; the second, to the covariation of the mean values for each Z. The special case $X = Y$ yields the decomposition

$$\mathrm{Var}\,(X) = \mathcal{E}[\mathrm{Var}\,(X|Z)] + \mathrm{Var}\,[\mathcal{E}(X|Z)] \tag{9.174}$$

for the variance.

Setting

$$V_{ij}(t) = \mathrm{Cov}\,[p_i(t),\, p_j(t)] \ , \tag{9.175}$$

identifying $X = p'_i$, $Y = p'_j$, and $Z = \mathbf{p}$ in (9.173), and utilizing (9.165b) and (9.170) we find

$$V'_{ij} = \mathcal{E}[(2N)^{-1}p_i(\delta_{ij} - p_j)] + \mathrm{Cov}\,(p_i, p_j)\ . \tag{9.176}$$

Inserting (9.172) and (9.175) into (9.176) leads to the difference equation

$$V'_{ij} = (2N)^{-1}\overline{p}_i(\delta_{ij} - \overline{p}_j) + [1 - (2N)^{-1}]V_{ij}\ , \tag{9.177}$$

with the solution

$$V_{ij}(t) = \hat{V}_{ij} + [V_{ij}(0) - \hat{V}_{ij}][1 - (2N)^{-1}]^t\ , \tag{9.178a}$$

which converges to the equilibrium covariance

$$\hat{V}_{ij} = \overline{p}_i(\delta_{ij} - \overline{p}_j)\ . \tag{9.178b}$$

In particular, at equilibrium the variance of p_i between populations is just $\overline{p}_i(1 - \overline{p}_i)$.

If we suppose that the initial allelic frequencies are fixed, then $V_{ij}(0) = 0$ and (9.178a) reduces to

$$V_{ij}(t) = \hat{V}_{ij}\left[1 - \left(1 - \frac{1}{2N}\right)^t\right] = \hat{V}_{ij}F(t)\ , \tag{9.179}$$

where the connection with the inbreeding coefficient comes from (9.96). We owe the diallelic solution to Wright (1942).

Since in our model adults are sampled from an infinite pool of zygotes, it is more consistent to start with the multinomial covariances

$$V_{ij}(0) = (2N)^{-1}\overline{p}_i(\delta_{ij} - \overline{p}_j) = (2N)^{-1}\hat{V}_{ij}\ , \tag{9.180}$$

in which case (9.178a) specializes to

$$V_{ij}(t) = \hat{V}_{ij}\left[1 - \left(1 - \frac{1}{2N}\right)^{t+1}\right] = \hat{V}_{ij}F(t+1)\ . \tag{9.181}$$

This agrees with the rather different approaches of Cockerham (1969) and Pollak (1977). The lag of one generation occurs in (9.181) because, although the initial generation of adults is not inbred, its gene frequencies are subject to sampling variation.

Our results allow us to deduce the expected genotypic proportions. From (9.151) we get

$$\mathcal{E}(P'_{ij}) = \mathcal{E}[(2 - \delta_{ij})p_i p_j] = (2 - \delta_{ij})(\overline{p}_i\overline{p}_j + V_{ij})\ . \tag{9.182}$$

Substituting (9.178b) and (9.181) into (9.182) gives the mean unordered frequencies

$$\mathcal{E}[P_{ij}(t)] = \overline{p}_i\delta_{ij}F(t) + \overline{p}_i\overline{p}_j(2 - \delta_{ij})[1 - F(t)]\ , \tag{9.183}$$

in agreement with the general formula (9.80) for the expected value of the ordered frequencies. From (9.183) we recover a special case of the expression (9.81) for the expected heterozygosity:

$$h(t) = h(0)H(t) = \left(1 - \sum_i \bar{p}_i^2\right)\left(1 - \frac{1}{2N}\right)^t .$$ (9.184)

We have already remarked that some allele is eventually fixed. According to (9.181) and (9.184), as a result of this process, the variance in gene frequency between lines increases at the same rate as the expected homozygosity.

Our formulae enable us to investigate the variance of a quantitative character within and between subpopulations. We consider infinitely many panmictic lines of N monoecious individuals and assume that the character X is controlled by a single locus without dominance, so that the trait value of A_iA_j is $c_i + c_j$. With $Z = \mathbf{P}$ in (9.174), the first term represents the expected variance within a subpopulation, V_w, whereas the second is the variance between the subpopulation means, V_b.

Within lines, we obtain

$$\mathcal{E}(X|\mathbf{P}) = \sum_{ij} \tilde{P}_{ij}(c_i + c_j) = 2\sum_i p_i c_i ,$$ (9.185a)

$$\mathcal{E}(X^2|\mathbf{P}) = \sum_{ij} \tilde{P}_{ij}(c_i + c_j)^2 = 2\sum_i p_i c_i^2 + 2\sum_{ij} \tilde{P}_{ij}c_i c_j ,$$ (9.185b)

whence

$$\text{Var}(X|\mathbf{P}) = 2\left(\sum_i p_i c_i^2 + \sum_{ij} \tilde{P}_{ij}c_i c_j - 2\sum_{ij} p_i p_j c_i c_j\right) .$$ (9.186)

Using successively (9.172), (9.183), (9.181), (9.178b), and (9.95a) to compute the expectation of (9.186) leads to

$$V_w(t) = (1 - N^{-1})V_r[1 - F(t)] ,$$ (9.187a)

where

$$V_r = 2\sum_i \bar{p}_i c_i^2 - 2\left(\sum_i \bar{p}_i c_i\right)^2$$ (9.187b)

would be the variance in the entire infinite population with random mating.

According to (9.174), the variance between lines is just the variance of (9.185a); with the help of (9.178b) and (9.181), we find

$$V_b(t) = 2V_r F(t+1) .$$ (9.188)

From (9.174) and (9.95a) we deduce the total variance in the population:

$$V(t) = V_w(t) + V_b(t) = V_r[1 + F(t)] .$$ (9.189)

This agrees with (9.39) because $\overline{P}_{ij}$ is the frequency of A_iA_j in the entire infinite population.

Thus, the variance within subpopulations is proportional to the panmictic index, and therefore approaches zero at the same rate as each line tends to fixation. The variance between the subpopulation means increases monotonically to twice the panmictic variance. The same holds for the total variance.

Our results are consistent with those of Lush (1948) and Cockerham (1969). Wright (1951) and Crow and Kimura (1970, pp. 339–340) neglect the stochastic deviation from Hardy-Weinberg proportions within lines. Therefore, they do not have the factor $1 - N^{-1}$ in (9.187a) and obtain $F(t)$ instead of $F(t+1)$ in (9.188). Note, however, that we must have $V_w = 0$ for $N = 1$. The discrepancies are small for large N.

With dominance, the third and fourth moments of the gene frequency are also required. Robertson (1952) and, more explicitly, Crow and Kimura (1970, pp. 331–344) calculated these moments for two alleles, and hence computed the variances within and between lines in the Hardy-Weinberg approximation.

We have already remarked that without mutation (excluding some rather special cases of lethality), some allele is eventually fixed. For pure random drift, we can easily determine the probability, $v_i(\mathbf{p})$, that this allele is A_i. Since $\{\mathbf{p}(t)\}$ is a finite Markov chain, (9.170) implies that $\{p_i(t)\}$ is a martingale for each i (Karlin and Taylor, 1975, Chap. 6), and the Optional Stopping Theorem (Karlin and Taylor, 1975, p. 261) yields

$$\mathcal{E}[p_i(T)] = \mathcal{E}[p_i(0)] = \overline{p}_i \; , \tag{9.190}$$

where T is the (random) absorption time. But $p_i(T)$ is 1 or 0 with respective probabilities v_i and $1 - v_i$; hence, we infer from (9.190)

$$v_i(\mathbf{p}) = \overline{p}_i \; . \tag{9.191}$$

So the fixation probability of an allele is just its initial frequency.

As pointed out by Ewens (1979, p. 17), we can also derive (9.191) by noting that ultimately all genes in the population are identical by descent. By symmetry, the probability that the allele fixed is descended from an initial A_i gene is evidently the initial frequency of A_i.

Observe that (9.191) enables us to confirm (9.178b).

Wright (1942) and Malécot (1948) obtained (9.191) for two alleles. Kimura (1957) deduced (9.191) for two alleles in the diffusion approximation, which produces the exact result in this case. Since all alleles are equivalent, we may assign the alleles other than A_i to one group and thereby obtain the multiallelic result from the diallelic one.

From (9.191) we conclude that a single mutant is fixed with probability $1/(2N)$, a formula implicit in work of Fisher (1930, 1930a) and Wright (1931) concerning two alleles with selection. Thus, the probability that a single neutral mutant is fixed in a large population is extremely small.

9.11 Gene-Frequency Change due to Mutation and Random Drift

Here we shall calculate the mean and variance of the gene frequency for the neutral, diallelic Wright-Fisher model with mutation. Changing the notation slightly, we designate the frequencies of A_1 and A_2 by p_t and q_t, respectively, and assume that A_1 mutates to A_2 at rate u, the reverse mutation rate being v.

From (9.147) and (9.157) we find

$$p^{\star\star} = v + \lambda p, \qquad \lambda = 1 - u - v . \tag{9.192}$$

The expectation of (9.165a) yields the recursion relation

$$\overline{p}' = v + \lambda \overline{p} \tag{9.193}$$

for the average frequency of A_1. Owing to the linearity of mutation, exhibited in (9.192), the mean allelic frequency is just the deterministic one. We posit that $u > 0$ or $v > 0$; then the solution of (9.193) reads

$$\overline{p}_t = \hat{p} + (\overline{p}_0 - \hat{p})\lambda^t , \tag{9.194a}$$

where

$$\hat{p} = v/(u + v) \tag{9.194b}$$

represents the usual deterministic equilibrium gene frequency, and therefore also the mean equilibrium frequency.

To evaluate

$$V_t = \mathrm{Var}\,(p_t) , \tag{9.195}$$

note first that (9.165b) yields

$$\mathrm{Var}\,(p'|\mathbf{p}) = (2N)^{-1}p^{\star\star}(1 - p^{\star\star}) . \tag{9.196}$$

Substituting (9.192) and (9.196) into (9.174), we infer

$$V' = \kappa V + (2N)^{-1}[v(1 - v) + \lambda(1 - 2v)\overline{p} - \lambda^2\overline{p}^2] , \tag{9.197a}$$

with

$$\kappa = \lambda^2[1 - (2N)^{-1}] = (1 - u - v)^2[1 - (2N)^{-1}] . \tag{9.197b}$$

Inserting (9.194) into (9.197a) leads to

$$V' - \kappa V = (2N)^{-1}[\hat{p}\hat{q} + b(\hat{q} - \hat{p})\lambda^t - b^2\lambda^{2t}] , \tag{9.198}$$

in which $b = \lambda(\overline{p}_0 - \hat{p})$. With the method applied to (5.18), we deduce that (9.198) has the solution (Nagylaki, 1979c)

$$V_t = \hat{V} + (V_0 - \hat{V})\kappa^t + \left[\frac{(\overline{p}_0 - \hat{p})(\hat{q} - \hat{p})}{2N - \lambda(2N - 1)}\right](\lambda^t - \kappa^t) - (\overline{p}_0 - \hat{p})^2(\lambda^{2t} - \kappa^t) , \tag{9.199a}$$

where

$$\hat{V} = \frac{\hat{p}\hat{q}}{2N(1-\kappa)} = \frac{\hat{p}\hat{q}}{2N - (1 - u - v)^2(2N-1)} \tag{9.199b}$$

is the variance of the gene frequency at equilibrium.

As a check, observe that V_t is unaltered by the simultaneous interchanges $u \leftrightarrow v$ and $\overline{p}_0 \leftrightarrow \overline{q}_0$. Since $1 > \lambda > \lambda^2 > \kappa > 0$, therefore, unless $\overline{p}_0 = \hat{p}$ or $u = v$, the variance converges at the same slow asymptotic rate, λ^t, as the mean. If $\overline{p}_0 \neq \hat{p}$ but $u = v$, convergence occurs at the asymptotic rate λ^{2t}. If $\overline{p}_0 = \hat{p}$, the rate is κ^t.

For weak mutation ($u, v \ll 1$), $\hat{V}$ is well approximated by Wright's (1931) formula,

$$\hat{V} \approx \frac{\hat{p}\hat{q}}{1 + 4N(u+v)} \ . \tag{9.200}$$

In fact, (9.200) is a slight underestimate: with $u + v \leq 1$, a little algebra demonstrates

$$\frac{\hat{p}\hat{q}}{1 + 4N(u+v)} < \hat{V} < \frac{[1 + 2(u+v)]\hat{p}\hat{q}}{1 + 4N(u+v)} \ . \tag{9.201}$$

Thus, the equilibrium gene frequency is very likely to be close to $\hat{p}$ if $4N(u+v) \gg 1$; significant departures from $\hat{p}$ are probable otherwise. We can approximate (9.199a) for $u, v, N^{-1} \ll 1$ by

$$\begin{aligned}
V_t &\approx \hat{V} + (V_0 - \hat{V})e^{-[2(u+v)+(2N)^{-1}]t} \\
&+ \left[\frac{(\overline{p}_0 - \hat{p})(\hat{q} - \hat{p})}{1 + 2N(u+v)}\right] \left[e^{-(u+v)t} - e^{-[2(u+v)+(2N)^{-1}]t}\right] \\
&- (\overline{p}_0 - \hat{p})^2 \left[e^{-2(u+v)t} - e^{-[2(u+v)+(2N)^{-1}]t}\right] \ . \tag{9.202}
\end{aligned}$$

Setting $u = 0$ in (9.199), we derive the variance

$$V_t = V_0\kappa^t + \frac{\overline{q}_0(\lambda^t - \kappa^t)}{2N - \lambda(2N-1)} - \overline{q}_0^2(\lambda^{2t} - \kappa^t) \tag{9.203}$$

for irreversible mutation. This tends to zero because A_1 is ultimately fixed. For $v \ll 1$ and $N \gg 1$, (9.203) gives

$$\begin{aligned}
V_t &\approx V_0 e^{-[2v+(2N)^{-1}]t} + \left[\frac{\overline{q}_0}{1 + 2Nv}\right] \left[e^{-vt} - e^{-[2v+(2N)^{-1}]t}\right] \\
&- \overline{q}_0^2 \left[e^{-2vt} - e^{-[2v+(2N)^{-1}]t}\right] \ . \tag{9.204}
\end{aligned}$$

If we let $v \to 0$ in (9.203), we recover the diallelic special case of (9.178).

Our results enable us to analyze approximately *mutation-selection balance* in a finite population. With strong selection against the mutant A_1, the mutant homozygotes A_1A_1 are so rare that their fitness is irrelevant; let A_2A_2 and A_1A_2 have fitnesses 1 and $1 - s$, respectively, where $0 < s \leq 1$. From (9.165b) we obtain

$$\text{Var}(p'|\mathbf{p}) = (2N)^{-1}[p^{**}(1 - p^{**}) + \tilde{P}_{11}^{**} - (p^{**})^2] \ , \tag{9.205}$$

which indicates that random drift has intensity N^{-1}. To ensure that high values of p have very low probability, we posit that $v \ll s$ and $N^{-1} \ll s$; with $v \ll s$, we lose little biological generality by assuming also that $u \ll s$. Then selection is much stronger than mutation and random drift.

From (9.157) and (9.158) we find

$$p^{\star\star} \approx v + (1 - s)p \ . \tag{9.206}$$

Since $p \ll 1$ with high probability, we may neglect the last two terms in (9.205), which consequently reduces to (9.196). Therefore, comparing (9.206) with (9.192), we conclude that the formulae for pure mutation apply with the replacement $u + v \to s$. In particular, (9.194) informs us that

$$\overline{p}_t \to \hat{p} \approx v/s \tag{9.207}$$

as $t \to \infty$. Thus, the expected gene frequency at equilibrium is close to its deterministic value.

From (9.199b) we deduce the approximate variance of p at equilibrium for $u, v, N^{-1} \ll s$:

$$\hat{V} \approx \frac{v}{2Ns^2(2 - s)} \ . \tag{9.208}$$

Since $\hat{V} \lesssim v/(2Ns^2) \ll 1$, (9.208) is consistent with our approximation that $p \ll 1$. If $A_1 A_2$ is lethal ($s = 1$), (9.208) becomes

$$\hat{V} \approx v/(2N) \ . \tag{9.209}$$

If selection is slow ($s \ll 1$), (9.208) reduces to

$$\hat{V} \approx v/(4Ns^2) \ . \tag{9.210}$$

The coefficient of variation may be quite large: for (9.210), this reads

$$\hat{V}^{1/2}/\hat{p} \approx (4Nv)^{-1/2} \ , \tag{9.211}$$

independent of the selection intensity.

The observation (9.207) is due to Wright (1931). Haldane (1949a) and Skellam (1949) obtained (9.208) from branching processes. Bartlett (1955, p. 42) and Ewens (1969, pp. 87–89) generalized the branching-process analysis to an arbitrary heterozygote offspring distribution; their formula reduces to (9.208) if this distribution is Poisson with mean $1 - s$. Nei (1968) deduced (9.210) from the diffusion approximation.

9.12 The Island Model

We can investigate the island model (Wright, 1931) by modifying the life cycle below (9.149). We consider a population of N monoecious individuals in the absence of selection and mutation. Before population regulation, a fraction m

$(0 \leq m \leq 1)$ of the population is replaced by migrants with fixed ordered genotypic frequencies $\tilde{Q}_{ij}$. Evidently, (9.149) holds with

$$\tilde{P}_{ij}^{\star\star} = (1 - m)p_i p_j + m\tilde{Q}_{ij} \ . \tag{9.212}$$

This model applies equally well to a population composed of infinitely many demes of N individuals, provided the demes exchange migrants wholly at random. Then m represents the proportion of migrants any particular deme exchanges with all the other demes, and $\tilde{Q}_{ij}$ signifies the ordered genotypic frequencies in the entire population.

From (9.212) we obtain

$$p_i^{\star\star} = (1 - m)p_i + mq_i \ , \tag{9.213}$$

where q_i designates the frequency of A_i in immigrants. The expectation of (9.165a) yields the recursion relation satisfied by the average allelic frequencies $\bar{p}_i$:

$$\bar{p}_i' = (1 - m)\bar{p}_i + mq_i \ ; \tag{9.214}$$

this has the solution

$$\bar{p}_i(t) = q_i + [\bar{p}_i(0) - q_i](1 - m)^t \ . \tag{9.215}$$

Although (9.214) can be obtained by specializing mutation to migration as explained in Sect. 2.3, this observation does not extend to the higher moments, because (9.212) does not satisfy (9.166). Notice that the population on the island is not in Hardy-Weinberg proportions after migration even if the immigrants are; in contrast, as we have seen, mutation preserves Hardy-Weinberg proportions.

Inserting (9.212) and (9.213) into (9.165b) leads to

$$\begin{aligned}
\mathrm{Cov}\,(p_i', p_j'|\mathbf{p}) = (2N)^{-1}\{&(1 - m)[(2m - 1)p_i p_j + (\delta_{ij} - 2mq_j)p_i \\
&- 2mq_i p_j] + m[\tilde{Q}_{ij} + q_i(\delta_{ij} - 2mq_j)]\} \ .
\end{aligned} \tag{9.216}$$

We substitute (9.165a), (9.213), and (9.216) into the identity (9.173) to derive a difference equation for the absolute covariances V_{ij}. After we replace $\bar{p}_i$ by (9.215), this equation becomes

$$\begin{aligned}
V_{ij}' - \mu V_{ij} = (2N)^{-1}\{&q_i(\delta_{ij} - q_j) + m(\tilde{Q}_{ij} - q_i q_j) \\
&+ [\delta_{ij}(\bar{p}_i(0) - q_i) + 2q_i q_j - q_i \bar{p}_j(0) - q_j \bar{p}_i(0)](1 - m)^{t+1} \\
&- (1 - 2m)[\bar{p}_i(0) - q_i][\bar{p}_j(0) - q_j](1 - m)^{2t+1}\} \ ,
\end{aligned} \tag{9.217}$$

where

$$\mu = (1 - m)[1 - m - (2N)^{-1}(1 - 2m)] \ . \tag{9.218}$$

Therefore,

$$\begin{aligned}
V_{ij}(t) = \hat{V}_{ij} + [V_{ij}(0) - \hat{V}_{ij} - b_{ij} - c_{ij}]\mu^t + b_{ij}(1 - m)^{2t} \\
+ c_{ij}(1 - m)^t \ ,
\end{aligned} \tag{9.219a}$$

in which

$$b_{ij} = -[\bar{p}_i(0) - q_i][\bar{p}_j(0) - q_j] \ , \tag{9.219b}$$

$$c_{ij} = \frac{\delta_{ij}[\bar{p}_i(0) - q_i] + 2q_iq_j - q_i\bar{p}_j(0) - q_j\bar{p}_i(0)}{1 + 2m(N-1)} \ , \tag{9.219c}$$

and the equilibrium covariances read

$$\hat{V}_{ij} = \frac{q_i(\delta_{ij} - q_j) + m(\tilde{Q}_{ij} - q_iq_j)}{(1-m)(1-2m) + 2Nm(2-m)} \ . \tag{9.219d}$$

Simple rearrangements of (9.218) show that $0 \le \mu < 1 - m$. Hence, unless $c_{ij} = 0$, the rate of convergence to (9.219d) is $(1-m)^t$. Note that $\tilde{Q}_{ij} = q_iq_j$ if the immigrants are in Hardy-Weinberg ratios. With the interpretation of infinitely many demes, $\hat{V}_{ii}$ is the equilibrium variance of the frequency of A_i between subpopulations. For $m \ll 1$, (9.219d) can be approximated as

$$V_{ij} \approx \frac{q_i(\delta_{ij} - q_j)}{1 + 4Nm} \ . \tag{9.220}$$

We owe the diallelic special case of (9.220) to Wright (1931). As suggested by Sect. 2.3, the diallelic form of (9.220) agrees with (9.200) if we identify $m = u + v$. With $m \ll 1$ and $N \gg 1$, (9.218) yields

$$\mu \approx 1 - 2m - (2N)^{-1} \ , \tag{9.221}$$

and for two alleles (9.219) reduces to (9.202).

The method of this section remains applicable even if the migration rate and migrant gene frequencies fluctuate at random. The diallelic case was analyzed in the Hardy-Weinberg approximation by Nagylaki (1979c).

9.13 The Variance Effective Population Number

In Sect. 9.8 we showed how to evaluate the influence of arbitrary variation of the number of progeny on inbreeding (and hence on the expected heterozygosity) by introducing an inbreeding effective population number. For the analysis of the random drift of gene frequencies, especially by diffusion methods, we require the consequences of an arbitrary reproductive variance on the stochastic change of gene frequencies. This leads to the variance effective population number.

To avoid tedious calculations, we restrict ourselves to monoecious populations. We allow the number of individuals, N_t, to vary deterministically and abbreviate the parental and offspring population numbers, N_{t-1} and N_t, by N and N'. We assume that mutation, selection, and migration are absent.

First, we examine the reproduction of the population. Let K_i represent the (random) number of successful gametes from parent i. Since no deterministic evolutionary forces are acting, reproduction must be completely symmetric, i.e., the joint distribution of the N random variables K_i is symmetric under all pairwise interchanges of the K_i. We call the random variables K_i *exchangeable*. Define the vectors $\boldsymbol{\xi} = (\xi_1, \ldots, \xi_N)$ and $\boldsymbol{\eta} = (\eta, \ldots, \eta)$ and note that the total number of successful gametes is $2N'$. Therefore, the joint distribution of the K_i has a completely symmetric probability-generating function $\Phi(\boldsymbol{\xi})$ that satisfies

$$\Phi(\boldsymbol{\eta}) = \eta^{2N'} \ . \tag{9.222}$$

Exchangeability is the only assumption imposed on reproduction; in particular, we do not posit panmixia.

The symmetry of Φ implies

$$\frac{\partial \Phi}{\partial \xi_i}(\boldsymbol{\eta}) = \frac{\partial \Phi}{\partial \xi_1}(\boldsymbol{\eta}) \ , \tag{9.223a}$$

$$\frac{\partial^2 \Phi}{\partial \xi_i^2}(\boldsymbol{\eta}) = \frac{\partial^2 \Phi}{\partial \xi_1^2}(\boldsymbol{\eta}) \ , \tag{9.223b}$$

$$\frac{\partial^2 \Phi}{\partial \xi_i \partial \xi_j}(\boldsymbol{\eta}) = \frac{\partial^2 \Phi}{\partial \xi_1 \partial \xi_2}(\boldsymbol{\eta}), \quad i \neq j \ . \tag{9.223c}$$

The number of gametes from any specified individual has the probability-generating function

$$\psi(\xi) = \Phi(\xi, 1, \ldots, 1) \ . \tag{9.224}$$

The mean number of gametes from a given individual is

$$\mu = \mathcal{E}(K_i) = \frac{d\psi}{d\xi}(1) = \frac{\partial \Phi}{\partial \xi_1}(\mathbf{u}) \ , \tag{9.225}$$

where $\mathbf{u}$ denotes the vector of N ones. But (9.222) and (9.223a) yield

$$2N'\eta^{2N'-1} = \frac{d}{d\eta}\Phi(\boldsymbol{\eta}) = \sum_{i=1}^{N} \frac{\partial \Phi}{\partial \xi_i}(\boldsymbol{\eta}) = N\frac{\partial \Phi}{\partial \xi_1}(\boldsymbol{\eta}) \ . \tag{9.226}$$

Evaluating (9.226) at $\boldsymbol{\eta} = \mathbf{u}$ and employing (9.225) reveals that

$$\mu = 2N'/N \ , \tag{9.227}$$

in agreement with (9.128).

The reproductive variance reads

$$\sigma^2 = \mathrm{Var}\,(K_i) = \frac{d^2\psi}{d\xi^2}(1) - \left[\frac{d\psi}{d\xi}(1)\right]^2 + \frac{d\psi}{d\xi}(1) = \frac{\partial^2 \Phi}{\partial \xi_1^2}(\mathbf{u}) - \mu^2 + \mu \ ,$$

whence

$$\frac{\partial^2 \Phi}{\partial \xi_1^2}(\mathbf{u}) = \sigma^2 + \mu^2 - \mu \ . \tag{9.228}$$

We can calculate the mixed partial derivatives by differentiating (9.226):

$$2N'(2N'-1)\eta^{2N'-2} = \frac{d^2}{d\eta^2}\Phi(\boldsymbol{\eta}) = N\sum_{i=1}^{N}\frac{\partial^2\Phi}{\partial\xi_i\partial\xi_1}(\boldsymbol{\eta}) \ . \tag{9.229}$$

Setting $\boldsymbol{\eta} = \mathbf{u}$ in (9.229) and using (9.223c) and (9.227) leads to

$$\mu(N\mu-1) = \frac{\partial^2\Phi}{\partial\xi_1^2}(\mathbf{u}) + (N-1)\frac{\partial^2\Phi}{\partial\xi_1\partial\xi_2}(\mathbf{u}) \ . \tag{9.230}$$

Combining (9.228) and (9.230), we obtain

$$\frac{\partial^2\Phi}{\partial\xi_1\partial\xi_2}(\mathbf{u}) = \mu^2 - \frac{\sigma^2}{N-1} \ . \tag{9.231}$$

To incorporate genetics, we derive an identity of wide usefulness, whose validity does not require exchangeability. We suppose that there are n alleles, A_j, and let L_{ij} denote the number of gametes produced by individual i that carry A_j. We designate by r_{ij} the probability that a gamete chosen at random from individual i carries A_j. The vector $\mathbf{L}_i = (L_{i1},\ldots,L_{in})$ specifies the gametic output of individual i and satisfies

$$\sum_{j=1}^{n}L_{ij} = K_i \ . \tag{9.232}$$

The Mendelian mechanism informs us that with the vector of gametic numbers $(\mathbf{K})$ fixed, the random vectors $\mathbf{L}_1,\mathbf{L}_2,\ldots,\mathbf{L}_N$ are mutually independent and $\mathbf{L}_i$ is multinomially distributed with index K_i and parameters r_{ij} for each i.

The above remarks permit us to compute the joint probability-generating function of the L_{ij}:

$$\mathcal{E}\left(\prod_{i=1}^{N}\prod_{j=1}^{n}w_{ij}^{L_{ij}}\right) = \mathcal{E}\left[\mathcal{E}\left(\prod_{i=1}^{N}\prod_{j=1}^{n}w_{ij}^{L_{ij}}\Big|\mathbf{K}\right)\right]$$

$$= \mathcal{E}\left[\prod_{i=1}^{N}\mathcal{E}\left(\prod_{j=1}^{n}w_{ij}^{L_{ij}}\Big|K_i\right)\right]$$

$$= \mathcal{E}\left[\prod_{i=1}^{N}\left(\sum_{j=1}^{n}r_{ij}w_{ij}\right)^{K_i}\right]$$

$$= \Phi(\boldsymbol{\xi}^\star) \ , \tag{9.233a}$$

where

$$\xi_i^\star = \sum_{j=1}^{n}r_{ij}w_{ij} \ . \tag{9.233b}$$

Let M_j be the total number of A_j alleles in the offspring. We can deduce the joint probability-generating function, $\phi(\zeta)$, of the M_j by setting $w_{ij} = \zeta_j$ in (9.233):

$$\phi(\zeta) = \Phi(\xi^{**}), \qquad \xi_i^{**} = \sum_{j=1}^{n} r_{ij}\zeta_j . \tag{9.234}$$

To apply (9.234) to our problem, we change our enumeration of the parents. We designate by ν_{ij}, where $1 \leq i \leq j \leq n$, the number of unordered $A_i A_j$ parents and assign the dummy variable $z_{ij,k}$, where $1 \leq i \leq j \leq n$ and $1 \leq k \leq \nu_{ij}$, to the kth $A_i A_j$ parent. Then with any fixed, convenient ordering of the N variables $z_{i,jk}$, (9.234) becomes

$$\phi(\zeta) = \Phi(\mathbf{z}), \qquad z_{ij,k} = \tfrac{1}{2}(\zeta_i + \zeta_j) . \tag{9.235}$$

We calculate the first two conditional moments of the gene frequencies by differentiating (9.235). We have

$$\frac{\partial \phi}{\partial \zeta_i} = \tfrac{1}{2}\sum_{k=i}^{n}\sum_{l=1}^{\nu_{ik}} \frac{\partial \phi}{\partial z_{ik,l}} + \tfrac{1}{2}\sum_{k=1}^{i}\sum_{l=1}^{\nu_{ki}} \frac{\partial \phi}{\partial z_{ki,l}} . \tag{9.236}$$

Successive utilization of (9.223a), (9.225),

$$\nu_{ik} = NP_{ik} , \tag{9.237}$$

and (9.145) in (9.236) yields

$$\mathcal{E}(2N'p_i'|\mathbf{P}) = \frac{\partial \phi}{\partial \zeta_i}(\mathbf{u}) = 2N'p_i , \tag{9.238}$$

and hence

$$\mathcal{E}(p_i'|\mathbf{P}) = p_i , \tag{9.239}$$

as is obvious by exchangeability.

Differentiating (9.236) and using (9.223b), (9.223c), (9.237), and (9.145) tells us

$$\frac{\partial^2 \phi}{\partial \zeta_i \partial \zeta_j}(\mathbf{u}) = N^2 p_i p_j \frac{\partial^2 \Phi}{\partial \xi_1 \partial \xi_2}(\mathbf{u}) + \tfrac{1}{2}N(\delta_{ij}p_i + \tilde{P}_{ij})\left[\frac{\partial^2 \Phi}{\partial \xi_1^2}(\mathbf{u}) - \frac{\partial^2 \Phi}{\partial \xi_1 \partial \xi_2}(\mathbf{u})\right] , \tag{9.240}$$

where $\tilde{P}_{ij}$ is the ordered genotypic frequency defined by (9.153). The conditional covariances are given by

$$(2N')^2 \operatorname{Cov}(p_i', p_j'|\mathbf{P}) = \frac{\partial^2 \phi}{\partial \zeta_i \partial \zeta_j}(\mathbf{u}) - \frac{\partial \phi}{\partial \zeta_i}(\mathbf{u})\frac{\partial \phi}{\partial \zeta_j}(\mathbf{u}) + \delta_{ij}\frac{\partial \phi}{\partial \zeta_i}(\mathbf{u}) . \tag{9.241}$$

Inserting (9.238) and (9.240) into (9.241), then substituting (9.228) and (9.231), recalling (9.227), and writing the genotypic frequencies in terms of the parental deviations from Hardy-Weinberg proportions,

$$\tilde{Q}_{ij} = \tilde{P}_{ij} - p_i p_j , \tag{9.242}$$

we find

$$\text{Cov}\,(p'_i, p'_j|\mathbf{P}) = \frac{p_i(\delta_{ij} - p_j)}{4N'} \left\{ 1 + \frac{N\sigma^2}{\mu(N-1)} \right.$$

$$\left. - \left[1 - \frac{N\sigma^2}{\mu(N-1)} \right] \frac{\tilde{Q}_{ij}}{p_i(\delta_{ij} - p_j)} \right\} \,. \qquad (9.243)$$

Now, for our basic multinomial neutral model of random drift,

$$\text{Cov}\,(p'_i, p'_j|\mathbf{P}) = (2N)^{-1} p_i(\delta_{ij} - p_j) \,,$$

which suggests that we define the variance effective population number, $N_e^{(v)}$, through

$$\text{Cov}\,(p'_i, p'_j|\mathbf{P}) = (2N_e^{(v)})^{-1} p_i(\delta_{ij} - p_j) \,. \qquad (9.244)$$

Comparing (9.243) with (9.244), we obtain

$$N_e^{(v)} = 2N_t \left\{ 1 + \frac{N_{t-1}\sigma^2}{\mu(N_{t-1} - 1)} - \left[1 - \frac{N_{t-1}\sigma^2}{\mu(N_{t-1} - 1)} \right] \frac{\tilde{Q}_{ij}}{p_i(\delta_{ij} - p_j)} \right\}^{-1} \,. \qquad (9.245)$$

Kimura and Crow (1963a) derived the diallelic special case of (9.245) by a different method. With the approximation $N_{t-1}/(N_{t-1} - 1) \approx 1$ for large population number, their formula agrees with Crow's (Crow, 1954; Crow and Morton, 1955). Ethier and Nagylaki (1980) established (9.245) for constant population size and multiple alleles in a particular panmictic model. See also Crow and Denniston (1988).

In the form (9.245), the dependence of the variance effective population number on the parental genotypic frequencies precludes its dynamical usefulness. For multinomial sampling of gametes, however, this dependence disappears. Then the joint probability-generating function of the gametic distribution is

$$\Phi(\boldsymbol{\xi}) = \left(\frac{1}{N_{t-1}} \sum_{i=1}^{N_{t-1}} \xi_i \right)^{2N_t} \,, \qquad (9.246)$$

whence (9.224) confirms that the number of gametes from any given individual is binomially distributed:

$$\psi(\xi) = (1 - N_{t-1}^{-1} + N_{t-1}^{-1}\xi)^{2N_t} \,. \qquad (9.247)$$

With the aid of (9.235), (9.237), and (9.145), we deduce from (9.246) that genes are also sampled multinomially:

$$\phi(\boldsymbol{\zeta}) = \left(\sum_{i=1}^{n} p_i \zeta_i \right)^{2N_t} \,. \qquad (9.248)$$

In view of (9.247), we again have (9.131), and hence (9.245) reduces to $N_e^{(v)} = N_t$ (Kimura and Crow, 1963a). Thus, as is reasonable from considerations of allelic identity *versus* gene-frequency variation, in the ideal case the inbreeding effective number is the number of parents, whereas the variance effective number equals the number of offspring.

Large panmictic populations will be close to Hardy-Weinberg proportions, so that $\tilde{Q}_{ij} \approx 0$. In our present notation, (3.40) becomes

$$\mathcal{E}(\tilde{Q}_{ij}|\mathbf{p}) = -(2N_{t-1} - 1)^{-1}p_i(\delta_{ij} - p_j) \ . \tag{9.249}$$

With random mating and large population number, following the approximation used by Kimura and Crow (1963a) for two alleles, we may replace $\tilde{Q}_{ij}$ by (9.249) in (9.245):

$$N_e^{(v)} = \frac{\mu(2N_{t-1} - 1)}{2(1 + \mu^{-1}\sigma^2)} \ . \tag{9.250}$$

Consult Pollak (1977) for a different derivation of (9.250) in the diallelic case.

With no reproductive variance, we have $\sigma^2 = 0$, and (9.250) simplifies to $N_e^{(v)} = 2N_t - \frac{\mu}{2}$. Thus, in large populations $N_e^{(v)} \approx 2N_t$. Since $N_e^{(v)} = N_t$ for multinomial sampling, about one-half of the variance in gene frequencies in an ideal population comes from Mendelian segregation and the rest from variation in the number of progeny.

The variance effective population number (9.250) is clearly a decreasing function of the reproductive variance. The empirical result $\sigma^2 > \mu$ (Crow and Morton, 1955) implies that $N_e^{(v)} < N_t - \frac{\mu}{4} < N_t$. Below (9.129) we saw that $\sigma^2 \leq \mu^2(N_{t-1} - 1)$. Inserting this into (9.250) informs us

$$N_e^{(v)} \geq 1 + \frac{\mu - 2}{2[1 + \mu(N_{t-1} - 1)]} \ ;$$

noting that $\mu \geq 2/N_{t-1}$ simplifies this to

$$N_e^{(v)} \geq 1 - \frac{N_{t-1} - 1}{3N_{t-1} - 2} > \tfrac{2}{3} \ .$$

Solving (9.129) for N_{t-1} and substituting into (9.250) leads to (Pollak, 1977)

$$N_e^{(v)} = N_e^{(i)} + \frac{\mu - 2}{1 + \mu^{-1}\sigma^2}(N_e^{(i)} - \tfrac{1}{2}) \ . \tag{9.251}$$

But below (9.129) we proved that $N_e^{(i)} \geq 1$. Therefore, (9.251) demonstrates that $N_e^{(v)} > N_e^{(i)}$ in an increasing population ($\mu > 2$), $N_e^{(v)} < N_e^{(i)}$ in a decreasing population ($\mu < 2$), and $N_e^{(v)} = N_e^{(i)}$ if the population number is constant ($\mu = 2$).

For a population of constant size N, (9.250) reduces to

$$N_e^{(v)} = \frac{2(2N - 1)}{\sigma^2 + 2} \ , \tag{9.252}$$

Wright's (1938, 1939) formula for $N_e^{(i)}$. By a different method, Haldane (1939) derived the approximation

$$N_e^{(v)} \approx \frac{4N}{\sigma^2 + 2} \tag{9.253}$$

for a large population.

The calculation of the variance effective population number for *dioecious populations* is much more difficult than for monoecious ones. Assuming that the population is large and genes in the parents are combined at random, Pollak (1977) found for a diallelic autosomal locus that $N_e^{(v)}$ is given by the inbreeding effective number (9.139) with t replaced by $t+1$. The approximation

$$N_e^{(v)} \approx \frac{4N^2}{N_1(\sigma_1^2 + \mu_1^2 - \mu_1) + N_2(\sigma_2^2 + \mu_2^2 - \mu_2)} \tag{9.254}$$

($\mu_j = N/N_j$) for constant population size had been derived earlier by Moran and Watterson (1959) and Latter (1959). It can also be extracted from Feldman's (1966) analysis of a model of random mating; Ethier and Nagylaki (1980) validated (9.254) in the multiallelic generalization of Feldman's model. See also Crow and Denniston (1988).

Moran and Watterson (1959) established also the formula corresponding to (9.254),

$$N_e^{(v)} \approx \frac{9N^2}{2[N_1(\sigma_1^2 + \mu_1^2 - \mu_1) + 2N_2(\sigma_2^2 + \mu_2^2 - \mu_2)]} \, , \tag{9.255}$$

for a diallelic X-linked locus. This was verified by Ethier and Nagylaki (1980) in a multiallelic model. With multinomial sampling, (9.140) applies, and consequently (9.255) is approximated for large populations by (Wright, 1939)

$$N_e^{(v)} \approx \frac{9N_1 N_2}{2(2N_1 + N_2)} \, . \tag{9.256}$$

The right-hand side of (9.256) cannot exceed $\frac{1}{2}N_1 + N_2$; equality occurs if and only if $N_1 = N_2 = \frac{1}{2}N$, in which case $N_e^{(v)} \approx \frac{3}{4}N$. In the limits $N_1 \ll N_2$ and $N_1 \gg N_2$, (9.256) reduces to $\frac{9}{2}N_1$ and $\frac{9}{4}N_2$, respectively.

9.14 Problems

9.1   Calculate F_I for the pedigree in Fig. 9.18. All inbreeding is shown.

9.2   Is the pedigree in Fig. 9.19 possible for a dioecious organism? A monoecious one? Why?

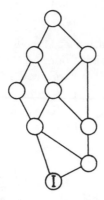

Fig. 9.18. A pedigree for computing the inbreeding coefficient.

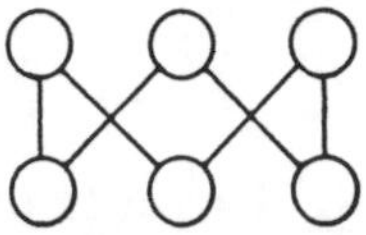

Fig. 9.19. Can this pedigree be sex-labelled?

9.3   If an ancestor who has at least one ancestor in the pedigree is assigned an inbreeding coefficient, recourse must be had to basic principles and the information may not suffice for the computation. Establish the following results for the pedigrees in Fig. 9.20.

(a) $F_I = \frac{1}{4}F_B + \frac{1}{8}(1 + F_A)$.

Note that $F_B > 0$ implies that A is not the only common ancestor of I.

(b) $F_I = \frac{1}{4}[F_B + F_C + \frac{1}{2}(1 + F_A) + \mathcal{P}(a \equiv b)]$.

(c) $F_I = \frac{1}{2}\mathcal{P}(a \equiv b) + \frac{1}{16}(1 + F_A)$.

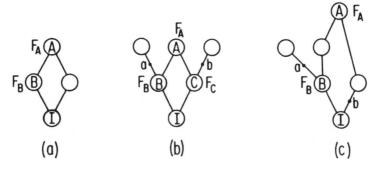

Fig. 9.20. Inbred ancestors with ancestors in the pedigree.

9.4   Calculate F_I for an X-linked locus for Fig. 9.21.

9.5   Evaluate the k-coefficients for the individuals I and J in Fig. 9.22.

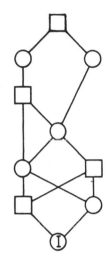

Fig. 9.21. A pedigree for calculating the inbreeding coefficient with X-linkage.

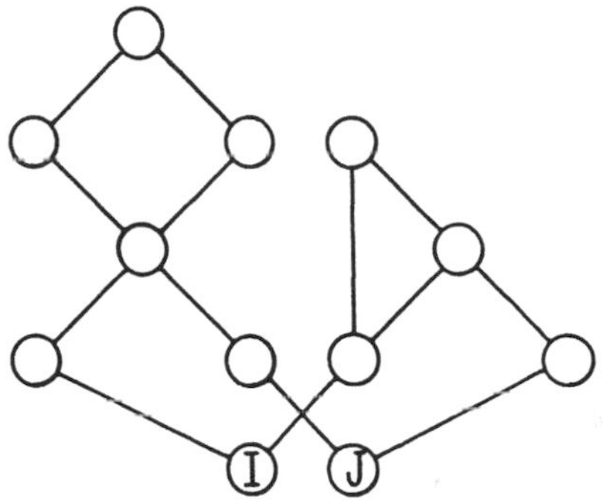

Fig. 9.22. A pedigree for calculating the k-coefficients.

9.6   Calculate S_{RR}, Q_{RD}, and $b(q)$ for double first cousins.

9.7   Use the numerical values in Sect. 9.3 to compute the probability, $c(q)$, that a schizophrenic individual has a schizophrenic niece. Calculate also the probability, $d_2(q)$, that a normal individual with a schizophrenic uncle is heterozygous.

9.8   Let an individual in generation t have n_τ ancestors in generation $t - \tau$ and define $n_\tau = 0$ for $\tau < 0$. Show that for half-first-cousin mating, displayed in Fig. 9.15,

$$n_\tau = 2n_{\tau-1} - n_{\tau-3} , \qquad \tau \geq 1 .$$

Deduce that for $\tau \geq 0$

$$n_\tau = -1 + \tfrac{1}{\sqrt{5}}[(\sqrt{5} + 2)\lambda_+^\tau + (\sqrt{5} - 2)\lambda_-^\tau]$$
$$\approx -1 + 1.89(1.62)^\tau , \qquad \tau \gg 1 ,$$

where $\lambda_\pm = \tfrac{1}{2}(1 \pm \sqrt{5})$.

9.9 Suppose a male S, with homozygosity f_S, mates successively with his daughter, granddaughter, etc., as exhibited in Fig. 9.23. Prove that for $t \geq 0$

$$f_{t+1} = \hat{f} + (g_0 - \hat{f})(\tfrac{1}{2})^t \,, \tag{9.257}$$

with $\hat{f} = \tfrac{1}{2}(1 + f_S)$. Note that by setting $f_S = 1$ we obtain from (9.257) the heterozygosity

$$h_{t+1} = (1 - g_0)(\tfrac{1}{2})^t \tag{9.258}$$

for a homozygous male or an X-linked locus. These results are due to Jennings (1916).

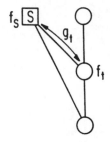

Fig. 9.23. Repeated backcrossing.

9.10 Demonstrate that the alternating parent-offspring mating system in Fig. 9.24 leads to the recursion relation (9.44) for sib mating. This system was studied numerically by Jennings (1916) and solved analytically by Robbins (1918).

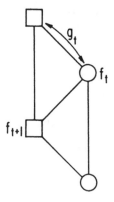

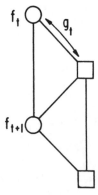

Fig. 9.24. Mating with the younger parent.

Fig. 9.25. Mating with the younger parent for an X-linked locus.

9.11 Relabel Fig. 9.24 for an X-linked locus as shown in Fig. 9.25. Prove that (9.258) holds.

9.12 Establish that sib mating for an X-linked locus leads to the same recursion relation, (9.44), as for an autosomal one. We owe the numerical investigation to Jennings (1916) and the analytic work to Robbins (1918).

9.13 Figure 9.26 displays the mating of a male with his two half sisters, full sisters of each other. Show that the ultimate rate of approach to homozygosis is given by the dominant root of (Wright, 1921; 1969, p. 189)

$$16\lambda^3 - 8\lambda^2 - 4\lambda - 1 = 0 \ . \tag{9.259}$$

This root is approximately 0.870, between the values for full and half-sibs, as expected.

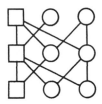

Fig. 9.26. Mating of a male with his two half sisters, full sisters of each other.

9.14 Prove that (9.259) holds also for an X-linked locus. Demonstrate that with the sexes interchanged in Fig. 9.26, the recursion relations are essentially the same as for sib mating.

9.15 Establish that (9.48) applies to circular mating of three individuals (Kimura and Crow, 1963).

9.16 The method of mating types (Pearl, 1914a; Robbins, 1917; Bartlett and Haldane, 1934; Fisher, 1949; Kempthorne, 1957, Chap. 6; Karlin, 1968) yields more information, with more labor, than techniques based on probabilities of identity. Mating types are essential for the inclusion of selection in regular systems of inbreeding (Karlin, 1968; Wright, 1969, Chap. 10).

As a simple example, consider sib mating at a diallelic autosomal locus without selection. With alleles A and a, let x_i, where $i = 1, 2, \ldots, 6$, represent the probabilities that the sibship is $AA \times AA$, $AA \times Aa$, $AA \times aa$, $Aa \times Aa$, $Aa \times aa$, and $aa \times aa$, respectively. (If there were infinitely many lines, the x_i would be the frequencies of the six mating types.) Show that

$$x_1' = x_1 + \tfrac{1}{4}x_2 + \tfrac{1}{16}x_4 \ , \tag{9.260a}$$

$$x_2' = \tfrac{1}{2}x_2 + \tfrac{1}{4}x_4 \ , \tag{9.260b}$$

$$x_3' = \tfrac{1}{8}x_4 \ , \tag{9.260c}$$

$$x_4' = \tfrac{1}{4}x_2 + x_3 + \tfrac{1}{4}x_4 + \tfrac{1}{4}x_5 \ , \tag{9.260d}$$

$$x_5' = \tfrac{1}{4}x_4 + \tfrac{1}{2}x_5 \ , \tag{9.260e}$$

$$x_6' = \tfrac{1}{16}x_4 + \tfrac{1}{4}x_5 + x_6 \ . \tag{9.260f}$$

Deduce from (9.260) that the expected frequency of A,

$$\bar{p} = x_1 + \tfrac{3}{4}x_2 + \tfrac{1}{2}x_3 + \tfrac{1}{2}x_4 + \tfrac{1}{4}x_5 \ ,$$

is conserved and the expected heterozygosity,

$$h = \tfrac{1}{2}x_2 + x_4 + \tfrac{1}{2}x_5 \ ,$$

satisfies

$$4h'' - 2h' - h = 0 \ . \tag{9.261}$$

Prove also that (9.261) follows from (9.45).

We owe (9.260) to Robbins (1917); a special case is due to Pearl (1914a). Consult Schäfer (1936) and Haldane (1937, 1955) for other results.

9.17 Investigate mixed selfing and panmixia in an infinite population for two loci. Let r and c denote the proportion of the population produced by selfing and the recombination frequency, respectively. Assume that $0 < r < 1$ and $0 < c \le \tfrac{1}{2}$.

(a) Designate by p_i and q_j the frequencies of the alleles A_i and B_j. Let P_{ij} represent the probability that a chromosome chosen at random carries A_i and B_j, Q_{ij} represent the probability that a chromosome chosen at random carries A_i and the other chromosome in the same individual carries B_j, and R_{ij} represent the probability that a chromosome chosen at random carries A_i and another chromosome chosen at random from another individual carries B_j. Show that (Kimura, 1963)

$$P_{ij}' = (1 - c)P_{ij} + cQ_{ij} \ , \tag{9.262a}$$

$$Q_{ij}' = r\tfrac{1}{2}(P_{ij} + Q_{ij}) + (1 - r)R_{ij} \ , \tag{9.262b}$$

$$R_{ij}' = R_{ij} \ . \tag{9.262c}$$

Infer that (Kimura, 1963) $R_{ij}(t) = p_i q_j$, and $P_{ij}(t) \to p_i q_j$ and $Q_{ij}(t) \to p_i q_j$ as $t \to \infty$. Thus, *genes* at different loci are stochastically independent at equilibrium.

(b) In spite of the last result, *genotypes* at different loci are mutually dependent at equilibrium. To see this, examine the evolution of the probabilities of identity by descent. Let F and Φ signify the inbreeding coefficient and the probability that an individual chosen at random is autozygous at both loci. Demonstrate that (Narain, 1966)

$$F' = r\tfrac{1}{2}(1 + F) \, , \tag{9.263a}$$

$$\Phi' = r\{[(1-c)^2 + c^2]\tfrac{1}{2}(1+\Phi) + 2c(1-c)F\} \, . \tag{9.263b}$$

Conclude that (Narain, 1966)

$$F(t) \to \frac{r}{2-r}, \qquad \Phi(t) \to \frac{r[2b + r(2 - 3b)]}{(2 - r)(2 - rb)} \tag{9.264}$$

as $t \to \infty$, where $b = (1-c)^2 + c^2$. Measuring the association between the two loci by $\Delta = \Phi - F^2$, we find (Weir and Cockerham, 1973)

$$\Delta(t) \to \hat{\Delta} = \frac{2r(1-r)[r + 2b(1-r)]}{(2-r)^2(2-rb)} \tag{9.265}$$

as $t \to \infty$. As is intuitively reasonable, (9.265) reveals that the association between the loci weakens as the recombination frequency increases, and hence $\hat{\Delta}$ must be no less than the value for unlinked loci:

$$\hat{\Delta} \ge \frac{4r(1-r)}{(2-r)^2(4-r)} > 0 \, . \tag{9.266}$$

(c) Assume that the population is initially panmictic and posit that $P_{ij} = Q_{ij} = R_{ij} = p_i q_j$. According to Part a, linkage equilibrium in this form holds for all time if it does so initially, and it always holds at equilibrium. Show that the frequency, $T_{ij,kl}$, of the ordered genotype $A_i B_j / A_k B_l$ is given by the general formula (Haldane, 1949; Cockerham and Weir, 1973; Weir and Cockerham, 1973)

$$T_{ij,kl} = \Phi p_i q_j \delta_{ik} \delta_{jl} + (F - \Phi) p_i q_j (q_l \delta_{ik} + p_k \delta_{jl})$$
$$+ (1 + \Phi - 2F) p_i q_j p_k q_l \tag{9.267a}$$
$$= [F p_i \delta_{ik} + (1 - F) p_i p_k][F q_j \delta_{jl} + (1 - F) q_j q_l]$$
$$+ \Delta p_i q_j (\delta_{ik} - p_k)(\delta_{jl} - q_l) \, . \tag{9.267b}$$

Bennett and Binet (1956; there should be no $p(a)$ in the penultimate equation on p. 53) and Kimura (1958a) presented the equilibrium genotypic frequencies for partial selfing with two and multiple alleles, respectively.

9.18 A population of N monoecious individuals selfs for $T - 1$ generations, then mates at random, including selfing with probability $1/N$; it repeats this process every T generations. Prove that the heterozygosity satisfies

$$h_t = h_T \mu^{-1} \lambda^t \tag{9.268a}$$

for $t = nT$, where $n = 1, 2, \ldots$ and

$$\mu = 1 - N^{-1}[1 - (\tfrac{1}{2})^T], \qquad \lambda = \mu^{1/T} . \tag{9.268b}$$

Observe that by choosing T sufficiently large, we can make λ arbitrarily close to one, and hence the rate of decay of heterozygosity arbitrarily slow. The interpretation is the same as the one below (9.79) for circular mating; by relaxing the requirement that the mating system have a period of one generation, we obtain here even slower rates of convergence. The above model is a modification of one due to Cockerham (1970).

9.19 Calculate the inbreeding effective population number for an X-linked locus as follows (Nagylaki, 1981). Assume that a population of N_1 males and N_2 females mates at random and let P_1 (P_2) denote the probability that two distinct, homologous, randomly chosen paternal (maternal) genes in generation t come from the same individual in generation $t - 1$. Designate by f_t the probability that a female chosen at random in generation t is homozygous; g_t signifies the probability that two homologous genes, one chosen at random from a male and one from a female, are the same allele. Let k_t represent the probability that two homologous genes chosen at random from distinct males are the same allele; l_t is the corresponding probability for two females.

(a) Establish that (Karlin, 1968; Nagylaki, 1981)

$$f' = g , \tag{9.269a}$$

$$g' = \tfrac{1}{2}g + \tfrac{1}{2}[P_2\tfrac{1}{2}(1 + f) + (1 - P_2)l] , \tag{9.269b}$$

$$k' = P_2\tfrac{1}{2}(1 + f) + (1 - P_2)l , \tag{9.269c}$$

$$l' = \tfrac{1}{4}[P_1 + (1 - P_1)k] + \tfrac{1}{4}[P_2\tfrac{1}{2}(1 + f) + (1 - P_2)l] + \tfrac{1}{2}g . \tag{9.269d}$$

(b) Show that the eigenvalues that control the rate of decay of genetic variability are zero and the three roots of

$$8\lambda^3 - 2(2 + a)\lambda^2 - (2 - a + 2b)\lambda + b = 0 , \tag{9.270a}$$

where

$$a = 1 - P_2, \qquad b = (1 - P_1)(1 - P_2) . \tag{9.270b}$$

(c) Writing the dominant root of (9.270) as

$$\lambda_0 = 1 - (2N_e^{(i)})^{-1} + O(P_1^2 + P_2^2) \tag{9.271}$$

for a large population ($P_1 \to 0$ and $P_2 \to 0$), deduce

$$N_e^{(i)} = \frac{9}{2(P_1 + 2P_2)} . \tag{9.272}$$

(d) Conclude that, in the notation of (9.136) and (9.138),

$$N_e^{(i)} = \frac{9N(N-1)}{2[N_1(\sigma_1^2 + \mu_1^2 - \mu_1) + 2N_2(\sigma_2^2 + \mu_2^2 - \mu_2)]} , \qquad (9.273)$$

in agreement with the asymptotic form (9.255) of $N_e^{(v)}$ for $N \gg 1$.

(e) Demonstrate that in the ideal case of reproduction by sampling with replacement, (9.273) reduces to (9.256). Wright (1933a), Malécot (1951), and Kimura (1963) used path coefficients, identity by descent, and identity in state, respectively, to derive our last result.

9.20 A mating system has a period of one generation, and mating occurs only within generations. Every allele mutates to a novel type at rate u. In the notation of (9.82) to (9.85), prove that the probabilities of allelic nonidentity converge to

$$\hat{z} = 2u(I - B)^{-1}V + O(u^2) , \qquad (9.274)$$

where V is the vector of ones (cf. Boucher and Nagylaki, 1988). As an example, deduce from (9.274) that if mutation is introduced into Problem 9.13, the heterozygosity at equilibrium is very close to $52u/3$.

9.21 Show that if mutation to new alleles at rate u is incorporated into the three models of Sect. 9.8, then all the results of Sect. 9.7 remain valid, provided we replace N by $N_e^{(i)}$. In particular, the homozygosity at equilibrium reads (Kimura and Crow, 1964; Crow and Kimura, 1970, pp. 322–327)

$$\hat{f} = \frac{1 + O(u)}{1 + 4N_e^{(i)}u} . \qquad (9.275)$$

9.22 Include mutation to novel alleles at rate u in Problem 9.19. Demonstrate that if $u \ll 1$, all the probabilities of identity converge to (9.275) at the approximate rate (assuming also that $N \gg 1$)

$$\exp\{-[2u + (2N_e^{(i)})^{-1}]t\} . \qquad (9.276)$$

For multinomial sampling and identity by descent, these results are due to Malécot (1951); they were generalized by Nagylaki (1981).

9.23 A monoecious population mates entirely at random, including selfing. Every allele mutates at total rate u and is equally likely to mutate to any one of the $k - 1$ ($k \geq 2$) other alleles.

(a) Show that after one generation the homozygosity satisfies the recursion relation

$$f_{t+1} = b + c(2N_e^{(i)})^{-1} + c[1 - (2N_e^{(i)})^{-1}]f_t , \qquad (9.277a)$$

where

$$b = \frac{u(2k - 2 - ku)}{(k-1)^2}, \qquad c = \left(\frac{k - 1 - ku}{k - 1}\right)^2 . \qquad (9.277b)$$

(b) Infer that if $u \ll 1$, the heterozygosity converges to (Kimura, 1968a; Crow and Kimura, 1970, pp. 324–325)

$$\hat{h} = 4N_e^{(i)}u \left(1 + \frac{4N_e^{(i)}uk}{k - 1}\right)^{-1} [1 + O(u)] \qquad (9.278)$$

at the approximate rate (assuming also that $N \gg 1$)

$$\exp\left[-\left(\frac{2uk}{k - 1} + \frac{1}{2N_e^{(i)}}\right)t\right] . \qquad (9.279)$$

Note the correct reduction to the model of infinitely many alleles as $k \to \infty$.

9.24 A time-homogeneous Markov chain has states $0, 1, \ldots, M$ and transition probabilities T_{ij}. There are two absorbing states, 0 and M; v_i (> 0) denotes the probability of ultimately reaching M, starting at i ($1 \leq i \leq M$).

(a) Prove that the process conditioned on fixation at M is a Markov chain with transition probabilities (Ewens, 1973)

$$T_{ij}^\star = T_{ij}v_j/v_i, \qquad 1 \leq i, j \leq M . \qquad (9.280)$$

(b) For the diallelic case of the pure random-drift model of Sect. 9.10, let i and j represent the numbers of A_1 genes in successive generations and set $x = i/(2N)$. Establish that (Ewens, 1979, p. 76)

$$T_{ij}^\star = \binom{2N - 1}{j - 1} x^{j-1}(1 - x)^{2N-j}, \qquad 1 \leq i, j \leq 2N . \qquad (9.281)$$

(c) Designate the frequency of A_1 in generation t by X_t and assume $X_0 = p$ and $q = 1 - p$. Show that the mean and variance of the conditional process read

$$\mathcal{E}(X_t) = 1 - q\kappa^t , \qquad (9.282a)$$
$$\text{Var}(X_t) = q[(q - \tfrac{1}{2})\mu^t + \tfrac{1}{2}\kappa^t - q\kappa^{2t}] , \qquad (9.282b)$$

where

$$\kappa = 1 - (2N)^{-1}, \qquad \mu = \kappa(1 - N^{-1}) . \qquad (9.282c)$$

Observe that the mean tends to one and the variance approaches zero at the asymptotic rate κ^t.

9.25 A dioecious model corresponding to the monoecious one of Sect. 9.10 may be constructed as follows (Moran, 1958; 1962, pp. 144–152; Ethier and Nagylaki, 1980). Let N_s, $P_{ij}^{(s)}$, and $p_i^{(s)}$ denote the number of individuals, the unordered frequency of $A_i A_j$ ($i \leq j$), and the frequency of A_i, respectively, in sex s ($s = 1, 2$) just before reproduction. To specify the transition probabilities, we suppose that, given $\mathbf{P}^{(1)}$ and $\mathbf{P}^{(2)}$, the genotypic frequencies $\mathbf{P}^{(1)\prime}$ in males and $\mathbf{P}^{(2)\prime}$ in females in the next generation are independent. The genotypic numbers $N_s P_{ij}^{(s)\prime}$ in each sex have a multinomial distribution with index N_s and parameters

$$\tfrac{1}{2}(2 - \delta_{ij})(p_i^{(1)} p_j^{(2)} + p_j^{(1)} p_i^{(2)}) \; . \tag{9.283}$$

(a) Prove that

$$\mathcal{E}[\tfrac{1}{2}(p_i^{(1)\prime} + p_i^{(2)\prime}) | \mathbf{p}^{(1)}, \mathbf{p}^{(2)}] = \tfrac{1}{2}(p_i^{(1)} + p_i^{(2)}) \; . \tag{9.284}$$

(b) Conclude that the unweighted average expectation $\tfrac{1}{2}(\overline{p}_i^{(1)} + \overline{p}_i^{(2)})$ is constant and equal to the probability of fixation of A_i.

9.26 Repeat Problem 9.25 for an X-linked locus. Let N_s, $P_{ij}^{(2)}$, and $p_i^{(s)}$ have the same meaning as above. The genotypic frequencies $\mathbf{p}^{(1)\prime}$ and $\mathbf{P}^{(2)\prime}$ are again conditionally independent, and $P_{ij}^{(2)\prime}$ has the same distribution as in the autosomal case. The male genotypic numbers $N_1 p_i^{(1)\prime}$ are multinomially distributed with index N_1 and parameters $p_i^{(2)}$ (Ethier and Nagylaki, 1980).

(a) Demonstrate that

$$\mathcal{E}(\tfrac{1}{3}p_i^{(1)\prime} + \tfrac{2}{3}p_i^{(2)\prime} | \mathbf{p}^{(1)}, \mathbf{p}^{(2)}) = \tfrac{1}{3}p_i^{(1)} + \tfrac{2}{3}p_i^{(2)} \; . \tag{9.285}$$

(b) Conclude that the average expectation $\tfrac{1}{3}\overline{p}_i^{(1)} + \tfrac{2}{3}\overline{p}_i^{(2)}$ is constant and equal to the fixation probability of A_i.

9.27 For the model of Sect. 9.11, show that the correlation between p_t and p_{t+1} is $\lambda(V_t/V_{t+1})^{1/2}$. Deduce the equilibrium value of this correlation in the three cases $u > 0$, $v > 0$; $u = 0$, $v > 0$; and $u = v = 0$.

9.28 In the island model of Sect. 9.12, assume that a fraction m of gametes, rather than zygotes, is replaced by migrants with gene frequencies q_i. Show that (9.215), (9.219a), (9.219b), (9.220), and (9.221) still hold, but now

$$\mu = (1 - m)^2 \left(1 - \frac{1}{2N} \right) \; ,$$

the denominator in (9.219c) is

$$1 + m(2N - 1) \; ,$$

and

$$\hat{V}_{ij} = \frac{q_i(\delta_{ij} - q_j)}{(1 - m)^2 + 2Nm(2 - m)} \ .$$

See Nagylaki (1979c) for the derivation of the diallelic special case.

10. Quantitative Genetics

Quantitative genetics is the study of inherited characters that exhibit continuous or almost continuous variation. Such traits are usually influenced by many loci and the environment. Since a very large proportion of anatomical, physiological, and behavioral characters are quantitative, the evolutionary, economic, and medical importance of inherited continuous variation is difficult to overestimate. Height, brain volume, hair and skin color, blood pressure, and various measures of personality and intelligence all fall into this category.

A biologically rigorous treatment of quantitative genetics must be based explicitly on the principles of Mendelism, recombination, and gene action. Since many loci are involved, such an approach usually leads to formidable mathematical problems. Evolutionary questions are particularly difficult; in the presence of epistasis or assortative mating, so is the analysis of genetic control by the estimation of variance components from the correlation between relatives. Hence, much of the literature is biologically or mathematically quite intuitive.

Weinberg (1909, 1910) was the first to attack the subject systematically from Mendelian principles. The majority of the results in this chapter were obtained by Fisher (1918) in one of the classic papers in population genetics. Moran and Smith (1966) and, in part, Kempthorne (1957, Chap. 22) have presented expositions of this notoriously difficult and occasionally intuitive paper. Wright (1921, 1922) used his path coefficients to derive many formulae with purely additive gene action. Malécot (1939) established rigorously almost all of Fisher's results in greater generality and later (Malécot, 1948) employed identity by descent to analyze random mating more elegantly.

We shall begin with the decomposition of the phenotypic variance under random mating and continue with the calculation of the correlation between relatives. In Sects. 10.3 and 10.4, we shall discuss the augmentation of the genetic variance and the correlation between relatives due to assortative mating. We shall examine the short-term effects of selection in Sect. 10.5. Mutation-selection balance is the subject of Sect. 10.6.

For topics not considered here, the reader should consult the books of Kempthorne (1957), Falconer (1960), Mather and Jinks (1971), and Bulmer (1980). The extensive literature on selection and mutation can be traced from Barton (1989), Bürger, Wagner, and Stettinger (1989), Gillespie and Turelli (1989), Gimelfarb (1989), Hastings and Hom (1989), Keightley and Hill

(1989), Nagylaki (1989a), Tachida and Cockerham (1989), Wagner (1989), Zeng (1989), Zeng, Tachida, and Cockerham (1989), Turelli and Barton (1990), Barton and Turelli (1991), and Lande (1991).

We shall assume throughout this chapter that the population is infinitely large and generations are discrete and nonoverlapping. Unless otherwise indicated, we shall not distinguish the sexes.

10.1 The Decomposition of the Variance with Panmixia

We consider a single character and denote the phenotypic, genotypic, and environmental values of each individual by Y, G, and E, respectively. These random variables are related by $Y = f(G, E)$, for some function f. We shall always suppose that there is no *genotype-environment interaction*, i.e., genotype and environment determine phenotype additively:

$$Y = G + E \ . \tag{10.1}$$

Observe that even if f is nonlinear, as long as the deviations of G and E from their respective mean values are small relative to the strength of the nonlinearity of f, Taylor's theorem reveals that phenotype is approximately a linear function of genotype and environment. Translating the origin and scaling then yield (10.1). The coefficient of variation, defined as the standard deviation divided by the mean, is often a useful measure of the amount of deviation from the mean. Sometimes a nonlinear function can be converted to a linear one by transformation of variables. For example, if $G, E > 0$ and β and γ are constants, then the relation $Y = G^\beta E^\gamma$, which may be more reasonable for some characters than the additive one, passes into the form (10.1) upon taking logarithms and redefining variables.

Empirically, (10.1) is generally satisfactory (Falconer, 1960, Chap. 8; Cavalli-Sforza and Bodmer, 1971, Chap. 9; Wright, 1978, Chap. 9; Bulmer, 1980, pp. 20–27). Notice that if $Y_1 = G_1 + E$, $Y_2 = G_2 + E$, and $G_1 > G_2$, then $Y_1 > Y_2$; thus, in the absence of genotype-environment interaction, the phenotypic ordering of genotypes is the same in all environments.

In (10.1), G and E have the same units as Y. Usually only the phenotype can be determined empirically, although at least a partial measurement of the environment is often possible. The genotypic values must be derived by analysis.

The expectation and variance of (10.1) yield

$$\overline{Y} = \overline{G} + \overline{E} \tag{10.2}$$

and

$$V_Y = V_G + V_E + 2\,\mathrm{Cov}\,(G, E) \ . \tag{10.3}$$

The terms in (10.3) represent the phenotypic, genotypic (or total genetic), and environmental variances and twice the covariance between heredity and environment. The *heritability in the broad sense*,

$$H^2 = V_G/V_Y \ , \tag{10.4}$$

is the proportion of the phenotypic variance due purely to genes. In general, for each trait it depends on the population and the environmental factors present. The proportion of the phenotypic variance that may be modified by manipulation of extant environmental factors within their existing range is $1 - H^2$. In a uniform environment ($V_E = 0$, $V_G > 0$), $H^2 = 1$; if the population is genetically homogeneous ($V_G = 0$, $V_E > 0$), then $H^2 = 0$.

In experiments, genotypes are usually randomized across environments, so the covariance term in (10.3) is zero. Even in natural populations, data can frequently be fitted by assuming that genotype and environment are uncorrelated (Cavalli-Sforza and Bodmer, 1971, Chap. 9; Wright, 1978, Chap. 9). With this simplification, two examples will illustrate the significance of the broad heritability.

Let σ_Y, σ_G, and σ_E designate the phenotypic, genotypic, and environmental standard deviations. Then (10.1), (10.3), and (10.4) inform us that improving the environment from conception by one standard deviation would raise the adult phenotype by $\sigma_E = (1 - H^2)^{1/2}\sigma_Y$.

Calculating the covariance of each side of (10.1) with G and taking into account that G and E are uncorrelated reveals that the correlation between genotype and phenotype is the square root of the broad heritability:

$$r_{GY} = H \ . \tag{10.5}$$

If the regression of genotype on phenotype is linear, then (10.5) allows us to calculate the expected genotypic deviation of an individual with phenotypic deviation Y:

$$\mathcal{E}(G|Y) = \overline{G} + (r_{GY}\sigma_G/\sigma_Y)(Y - \overline{Y}) \tag{10.6}$$

$$= \overline{G} + H^2(Y - \overline{Y}) \ . \tag{10.7}$$

If the regression of G on Y is nonlinear, then (10.7) is the best linear approximation to the true regression, in the sense of minimizing the mean squared error.

To analyze the genotypic variance in the absence of selection and mutation, we posit that the population has reached equilibrium under random mating and that there is no epistasis. Neglecting epistasis is often a good approximation, and the data frequently do not suffice to distinguish epistasis from dominance (Falconer, 1960, Chap. 8; Cavalli-Sforza and Bodmer, 1971, Chap. 9; Wright, 1978, Chap. 9). The decomposition of the variance with epistasis was carried out by Cockerham (1954) for two alleles at each locus and Kempthorne (1954; 1955; 1955a; 1957, pp. 413–419) in general. Even with epistasis, the decomposition is independent of the linkage map.

Denoting the contribution of the ith locus to the genotypic value by $G^{(i)}$, for any number of additive loci we have

$$G = \sum_i G^{(i)} \ . \tag{10.8}$$

In Sect. 3.3 we established that at equilibrium two-locus genotypic frequencies are products of single-locus genotypic frequencies, i.e., the loci are pairwise stochastically independent. Therefore,

$$\mathrm{Cov}\,(G^{(i)}, G^{(j)}) = 0\,, \qquad i \neq j \,, \tag{10.9}$$

and (10.8) yields

$$V_G = \sum_i V_G^{(i)} \,, \tag{10.10a}$$

where

$$V_G^{(i)} = \mathrm{Var}\,(G^{(i)}) \tag{10.10b}$$

represents the contribution of locus i to the total genetic variance.

We showed in Sect. 4.10 that the genotypic variance at a single locus is the sum of the genic and dominance variances; here we need only specialize (4.179) through (4.189) to Hardy-Weinberg proportions. Let the allele $A_j^{(i)}$ at locus i have frequency $p_j^{(i)}$ and designate the genotypic contribution of $A_j^{(i)} A_k^{(i)}$ by $G_{jk}^{(i)}$. Then the mean contribution of locus i reads

$$\overline{G}^{(i)} = \sum_{jk} G_{jk}^{(i)} p_j^{(i)} p_k^{(i)} \,. \tag{10.11}$$

We write the deviation of $G_{jk}^{(i)}$ from the mean as

$$g_{jk}^{(i)} = G_{jk}^{(i)} - \overline{G}^{(i)} = \alpha_j^{(i)} + \alpha_k^{(i)} + d_{jk}^{(i)} \,, \tag{10.12}$$

in which $\alpha_j^{(i)}$ and $d_{jk}^{(i)}$ signify the average effect of $A_j^{(i)}$ and the dominance deviation of $A_j^{(i)} A_k^{(i)}$, respectively. The average effect of $A_j^{(i)}$ equals the average excess, $g_j^{(i)}$, of $A_j^{(i)}$,

$$\alpha_j^{(i)} = g_j^{(i)} = \sum_k g_{jk}^{(i)} p_k^{(i)} \,, \tag{10.13}$$

and the means

$$\sum_j \alpha_j^{(i)} p_j^{(i)} = 0 \,, \tag{10.14a}$$

$$\sum_k d_{jk}^{(i)} p_k^{(i)} = 0 \tag{10.14b}$$

vanish. The genotypic, additive, and dominance variances due to locus i are given by

$$V_G^{(i)} = \sum_{jk} (g_{jk}^{(i)})^2 p_j^{(i)} p_k^{(i)} \,, \tag{10.15a}$$

$$V_A^{(i)} = 2 \sum_j (\alpha_j^{(i)})^2 p_j^{(i)} \,, \tag{10.15b}$$

$$V_D^{(i)} = \sum_{jk} (d_{jk}^{(i)})^2 p_j^{(i)} p_k^{(i)} \,, \tag{10.15c}$$

and these satisfy

$$V_G^{(i)} = V_A^{(i)} + V_D^{(i)} \ . \tag{10.16}$$

Inserting (10.16) into (10.10) produces the decomposition of the genotypic variance for the character (Fisher, 1918):

$$V_G = V_A + V_D \ , \tag{10.17a}$$

with

$$V_A = \sum_i V_A^{(i)} , \qquad V_D = \sum_i V_D^{(i)} \ . \tag{10.17b}$$

The dominance variance, V_D, is typically less than the genic variance, V_A, but for many traits V_D/V_A is sufficiently large that neglecting the dominance variance provides only a rough approximation (Falconer, 1960, Chap. 8; Cavalli-Sforza and Bodmer, 1971, Chap. 9; Wright, 1978, Chap. 9). Substituting (10.17) into (10.3) gives the partition

$$V_Y = V_A + V_D + V_E + 2\,\mathrm{Cov}\,(G,E) \tag{10.18}$$

of the phenotypic variance.

The proportions of the phenotypic variance due to purely additive gene action and dominance are

$$h^2 = V_A/V_Y \ , \qquad D^2 = V_D/V_Y \ ; \tag{10.19}$$

h^2 is called the *heritability in the narrow sense* or simply *heritability*. Like the broad heritability, for each trait it generally depends on the population and the environmental factors present. In Sect. 10.5 we shall see that the narrow heritability enables us to calculate approximately the response to a generation of selection. Values of the narrow heritability of sundry characters in various species range from essentially zero to almost one (Falconer, 1960, Chaps. 8, 10; Cavalli-Sforza, 1971, Chap. 9; Wright, 1978, Chaps. 8, 9).

Before concluding this section, we extend our results to an *inbred population*. We assume that there is no dominance and initially Hardy-Weinberg proportions obtain at all loci. In the second part of Sect. 9.4, we investigated the effect of inbreeding on a single locus. We saw that inbreeding does not alter the panmictic value, $\overline{G}^{(i)}$, of the mean contribution of locus i; summing over loci shows that the average genotypic value, $\overline{G}$, is unaffected by inbreeding.

The calculation of the variance for random mating does not generalize directly, because, as Problem 9.17 demonstrates, the *genotypes* at different loci may be mutually dependent. Using a tilde for maternal alleles and none for paternal ones and recalling from Sect. 9.1 that the correlation between $\alpha^{(i)}$ and $\tilde{\alpha}^{(i)}$ is the inbreeding coefficient, F, we find

$$V_G = \mathrm{Var}\left[\sum_i (\alpha^{(i)} + \tilde{\alpha}^{(i)})\right]$$

$$= (1+F)V_G^0 + \sum_{i \neq j} \mathrm{Cov}\,(\alpha^{(i)} + \tilde{\alpha}^{(i)}, \alpha^{(j)} + \tilde{\alpha}^{(j)}) \ , \tag{10.20}$$

where V_G^0, given by the sum of (10.15b) over all loci, is the genotypic variance with panmixia. For partial selfing, in Problem 9.17 we proved that *genes* at different loci are pairwise stochastically independent at equilibrium; furthermore, (9.262) informs us that if pairwise stochastic independence holds initially, then it does so in all subsequent generations. It is not difficult to establish the same qualitative results for a population that consists of infinitely many demes of N monoecious individuals, provided we posit panmixia without selfing in each deme and assume that a proportion m of individuals in each deme comes from the offspring of a mating pool chosen at random from the entire population (see Problem 10.3). For another example and more general results, consult Problems 10.4 to 10.6.

Whenever allelic effects at different loci are uncorrelated, (10.20) gives

$$V_G = (1 + F)V_G^0 \ , \tag{10.21}$$

which is just the single-locus formula (9.39). If genotype and environment are stochastically independent, then inbreeding does not alter the environmental variance, and therefore (10.3) and (10.21) yield

$$V_Y = (1 + F)V_G^0 + V_E \ . \tag{10.22}$$

The panmictic value of the heritability, h_0^2, is raised to

$$h^2 = \frac{(1 + F)V_G^0}{(1 + F)V_G^0 + V_E} = \frac{(1 + F)h_0^2}{1 + F h_0^2} \ .$$

10.2 The Correlation Between Relatives with Panmixia

To evaluate the correlation between relatives, we assume that the population is at equilibrium under random mating and that selection, mutation, genotype-environment interaction, and epistasis are absent. Calling the relatives I and J and using a star on all random variables that refer to J, we obtain from (10.1)

$$\text{Cov}\,(Y, Y^\star) = \text{Cov}\,(G, G^\star) + \text{Cov}\,(G, E^\star) + \text{Cov}\,(E, G^\star) + \text{Cov}\,(E, E^\star) \ . \tag{10.23}$$

Unless a relationship is symmetric (e.g., twins or litter-mates), we may not generally assume that $\text{Cov}\,(G, E^\star) = \text{Cov}\,(E, G^\star)$.

To reduce the genotypic covariance in (10.23), we appeal to (10.8). Let A be any common ancestor of I and J. According to Sect. 3.3, at equilibrium the genotypes at distinct loci of A are pairwise stochastically independent. It follows that the genotypes at distinct loci of I and J are pairwise stochastically independent, and hence their contributions are uncorrelated:

$$\text{Cov}\,(G^{(i)}, G^{(j)\star}) = 0\,, \qquad i \neq j \ . \tag{10.24}$$

From (10.8) and (10.24) we infer

$$\text{Cov}(G, G^\star) = \sum_i \text{Cov}(G^{(i)}, G^{(i)\star}) \ . \tag{10.25}$$

In calculating the single-locus covariances, we employ κ, λ, μ, and ν as integer-valued random variables that specify allelic state. Thus,

$$\mathcal{P}(A_\kappa^{(i)} = A_k^{(i)}) = p_k^{(i)} \ .$$

Suppose that I and J have genotypes $A_\kappa^{(i)} A_\lambda^{(i)}$ and $A_\mu^{(i)} A_\nu^{(i)}$, respectively, at locus i. For both genotypes, we assign the alleles to paternal and maternal gametes at random. Therefore, (10.12) yields

$$\begin{aligned} \text{Cov}(G^{(i)}, G^{(i)\star}) &= \mathcal{E}(g_{\kappa\lambda}^{(i)} g_{\mu\nu}^{(i)}) \\ &= 4\mathcal{E}(\alpha_\kappa^{(i)} \alpha_\mu^{(i)}) + 2\mathcal{E}(\alpha_\kappa^{(i)} d_{\mu\nu}^{(i)}) + 2\mathcal{E}(d_{\kappa\lambda}^{(i)} \alpha_\mu^{(i)}) + \mathcal{E}(d_{\kappa\lambda}^{(i)} d_{\mu\nu}^{(i)}) \ . \end{aligned}$$
$$\tag{10.26}$$

The k-coefficients of Sect. 9.3 enable us to compute these expectations. Note that (10.14) means

$$\mathcal{E}(\alpha_\kappa^{(i)}) = 0 \ , \tag{10.27a}$$
$$\mathcal{E}(d_{j\kappa}^{(i)}) = 0 \ , \tag{10.27b}$$

and the variance components (10.15) have the concise form

$$V_G^{(i)} = \mathcal{E}[(g_{\kappa\lambda}^{(i)})^2] \ , \tag{10.28a}$$
$$V_A^{(i)} = 2\mathcal{E}[(\alpha_\kappa^{(i)})^2] \ , \tag{10.28b}$$
$$V_D^{(i)} = \mathcal{E}[(d_{\kappa\lambda}^{(i)})^2] \ . \tag{10.28c}$$

Let $A_j^{(i)} \equiv A_k^{(i)}$ signify that $A_j^{(i)}$ and $A_k^{(i)}$ are identical by descent. From (10.27) and (10.28b) we deduce

$$\begin{aligned} \mathcal{E}(\alpha_\kappa^{(i)} \alpha_\mu^{(i)}) &= \mathcal{P}(A_\kappa^{(i)} \equiv A_\mu^{(i)}) \mathcal{E}(\alpha_\kappa^{(i)} \alpha_\mu^{(i)} | A_\kappa^{(i)} \equiv A_\mu^{(i)}) \\ &\quad + \mathcal{P}(A_\kappa^{(i)} \not\equiv A_\mu^{(i)}) \mathcal{E}(\alpha_\kappa^{(i)} \alpha_\mu^{(i)} | A_\kappa^{(i)} \not\equiv A_\mu^{(i)}) \\ &= F_{IJ} \mathcal{E}[(\alpha_\kappa^{(i)})^2] + (1 - F_{IJ}) \mathcal{E}(\alpha_\kappa^{(i)}) \mathcal{E}(\alpha_\mu^{(i)}) \\ &= \tfrac{1}{2} F_{IJ} V_A^{(i)} \ , \end{aligned} \tag{10.29}$$

in which we relied on the stochastic independence of genes that are not identical by descent. As in Chap. 9, F_{IJ} denotes the coefficient of consanguinity of I and J.

The second and third expectations in (10.26) have the same structure. For the second term, since J is not inbred, there are three mutually exclusive possibilities: $A_\kappa^{(i)} \equiv A_\mu^{(i)} \not\equiv A_\nu^{(i)}$, or $A_\kappa^{(i)} \equiv A_\nu^{(i)} \not\equiv A_\mu^{(i)}$, or $A_\kappa^{(i)} \not\equiv A_\mu^{(i)}$ and

$A_\kappa^{(i)} \not\equiv A_\nu^{(i)}$; the first two of these evidently contribute equally. Consequently, (10.27) leads to

$$
\begin{aligned}
\mathcal{E}(\alpha_\kappa^{(i)} d_{\mu\nu}^{(i)}) &= 2\mathcal{P}(A_\kappa^{(i)} \equiv A_\mu^{(i)} \not\equiv A_\nu^{(i)})\mathcal{E}(\alpha_\kappa^{(i)} d_{\mu\nu}^{(i)}|A_\kappa^{(i)} \equiv A_\mu^{(i)} \not\equiv A_\nu^{(i)}) \\
&\quad + \mathcal{P}(A_\kappa^{(i)} \not\equiv A_\mu^{(i)} \text{ and } A_\kappa^{(i)} \not\equiv A_\nu^{(i)}) \cdot \\
&\qquad \cdot \mathcal{E}(\alpha_\kappa^{(i)} d_{\mu\nu}^{(i)}|A_\kappa^{(i)} \not\equiv A_\mu^{(i)} \text{ and } A_\kappa^{(i)} \not\equiv A_\nu^{(i)}) \\
&= 2F_{IJ}\mathcal{E}(\alpha_\kappa^{(i)} d_{\kappa\nu}^{(i)}|A_\kappa^{(i)} \not\equiv A_\nu^{(i)}) + [k_0 + (2k_1)\tfrac{1}{2}]\mathcal{E}(\alpha_\kappa^{(i)})\mathcal{E}(d_{\mu\nu}^{(i)}) \\
&= 2F_{IJ}\mathcal{E}[\alpha_\kappa^{(i)}\mathcal{E}(d_{\kappa\nu}^{(i)})|\kappa)] \\
&= 0 \ .
\end{aligned}
\tag{10.30}
$$

We evaluate the last term in (10.26) by invoking (10.27b) and (10.28c):

$$
\begin{aligned}
\mathcal{E}(d_{\kappa\lambda}^{(i)} d_{\mu\nu}^{(i)}) &= k_2\mathcal{E}(d_{\kappa\lambda}^{(i)} d_{\mu\nu}^{(i)}| I \text{ and } J \text{ have both genes identical by descent}) \\
&\quad + 2k_1\mathcal{E}(d_{\kappa\lambda}^{(i)} d_{\mu\nu}^{(i)}| I \text{ and } J \text{ have exactly one gene identical} \\
&\hspace{8cm} \text{by descent}) \\
&\quad + k_0\mathcal{E}(d_{\kappa\lambda}^{(i)} d_{\mu\nu}^{(i)}| I \text{ and } J \text{ have no genes identical by descent}) \\
&= k_2\mathcal{E}[(d_{\kappa\lambda}^{(i)})^2] + 2k_1\mathcal{E}(d_{\kappa\lambda}^{(i)} d_{\kappa\nu}^{(i)}|A_\kappa^{(i)} \not\equiv A_\lambda^{(i)}, A_\kappa^{(i)} \not\equiv A_\nu^{(i)}, A_\lambda^{(i)} \not\equiv A_\nu^{(i)}) \\
&\quad + k_0\mathcal{E}(d_{\kappa\lambda}^{(i)})\mathcal{E}(d_{\mu\nu}^{(i)}) \\
&= k_2 V_D^{(i)} + 2k_1\mathcal{E}[d_{\kappa\lambda}^{(i)}\mathcal{E}(d_{\kappa\nu}^{(i)}|\kappa)|A_\kappa^{(i)} \not\equiv A_\lambda^{(i)}] \\
&= k_2 V_D^{(i)} \ .
\end{aligned}
\tag{10.31}
$$

Substituting (10.29), (10.30), and (10.31) into (10.26) gives

$$
\text{Cov}(G^{(i)}, G^{(i)\star}) = 2F_{IJ}V_A^{(i)} + k_2 V_D^{(i)} \ .
\tag{10.32}
$$

From (10.17b) and (10.25) we obtain the expansion of the genotypic covariance in terms of the additive genetic and dominance variances:

$$
\text{Cov}(G, G^\star) = 2F_{IJ}V_A + k_2 V_D \ .
\tag{10.33}
$$

Then (10.19) and (10.23) inform us that the phenotypic correlation between I and J reads

$$
r_{IJ} = 2F_{IJ}h^2 + k_2 D^2 + V_Y^{-1}[\text{Cov}(G, E^\star) + \text{Cov}(E, G^\star) + \text{Cov}(E, E^\star)] \ ,
\tag{10.34}
$$

where the phenotypic variance is given by (10.18).

As described in Sect. 9.1, Wright's rule, (9.8), permits routine calculation of coefficients of consanguinity; in Sect. 9.3 we reduced the computation of the k-coefficients to that of coefficients of consanguinity. Refer to Table 9.1 for coefficients of consanguinity and k-coefficients for sundry close relatives. According to (10.34), the dominance variance does not contribute to the correlation between unilineal relatives.

If I and J are in the same generation (e.g., siblings or cousins), it is often useful to decompose the phenotypic variance into components within (V_w) and between (V_b) such groups of relatives [cf. (9.174)]: $V_Y = V_w + V_b$. Then $r_{IJ} = V_b/V_Y$.

If the environments of individuals are uncorrelated with the genotypes and environments of their relatives, then (10.34) simplifies to

$$r_{IJ} = 2F_{IJ}h^2 + k_2 D^2 \ . \tag{10.35}$$

This allows estimation of the narrow heritability and the proportion of dominance variance. For example, the parent-offspring and sib-sib correlations are

$$r_{po} = \tfrac{1}{2}h^2 , \qquad r_s = \tfrac{1}{2}h^2 + \tfrac{1}{4}D^2 \ , \tag{10.36}$$

whence

$$h^2 = 2r_{po} , \qquad D^2 = 4(r_s - r_{po}) \ . \tag{10.37}$$

Observe that $r_s \geq r_{po}$.

Pearson (1904) obtained the three closest ancestral correlations (parent and offspring, grandparent and grandchild, and great grandparent and great grandchild) and the correlation between full siblings by neglecting environment and epistasis and positing that all loci contribute equally and are diallelic with complete dominance and gene frequency one half. Yule (1906) generalized Pearson's ancestral correlations by allowing the same arbitrary degree of dominance at each locus and including the effect of an uncorrelated environment. Weinberg (1909, 1910) derived the ancestral and fraternal correlations under the assumptions leading to (10.35) and the restriction that the degree of dominance be the same for all pairs of alleles. Fisher (1918) deduced (10.35) for the three closest ancestors, full siblings, uncle-niece, and single and double first cousins. We owe the formula $r_{IJ} = 2F_{IJ}$ for the correlation without environment or dominance to Wright (1922), who obtained it by path analysis. Using probabilistic arguments, Malécot (1948) extended Wright's result to incorporate an uncorrelated environment. Earlier (Malécot, 1939), he had included dominance for two classes of relatives: (i) those related only through one parent of one of them (e.g., parent-offspring and uncle-niece) and (ii) those for which no parent of either individual is related to both parents of the other (e.g., full siblings and double first cousins). Unfortunately, a false assumption (Malécot, 1948, p. 21) limits the validity of his subsequent, more elegant treatment to category ii, which excludes, *inter alia*, the above examples of category i. In any case, neither category includes relatives such as quadruple half-first cousins (Trustrum, 1961; van Aarde, 1975). The general result (10.35) is due to Kempthorne (1955; 1955a; 1957, pp. 330–332).

With epistasis, the joint effects of at least pairs of loci must be considered simultaneously in the two relatives, so the correlation between relatives depends on the linkage map of the loci that influence the trait. The analysis and results are simplest for unlinked loci. Fisher (1918) generalized his correlations in this case to include two-locus epistasis with diallelic loci. Cockerham (1954) extended Fisher's work to any relationship and arbitrary epistasis. We owe the general multiallelic formula without linkage, expressed in terms of the two probabilities of identity in (10.35), to Kempthorne (1954; 1955; 1955a; 1957, pp. 419–420). Cockerham (1956) calculated the correlation in some special cases with linkage. Schnell (1963) derived the covariances for ancestor-descendant and relationships for which no parent of either individual is related to both parents of the other. The general formula with linkage is due to van Aarde (1975).

Bohidar (1964) has deduced the decomposition of the variance and the correlation between relatives for a character influenced by both autosomal and X-linked loci. Consult James (1973) for a discussion of these results.

In human genetics, *twins* are often used to resolve the genetic and environmental contributions to the phenotypic variance. For two monozygotic twins adopted by different randomly chosen families, if we disregard environmental correlations due to prenatal environment, (10.23) yields

$$\text{Cov}\,(Y, Y^\star) = V_G \ . \tag{10.38}$$

Since for these twins genotype and environment are uncorrelated, therefore (10.3) shows that their phenotypic variance is

$$V_G + V_E = V_Y - 2\text{Cov}\,(G, E) \ , \tag{10.39}$$

where all terms refer to the general population, for which we wish to estimate parameters. Combining (10.38) and (10.39), we obtain

$$r_A = V_G/[V_Y - 2\text{Cov}\,(G, E)] \tag{10.40}$$

for the correlation between monozygotic twins reared apart. If genotype and environment are uncorrelated, (10.40) simplifies to the broad heritability:

$$r_A = H^2 \ . \tag{10.41}$$

For positively correlated genotype and environment, (10.40) informs us that $r_A > H^2$. However, the error in thus overestimating the broad heritability is probably reduced (if not sometimes reversed in direction) by special prenatal environmental factors that tend to make twins dissimilar (Nagylaki and Levy, 1973). The estimate (10.41) is accurate whenever the phenotypic variance of adopted twins reared apart is close to that of the general population.

The correlation between monozygotic twins reared together at home,

$$r_T = [V_G + 2\text{Cov}\,(G, E^\star) + \text{Cov}\,(E, E^\star)]/V_Y \ , \tag{10.42}$$

follows from (10.23). If genotype and environment are uncorrelated, i.e., $\text{Cov}\,(G, E) = \text{Cov}\,(G, E^\star) = 0$, then (10.41) and (10.42) inform us

$$r_T - r_A = \text{Cov}\,(E, E^\star)/V_Y \ . \tag{10.43}$$

Thus, $r_T - r_A$ specifies the proportion of the phenotypic variance due to environmental variation between families. Therefore, the proportion of the phenotypic variance due to environmental variation within families reads

$$(1 - H^2) - (r_T - r_A) \ . \tag{10.44}$$

Special postnatal environmental factors, however, may decrease this component for twins relative to its value for siblings. The comparison of the phenotypic variance of twins reared at home with that of the general population again provides a check.

For unrelated individuals reared together by a family chosen at random, (10.23) leads to the phenotypic correlation

$$r_U = \mathrm{Cov}\,(E, E^\star)/[V_Y - 2\mathrm{Cov}\,(G, E)] \ . \tag{10.45}$$

In the absence of genotype-environment correlation, this should agree with (10.43). In this case, adopted children have the phenotypic variance of the general population, and hence their correlation with their biological parents is one-half the narrow heritability.

Note that (10.38) to (10.45) follow directly from (10.1), and therefore hold even if mating is assortative.

Consult Falconer (1960, pp. 183–185) for a lucid discussion of comparisons between monozygotic and dizygotic twins.

Finally, we calculate the correlation between *inbred relatives*. We assume that there is no dominance and the population is initially in Hardy-Weinberg proportions at all loci. Since the *genotypes* at different loci may be mutually dependent, we proceed as in (10.20):

$$\mathrm{Cov}\,(G, G^\star) = \sum_{ij} \mathrm{Cov}\,(\alpha^{(i)} + \tilde{\alpha}^{(i)}, \alpha^{(j)\star} + \tilde{\alpha}^{(j)\star}) \ . \tag{10.46}$$

If *allelic effects* at different loci in all common ancestors of I and J are uncorrelated, as discussed below (10.20), then the covariances vanish for $i \neq j$. Invoking (10.17b) and (10.29) reduces (10.46) to

$$\mathrm{Cov}\,(G, G^\star) = 2F_{IJ}V_G^0 \ , \tag{10.47}$$

where V_G^0 designates the genetic variance with panmixia. Suppose that the environments are stochastically independent of each other and of the genotypes. In that case, (10.47) is also the phenotypic covariance, and from (10.22) and (10.47) we find

$$r_{IJ} = 2F_{IJ}h_0^2[(1 + F_I h_0^2)(1 + F_J h_0^2)]^{-1/2} \ , \tag{10.48}$$

where h_0^2 denotes the heritability with random mating, and I and J belong to classes of individuals with respective inbreeding coefficients F_I and F_J. According to (10.21) and (10.47), inbreeding increases the genetic variance but does not affect the covariance between relatives; consequently, it decreases the correlation between relatives, as exhibited by (10.48).

We owe (10.48) to Malécot (1948); the special case without environment was derived much earlier from path analysis by Wright (1922).

10.3 The Change in Variance due to Assortative Mating

Positive phenotypic correlation between mates has been thoroughly established for many physical and behavioral characters in human populations. Empirical correlations for many physical and personality traits are quite low, although the observed values range up to about 0.5. Correlations are significantly higher for various measures of intelligence: 0.5 is typical, but even correlations around 0.75 have been occasionally reported (Spuhler, 1968; Vandenberg, 1972; Jensen, 1978). As discussed by Wright (1978, pp. 367–369), several factors may contribute to such marital correlations. Since average inbreeding coefficients in human populations are at most a few percent and are usually one or two orders of magnitude smaller (Cavalli-Sforza and Bodmer, 1971, pp. 350-353), the effect of consanguineous mating in this regard is negligible. In many societies, socioeconomic classes are genetically and environmentally differentiated. Then the influence of class membership on mate choice can greatly enhance the genetic and environmental similarity of mates. Here we confine ourselves, however, to assortative mating determined solely by phenotype.

We assume that an individual's environment is stochastically independent of his genotype and that dominance and epistasis are absent. We assume also that an individual's total probability of mating is independent of his genotype. As shown in the introduction to Chap. 5, this implies the constancy of the allelic frequencies. Consult Wilson (1973) for a model of assortative mating with gene-frequency change.

According to (10.8), the mean genotypic value of the population is given in terms of the ordered single-locus genotypic frequencies $P_{jk}^{(i)}$ by

$$\overline{G} = \sum_{ijk} G_{jk}^{(i)} P_{jk}^{(i)} \ . \tag{10.49}$$

With no dominance,

$$G_{jk}^{(i)} = \beta_j^{(i)} + \beta_k^{(i)} \ , \tag{10.50}$$

for some constants $\beta_j^{(i)}$. Inserting (10.50) into (10.49) yields

$$\overline{G} = 2 \sum_{ij} \beta_j^{(i)} p_j^{(i)} \ , \tag{10.51}$$

which demonstrates that the mean genotype remains constant. Since with independent environments the average environmental value is obviously also constant, it follows from (10.2) that assortative mating does not change the average phenotype.

In Sects. 5.2 and 5.3, we saw that assortative mating increases homozygote frequencies and decreases heterozygote frequencies. Hence, (10.49) indicates that if dominance were present, the mean genotype would generally not be constant. If many loci of comparable effect contributed to the trait, however, we would expect the influence of assortative mating on each locus to be small. Then, as suggested by (5.43), the deviations from Hardy-Weinberg proportions would be slight, and we conclude from (10.49) that $\overline{G}$ would change very little.

We proceed to examine the influence of assortative mating on the variance. From (10.12), (10.50), and (10.51) we obtain

$$g_{jk}^{(i)} = \alpha_j^{(i)} + \alpha_k^{(i)} = \beta_j^{(i)} + \beta_k^{(i)} - 2\overline{\beta}^{(i)} \; , \qquad (10.52)$$

where

$$\overline{\beta}^{(i)} = \sum_j \beta_j^{(i)} p_j^{(i)} \; . \qquad (10.53a)$$

Therefore, the average efffects are given by

$$\alpha_j^{(i)} = \beta_j^{(i)} - \overline{\beta}^{(i)} \; . \qquad (10.53b)$$

The single-gene variance at locus i is

$$\sigma_i^2 = \mathcal{E}[(\alpha^{(i)})^2] = \sum_j (\alpha_j^{(i)})^2 p_j^{(i)} \; . \qquad (10.54)$$

Since the allelic frequencies are invariant, the *marginal* distributions of the random variables $\alpha^{(i)}$ are independent of time; in particular, (10.53) and (10.54) reveal that $\alpha_j^{(i)}$ and σ_i^2 are constants. The *joint* distribution of the average effects does change, however, because assortative mating induces correlations within individuals between the two genes at each locus and between genes at different loci.

Let individuals I and $I^\star$ mate to produce individual I'; stars and primes distinguish variables that pertain to $I^\star$ and I'. As before, tildes designate maternal gametes. In our definitions and calculations, we shall take repeated advantage of the symmetry between paternal and maternal gametes. Let k_{ij}, f_{ij}, and m_{ij} denote the correlations between a gene at locus i and one at locus j in the same gamete, uniting gametes, and mates, respectively. Thus, $k_{ii} = 1$ and f_{ii} is a measure of the deviation from Hardy-Weinberg proportions at locus i. In principle, m_{ij} can be calculated from the mating system and the determination of the character by genes and environment, but we shall do this only in the important special case at the end of this section. We display our variables in Fig. 10.1.

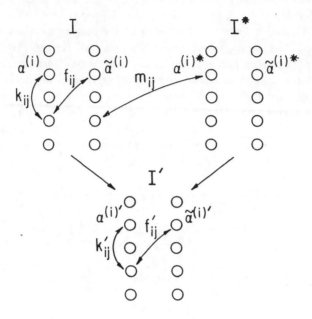

Fig. 10.1. Variables for assortative mating.

From (10.52) and (10.54) we compute the genetic variance:

$$V_G = \sum_{ij} \mathcal{E}[(\alpha^{(i)} + \tilde{\alpha}^{(i)})(\alpha^{(j)} + \tilde{\alpha}^{(j)})]$$

$$= 2\sum_{ij} \mathcal{E}(\alpha^{(i)}\alpha^{(j)} + \alpha^{(i)}\tilde{\alpha}^{(j)})$$

$$= 2\sum_{ij} (k_{ij} + f_{ij})\sigma_i\sigma_j \ . \tag{10.55}$$

Similarly, the covariance between the genotypic values of mates reads

$$C = \text{Cov}\,(G, G^\star)$$

$$= \sum_{ij} \mathcal{E}[(\alpha^{(i)} + \tilde{\alpha}^{(i)})(\alpha^{(j)\star} + \tilde{\alpha}^{(j)\star})]$$

$$= 4\sum_{ij} \mathcal{E}(\alpha^{(i)}\alpha^{(j)\star})$$

$$= 4\sum_{ij} m_{ij}\sigma_i\sigma_j \ . \tag{10.56}$$

Next, we derive recursion relations for f_{ij} and k_{ij}. Evidently,

$$\sigma_i \sigma_j f'_{ij} = \mathcal{E}(\alpha^{(i)\prime} \tilde{\alpha}^{(j)\prime}) = \mathcal{E}(\alpha^{(i)} \alpha^{(j)\star}) = \sigma_i \sigma_j m_{ij} \, ,$$

whence

$$f'_{ij} = m_{ij} \, . \tag{10.57a}$$

Denoting the recombination frequency, the occurrence of recombination, and the nonoccurrence of recombination between loci i and j by c_{ij} ($c_{ii} = 0$), $\mathcal{R}_{ij}$, and $\bar{\mathcal{R}}_{ij}$, respectively, we have

$$\begin{aligned}
\sigma_i \sigma_j k'_{ij} &= \mathcal{E}(\alpha^{(i)\prime} \alpha^{(j)\prime}) \\
&= P(\bar{\mathcal{R}}_{ij}) \mathcal{E}(\alpha^{(i)\prime} \alpha^{(j)\prime} | \bar{\mathcal{R}}_{ij}) + P(\mathcal{R}_{ij}) \mathcal{E}(u^{(i)\prime} \alpha^{(j)\prime} | \mathcal{R}_{ij}) \\
&= (1 - c_{ij}) \mathcal{E}(\alpha^{(i)} \alpha^{(j)}) + c_{ij} \mathcal{E}(\alpha^{(i)} \tilde{\alpha}^{(j)}) \\
&= (1 - c_{ij}) \sigma_i \sigma_j k_{ij} + c_{ij} \sigma_i \sigma_j f_{ij} \, ,
\end{aligned}$$

whence

$$k'_{ij} = (1 - c_{ij}) k_{ij} + c_{ij} f_{ij} \, . \tag{10.57b}$$

Since we do not know how m_{ij} depends on time, we confine ourselves to *equilibrium* and do not try to establish convergence. At equilibrium (indicated by carets throughout), (10.57) yields

$$\hat{f}_{ij} = \hat{m}_{ij} \, , \tag{10.58a}$$

$$\hat{k}_{ij} = \hat{m}_{ij} \, , \qquad i \neq j \, . \tag{10.58b}$$

Recalling that $k_{ii} = 1$ and substituting (10.58) into (10.55) leads to

$$\hat{V}_G = V_G^0 + 4 \sum_{ij} \hat{m}_{ij} \sigma_i \sigma_j - 2 \sum_i \hat{m}_{ii} \sigma_i^2 \, , \tag{10.59}$$

in which

$$V_G^0 = 2 \sum_i \sigma_i^2 \tag{10.60}$$

represents the genetic variance with random mating.

We define the effective number of loci as

$$n_e = \sum_{ij} \hat{m}_{ij} \sigma_i \sigma_j \Big/ \sum_i \hat{m}_{ii} \sigma_i^2 \, . \tag{10.61}$$

For a single locus, (10.61) yields $n_e = 1$. Separating the terms with $i = j$ in the numerator informs us that if the $\hat{m}_{ij}$ all have the same sign (as we expect for both assortative and disassortative mating), then $n_e \geq 1$. For assortative mating, from the assumption

$$\hat{m}_{ii} > 0, \qquad \hat{m}_{ij} \leq (\hat{m}_{ii} \hat{m}_{jj})^{1/2} \tag{10.62}$$

for all i and j, we deduce

$$\sum_{ij} \hat{m}_{ij}\sigma_i\sigma_j \leq \sum_{ij}(\sqrt{\hat{m}_{ii}}\sigma_i)(\sqrt{\hat{m}_{jj}}\sigma_j)$$

$$\leq \sum_{ij} \tfrac{1}{2}(\hat{m}_{ii}\sigma_i^2 + \hat{m}_{jj}\sigma_j^2)$$

$$= n \sum_i \hat{m}_{ii}\sigma_i^2 \ , \qquad (10.63)$$

where n represents the actual number of loci. We conclude from (10.61) and (10.63) that $n_e \leq n$. (Appropriate sign changes immediately extend our argument to disassortative mating.) With exchangeable loci, $\hat{m}_{ij} = \hat{m}$ and $\sigma_i = \sigma$ for all i and j, and therefore (10.61) gives $n_e = n$.

It would be desirable to prove from the underlying biological model, without extraneous assumptions, that $1 \leq n_e \leq n$. If $n_e \leq n$ always holds, the model must imply some restriction like (10.62); for otherwise, with $\hat{m}_{ij}$ fixed and positive for $i \neq j$, arbitrarily small positive values of $\hat{m}_{ii}$ would produce arbitrarily large n_e. Similarly, $n_e \geq 1$ entails restrictions on the signs of $\hat{m}_{ij}$: for instance, if $\hat{m}_{ii} > 0$ and $\hat{m}_{ij} < 0$ for all i and j such that $i \neq j$, then $n_e < 1$. We shall establish that $1 \leq n_e \leq n$ in the special case at the end of this section.

Evaluating (10.56) at equilibrium and employing (10.61) reduces (10.59) to

$$\hat{V}_G = V_G^0 + \gamma\hat{C} \ , \qquad (10.64)$$

in which

$$\gamma = 1 - (2n_e)^{-1} \ . \qquad (10.65)$$

We denote the equilibrium correlation between the genotypic values of mates by $\hat{\rho}$. Eliminating $\hat{C}$ from (10.64) with the aid of

$$\hat{C} = \hat{\rho}\hat{V}_G \ , \qquad (10.66)$$

we find

$$\hat{V}_G = V_G^0/(1 - \gamma\hat{\rho}) \ . \qquad (10.67)$$

Now, $1 \leq n_e < \infty$ corresponds to $\tfrac{1}{2} \leq \gamma < 1$, so $\hat{\rho} > 0$ implies

$$V_G^0 < V_G^0/(1 - \tfrac{1}{2}\hat{\rho}) \leq \hat{V}_G < V_G^0/(1 - \hat{\rho}) \ , \qquad (10.68)$$

where the upper limit is approached for $n_e \gg 1$.

We can also calculate the average correlation between homologous genes in an individual, provided we weight the single-locus correlations with the single-gene variances. Appealing successively to (10.58a), (10.61), (10.56), (10.60), (10.66), (10.67), and (10.65), we get

$$\overline{\hat{f}}_{ii} = \overline{\hat{m}}_{ii} = \frac{\sum_i \hat{m}_{ii}\sigma_i^2}{\sum_i \sigma_i^2} = \frac{\hat{\rho}}{\hat{\rho} + 2n_e(1 - \hat{\rho})} \ . \qquad (10.69)$$

From $1 \le n_e < \infty$ and $\hat{\rho} > 0$ we conclude

$$0 < \overline{\hat{f}}_{ii} \le \hat{\rho}/(2 - \hat{\rho}) < \hat{\rho} \; . \tag{10.70}$$

Observe that the genetic variance increases and the average correlation decreases as the effective number of loci increases. For $2n_e \gg \hat{\rho}/(1 - \hat{\rho})$, $\overline{\hat{f}}_{ii} \ll 1$; for any n_e, if $\hat{\rho}$ is not close to zero, the proportional increase in the variance is quite appreciable. So, with many loci, assortative mating can significantly increase the variance without altering the mean or causing high correlations between alleles. This can occur because of the large number of two-locus covariances that contribute to the genetic variance in (10.55)

For disassortative mating ($\hat{\rho} < 0$), (10.67) and (10.69) are unaltered, but all the inequalities in (10.68) and (10.70) must be reversed. Disassortative mating decreases the genetic variance and induces negative allelic correlations.

With independent environments, the environmental variance is obviously unchanged by assortative mating. Therefore, the phenotypic variance at equilibrium reads

$$\hat{V}_Y = \hat{V}_G + V_E \; , \tag{10.71}$$

and (10.67) and (10.71) yield the heritability

$$\hat{h}^2 = \frac{\hat{V}_G}{\hat{V}_Y} = \frac{h_0^2}{1 - \gamma\hat{\rho}(1 - h_0^2)} \; , \tag{10.72}$$

where

$$h_0^2 = V_G^0/V_Y^0 \tag{10.73}$$

($0 \le h_0 \le 1$) designates the panmictic heritability. Equation (10.72) exhibits the increase in heritability due to assortative mating.

To complete the analysis, we should like to evaluate $\hat{\rho}$ and n_e in terms of panmictic variables by appealing to the purely phenotypic determination of the mating frequencies. This has not been accomplished in general. We shall prove below that if the regressions of an individual's genotype on his phenotype and of his phenotype on that of his mate are linear, and if the phenotypic correlation between mates is the constant r ($-1 \le r \le 1$), then

$$\hat{\rho} = r\hat{h}^2 \; . \tag{10.74}$$

Inserting (10.72) into (10.74) leads to

$$\gamma(1 - h_0^2)\hat{\rho}^2 - \hat{\rho} + rh_0^2 = 0 \; . \tag{10.75}$$

With the aid of the assumption $n_e > \frac{1}{2}$, it is easy to show that (10.75) has the unique root

$$\hat{\rho} = [2\gamma(1 - h_0^2)]^{-1}\{1 - [1 - 4r\gamma h_0^2(1 - h_0^2)]^{1/2}\} \tag{10.76}$$

in $[-1, 1]$. In accordance with (10.74), $\operatorname{sgn} \hat{\rho} = \operatorname{sgn} r$ and $0 \le |\hat{\rho}| \le |r|$; as $h_0 \to 1$, $\hat{\rho} \to r$.

From (10.76) we infer that $\hat{\rho}$ increases if r or n_e increases and that $|\hat{\rho}|$ increases if h_0 increases. Therefore, (10.65), (10.67), and (10.72) reveal that the relative values of the genetic variance $(\hat{V}_G/V_G^0)$ and the heritability $(\hat{h}^2/h_0^2)$ are greater if assortative mating is stronger or the effective number of loci is larger. The greatest relative augmentation of the variance occurs when $h_0^2 = 1$; that of the heritability, when $h_0^2 = \frac{1}{2}$.

The effect of the increase in the genetic variance is greatest at the extremes of the phenotypic distribution. To illustrate this, suppose that at equilibrium under assortative mating with phenotypic correlation $r = 0.55$, a character Y has mean zero and standard deviation unity. Posit also that $\hat{h}^2 = 0.65$, the effective number of loci is sufficiently large to permit the approximation $\gamma = 1$, and Y is normally distributed. From (10.74) we obtain $\hat{\rho} \approx 0.358$; then (10.75) informs us that if random mating were instituted, at equilibrium the heritability would be $h_0^2 \approx 0.544$. By virtue of (10.67), the phenotypic variance would be reduced by panmixia to

$$V_Y^0 = h_0^{-2}V_G^0 = h_0^{-2}(1-\hat{\rho})\hat{h}^2\hat{V}_Y \approx 0.768 \ ,$$

which corresponds to a standard deviation $\sigma_Y^0 \approx 0.876$. The mean is still zero, and we again postulate normality of Y. Whereas at equilibrium under assortative mating, approximately 0.135% of the population has $Y > 3$, panmixia reduces this percentage more than four times, to approximately 0.031%.

Fisher (1918) was the first to treat assortative mating for a quantitative character. He posited stochastic independence of the environment, control of the character by many unlinked loci with contributions of the same order of magnitude, and the absence of epistasis. He evaluated approximately the genotypic variance at the equilibrium determined by assortative mating in terms of the initial, panmictic genotypic variance. Kempthorne (1957, Chap. 22) has presented a helpful exposition of this calculation. Fisher's (1918) conjecture that his results should be unaffected by linkage received some support in the useful exegesis of his paper by Moran and Smith (1966).

Malécot (1939) proved for diallelic loci that Fisher's formulae hold in the limit as the number of loci tends to infinity. Fisher's work had suggested that assortative mating increases the additive genetic variance, leaving the dominance variance approximately unaltered, and Fisher (1918) implicitly made this assumption in his study of the correlation between relatives. This result was also established by Malécot (1939). At least with unlinked loci and $n_e \gg 1$, so that $\gamma \approx 1$, his work allows us to apply our formulae to traits with dominance, provided we include the dominance variance V_D in V_E and everywhere interpret V_G as the genic variance V_A, ρ as the correlation between the genic values of mates, and h^2 as the narrow heritability.

Wright (1921) employed path analysis to derive our results in the special case of unlinked, diallelic, exchangeable (i.e., completely equivalent) loci with gene frequency one half. In this subsequent, much more elaborate investigation (Wright, 1969, Chap. 11), he used the same method without these restrictions.

Our approach above is based on the work of Crow and Felsenstein (1968), whose use of correlations was inspired by Wright (1921).

Consult Lande (1977) for the inclusion of selection and mutation.

The remainder of this section follows Nagylaki (1982).

We return now to the derivation of (10.74) and the determination of the effective number of alleles. Since we have seen that the means are constant, we may suppose without loss of generality that $\overline{G} = \overline{E} = \overline{Y} = 0$.

We assume that the regression of an individual's phenotype on that of his mate is linear. Since the phenotypic correlation is r, this means

$$\mathcal{E}(Y^*|Y) = rY \ . \tag{10.77}$$

Let

$$\phi(u, v) = \mathcal{E}(e^{i(uY+vY^*)})$$

($i = \sqrt{-1}$ here) represent the characteristic function of the joint distribution of Y and Y^*. Then (10.77) holds if and only if (Kendall and Stuart, 1973, p. 365)

$$\frac{\partial \phi}{\partial v}(u, 0) = r\frac{\partial \phi}{\partial u}(u, 0) \ . \tag{10.78}$$

Numerous bivariate distributions have linear regressions and (for a symmetric choice of parameters) identical marginals. For us, the most important example is the bivariate normal distribution, because we expect it to apply if either there are many alleles with approximately normally distributed effects at each of an arbitrary number of loci, or an arbitrary number of alleles at each of many loci, each of which makes only a small contribution to the genetic variance. Two bivariate gamma distributions, the inverted Dirichlet distribution, and the bivariate Pareto distribution furnish other examples of continuous distributions; these have positive regression coefficients (Johnson and Kotz, 1972, pp. 218, 222, 239, 285). The multinomial, negative multinomial, and bivariate Poisson are discrete distributions that satisfy our requirements; the regression coefficients are negative for the first of these and positive for the others (Johnson and Kotz, 1969, pp. 282, 295, 299).

We suppose also that the regression of an individual's genotypic value on his phenotype is linear. Then (10.7) yields

$$\mathcal{E}(G|Y) = h^2Y \ . \tag{10.79}$$

Let ψ and χ designate the characteristic functions of G and E, respectively. By the independence of G and E, the joint characteristic function of Y and G reads

$$\mathcal{E}(e^{i(uY+vG)}) = \psi(u + v)\chi(u) \ ,$$

whence the criterion (10.78) implies that (10.79) holds if and only if (Kendall and Stuart, 1973, pp. 430–431)

$$[\psi(u)]^{1-h^2} = [\chi(u)]^{h^2} \ . \tag{10.80}$$

The most important consequence of this condition is that if the environment is normally distributed, as may often be reasonably assumed, then the genotypic value must also have a normal distribution. From the characteristic functions in Lukacs (1970, pp. 6, 7, 18), it is easy to see that for suitable parameters (10.80) is also satisfied if G and E both have a Poisson, negative binomial, gamma, or Cauchy distribution. However, since the last of these has infinite variance, it is irrelevant to our problem. Thus, (10.79) appears to be a more severe restriction than (10.77). Excluding the Gaussian limit, it is unlikely that (10.77) and (10.79) both hold exactly, but they may provide good approximations.

With the aid of (10.77) and (10.79), we can easily deduce (10.74). For phenotypic assortative mating, if Y is fixed, then all other random variables that relate to I must be stochastically independent of all those that relate to I^*. An analogous result holds, of course, for fixed Y^*. Therefore, (10.79) and (10.77) give

$$
\begin{aligned}
\mathcal{E}(G^*|Y) &= \mathcal{E}[\mathcal{E}(G^*|Y,Y^*)|Y] \\
&= \mathcal{E}[\mathcal{E}(G^*|Y^*)|Y] \\
&= \mathcal{E}(h^2 Y^*|Y) \\
&= rh^2 Y \ .
\end{aligned}
\tag{10.81}
$$

From (10.79) and (10.81) we find

$$
\begin{aligned}
\rho V_G &= \mathcal{E}(GG^*) \\
&= \mathcal{E}[\mathcal{E}(GG^*|Y)] \\
&= \mathcal{E}[\mathcal{E}(G|Y)\mathcal{E}(G^*|Y)] \\
&= \mathcal{E}(h^2 Y r h^2 Y) \\
&= r h^4 V_Y \ ,
\end{aligned}
$$

whence

$$
\rho = r h^2 \ ,
\tag{10.82}
$$

which becomes (10.74) at equilibrium.

Finally, we consider the calculation of the effective number of loci. Let $\hat{l}_i$ signify the correlation at equilibrium between the allelic effect $\alpha^{(i)}$ at locus i and the phenotype Y. In addition to (10.77) and (10.79), we posit that the regression of $\alpha^{(i)}$ on Y is linear:

$$
\mathcal{E}(\alpha^{(i)}|Y) = \hat{l}_i \sigma_i \hat{\sigma}_Y^{-1} Y \ .
\tag{10.83}
$$

According to the discussion below (10.79), this linearity is likely to hold exactly only in the limit of normally distributed allelic effects.

Conditioning as in the derivation of (10.81) and employing (10.83) leads to

$$
\mathcal{E}(\alpha^{(j)*}|Y) = r \hat{l}_j \sigma_j \hat{\sigma}_Y^{-1} Y \ .
\tag{10.84}
$$

Conditioning as in the derivation of (10.82) and using (10.83) and (10.84) yields

$$
\hat{m}_{ij} \sigma_i \sigma_j = \mathcal{E}(\alpha^{(i)} \alpha^{(j)*}) = r \hat{l}_i \hat{l}_j \sigma_i \sigma_j \ ,
$$

whence

$$\hat{m}_{ij} = r\hat{l}_i\hat{l}_j \; . \tag{10.85}$$

To evaluate $\hat{l}_i$, recall that environment and genotype are independent and utilize (10.58):

$$\hat{l}_i\sigma_i\hat{\sigma}_Y = \mathcal{E}(\alpha^{(i)}Y)$$

$$= \mathcal{E}\left\{\alpha^{(i)}\left[\sum_j(\alpha^{(j)} + \tilde{\alpha}^{(j)}) + E\right]\right\}$$

$$= \sum_j(\hat{k}_{ij} + \hat{f}_{ij})\sigma_i\sigma_j$$

$$= \sigma_i\left[(1 - \hat{m}_{ii})\sigma_i + 2\sum_j \hat{m}_{ij}\sigma_j\right] \; . \tag{10.86}$$

Substituting (10.85) into (10.86) reveals

$$\hat{l}_i = (1 - r\hat{l}_i^2)b_i + 2r\hat{l}_i\sum_j \hat{l}_j b_j \; , \tag{10.87}$$

in which

$$b_i = \sigma_i/\hat{\sigma}_Y \; . \tag{10.88}$$

The hypothesis (10.83) implies the less detailed postulate (10.79): since (10.83) obviously holds with $\alpha^{(i)}$ replaced by $\tilde{\alpha}^{(i)}$, we have

$$\mathcal{E}(G|Y) = \sum_j \mathcal{E}(\alpha^{(j)} + \tilde{\alpha}^{(j)}|Y) = 2\sum_j \hat{l}_j\sigma_j\hat{\sigma}_Y^{-1}Y \; .$$

Comparing this with (10.79) at equilibrium and utilizing (10.88) informs us

$$\hat{h}^2 = 2\sum_j \hat{l}_j b_j \; . \tag{10.89}$$

Inserting (10.89) and then (10.74) into (10.87) produces the quadratic equation

$$rb_i\hat{l}_i^2 + (1 - \hat{\rho})\hat{l}_i - b_i = 0 \tag{10.90}$$

for $\hat{l}_i$.

To study the roots of (10.90), we need the following upper bound on b_i^2. By successive application of (10.88), (10.60), and (10.67) we obtain

$$b_i^2 \le \sum_j b_j^2 = \sum_j \sigma_j^2/\hat{V}_Y = V_G^0/(2\hat{V}_Y) = \tfrac{1}{2}\hat{h}^2(1 - \gamma\hat{\rho}) \; . \tag{10.91}$$

With the help of (10.91) and the assumption $n_e > \frac{1}{2}$, it is easy to prove that (excluding the spurious root $\hat{l}_i = -1$ for $r = h_0 = 1$) the unique solution of (10.90) in $[-1, 1]$ is the positive root

$$\hat{l}_i = (2rb_i)^{-1}\{-(1 - \hat{\rho}) + [(1 - \hat{\rho})^2 + 4rb_i^2]^{1/2}\} \; . \tag{10.92}$$

A little algebra shows that if locus i contributes more to the variance than locus j, i.e. $\sigma_i \geq \sigma_j$, then it is more highly correlated with the phenotype, i.e. $\hat{l}_i \geq \hat{l}_j$.

We proceed to derive an equation for n_e. From (10.71), (10.73), and (10.67) we deduce

$$\hat{V}_Y = [1 - \gamma\hat{\rho}(1 - h_0^2)](1 - \gamma\hat{\rho})^{-1}V_Y^0 \ . \tag{10.93}$$

Given r, h_0, σ_i for all i, and a value of n_e greater than $\frac{1}{2}$, (10.76) determines $\hat{\rho}(n_e)$ in $[-1, 1]$; then (10.60), (10.73), (10.93), (10.88), and (10.92) determine $\hat{l}_i(n_e)$ in $(0, 1]$. Therefore, substituting (10.85) into (10.61) gives

$$n_e = \Phi(n_e) = \left(\sum_i \hat{l}_i\sigma_i\right)^2 \Big/ \sum_i \hat{l}_i^2\sigma_i^2 \ . \tag{10.94}$$

As in the discussion of (10.61), by the positivity of $\hat{l}_i$, (10.94) implies that $1 \leq \Phi(n_e) \leq n$ for $n_e > \frac{1}{2}$. Hence, if we rewrite (10.94) in the form

$$\Psi(n_e) = n_e - \Phi(n_e) = 0 \ , \tag{10.95}$$

it follows that $\Psi(1) \leq 0$ and $\Psi(n) \geq 0$. We conclude that (10.95) has at least one root in $[1, n]$. Uniqueness is plausible but unproved.

Several aspects of our solution deserve comment.

At equilibrium all the correlations, and therefore the genetic variance, are independent of the linkage map.

We can cast (10.94) into the form

$$\frac{n_e}{n} = \left(\frac{1}{n}\sum_i \hat{l}_i\sigma_i\right)^2 \Big/ \left(\frac{1}{n}\sum_i \hat{l}_i^2\sigma_i^2\right) = [\mathcal{E}(\hat{l}\sigma)]^2 / \mathcal{E}[(\hat{l}\sigma)^2] \ ,$$

whence

$$(n - n_e)/n = \mathrm{Var}\,(\hat{l}\sigma)/\mathcal{E}[(\hat{l}\sigma)^2] \ . \tag{10.96}$$

Here the expectation and variance refer to the discrete distribution that assigns probability $1/n$ to each of the n loci that influence the character. From (10.96) it is again obvious that $n_e = n$ for exchangeable loci. In this case, $\hat{l}_i = \hat{l}$ and $\hat{m}_{ij} = \hat{m}$, so (10.69) and (10.85) tell us

$$r\hat{l}^2 = \hat{m} = \hat{\rho}[\hat{\rho} + 2n(1 - \hat{\rho})]^{-1} \ ; \tag{10.97}$$

(10.76) completes the solution.

If assortative mating and genetic determination are complete ($r = h_0 = 1$), then (10.75) yields $\hat{\rho} = 1$, and hence $\hat{l}_i = 1$ from (10.92). Thus, in this case linear regressions imply perfect correlation between the genotype and the allelic effects at each locus. With $\hat{l}_i = 1$, (10.94) simplifies to the explicit formula

$$n_e = \left(\sum_i \sigma_i\right)^2 \Big/ \sum_i \sigma_i^2 \ . \tag{10.98}$$

If $r < 1$ or $h_0 < 1$, as $b_i \to 0$ (10.92) leads to

$$\hat{l}_i \sim b_i/(1 - \hat{\rho}) \ll 1 \ . \tag{10.99}$$

Now, (10.88) enables us to express (10.94) as

$$n_e = \left(\sum_i \hat{l}_i b_i \right)^2 \Big/ \sum_i \hat{l}_i^2 b_i^2 \ . \tag{10.100}$$

Hence, if locus i contributes much less to the genetic variance than locus j, i.e. $b_i \ll b_j < 1$ [see (10.91)], then locus i also influences n_e much less than locus j.

As $r \to 0$, from (10.92) we infer that $\hat{l}_i \to b_i$, and consequently (10.94) reduces to Lande's (1981) panmictic formula,

$$n_e = \left(\sum_i \sigma_i^2 \right)^2 \Big/ \sum_i \sigma_i^4 \ . \tag{10.101}$$

Let us prove that the effective number (10.101) cannot exceed the one in (10.98) and that the two effective numbers are equal if and only if σ_i is independent of i. We must establish that

$$\left(\sum_i \sigma_i^2 \right)^3 \le \left(\sum_i \sigma_i \right)^2 \left(\sum_i \sigma_i^4 \right) \ . \tag{10.102}$$

By Hölder's inequality (Abramowitz, 1964, p.11),

$$\sum_i \sigma_i^2 = \sum_i \sigma_i^{2/3} \sigma_i^{4/3} \le \left[\sum_i (\sigma_i^{2/3})^{3/2} \right]^{2/3} \left[\sum_i (\sigma_i^{4/3})^3 \right]^{1/3} \ , \tag{10.103}$$

the cube of which is (10.102). Equality occurs in (10.103) if and only $\sigma_i^{4/3} = s(\sigma_i^{2/3})^{1/2} = s\sigma_i^{1/3}$, i.e., $\sigma_i = s$ for all i and some constant $s > 0$ (Abramowitz, 1964, p. 11).

The examples displayed in Table 10.1 were investigated numerically. The root n_e is unique in all cases. Since the function $\Phi(x)$ in (10.95) turns out to depend rather weakly on the value of x for $1 \le x \le n$, even a few iterations of

$$x_{i+1} = \Phi(x_i)$$

(starting with some intuitively reasonable value of x_0, such as 1, $\frac{1}{2}(n+1)$, or n) produce an accurate value of n_e. Note that Examples 7 to 12 have the same respective single-locus variances as Examples 1 to 6. The results agree with the remarks below (10.76) and (10.92). Since, in accordance with (10.100), minor loci have only a small effect on n_e, the effective number of loci is not a good guide to the actual number: compare Example 1 with Example 5, and Example 5 with Example 6. Examples 10 and 12 show that even for h_0^2 as low as $\frac{1}{2}$, if r is close to 1 and $n_e \gg 1$, the relative increase in the variance is quite high. This becomes obvious from (10.67) if one observes that in the limit $r = \gamma = 1$ (10.76) yields

$$\hat{\rho} = \begin{cases} h_0^2/(1 - h_0^2), & h_0^2 < \frac{1}{2} , \\ 1, & h_0^2 \geq \frac{1}{2} . \end{cases}$$

Table 10.1. The increase in variance due to assortative mating

Example	r	h_0^2	n	σ_1^2	σ_2^2	σ_i^2	n_e	$\hat{\rho}$	$\hat{V}_Y/V_Y^0$
1	$\frac{1}{2}$	$\frac{2}{3}$	2	0.80	0.20	−	1.530	0.363	1.216
2	$\frac{1}{2}$	$\frac{2}{3}$	3	0.80	0.10	0.10	1.608	0.364	1.223
3	$\frac{1}{2}$	$\frac{2}{3}$	3	0.98	0.01	0.01	1.052	0.355	1.153
4	$\frac{1}{2}$	$\frac{2}{3}$	10	0.55	0.05	0.05	3.397	0.373	1.311
5	$\frac{1}{2}$	$\frac{2}{3}$	10	0.91	0.01	0.01	1.259	0.359	1.184
6	$\frac{1}{2}$	$\frac{2}{3}$	10	0.15	0.05	0.10	9.548	0.379	1.373
7	1	$\frac{1}{2}$	2	0.80	0.20	−	1.610	0.642	1.397
8	1	$\frac{1}{2}$	3	0.80	0.10	0.10	1.763	0.653	1.439
9	1	$\frac{1}{2}$	3	0.98	0.01	0.01	1.069	0.594	1.231
10	1	$\frac{1}{2}$	10	0.55	0.05	0.05	4.228	0.745	1.964
11	1	$\frac{1}{2}$	10	0.91	0.01	0.01	1.357	0.622	1.324
12	1	$\frac{1}{2}$	10	0.15	0.05	0.10	9.643	0.815	2.696

10.4 The Correlation Between Relatives with Assortative Mating

As in the last section, we assume that an individual's environment is stochastically independent of his genotype and that selection, mutation, epistasis, and dominance are absent. We suppose also that the regressions of an individual's genotype on his phenotype and of his phenotype on that of his mate are linear. Since we may take $\overline{G} = \overline{E} = \overline{Y} = 0$ without loss of generality, these regression hypotheses have the form (10.77) and (10.79). Note that we do not require the more detailed and restrictive condition (10.83). Finally, we posit that equilibrium has been reached under phenotypic assortative mating with phenotypic correlation r.

The following simple, general results will serve vitally in the calculation of all the correlations. Let male B_1 mate with female B_2 to produce offspring B_3. Suppose that at locus i these individuals have the respective genotypes $A_j^{(i)} A_k^{(i)}$, $A_l^{(i)} A_m^{(i)}$, and $A_\kappa^{(i)} A_\lambda^{(i)}$, where the first gene is paternal, and Greek subscripts represent integer-valued random variables that specify allelic state. The script letters $\mathcal{G}_1$ and $\mathcal{G}_2$ signify the genotypes (as opposed to the genotypic *values*, G_1 and G_2) of individuals B_1 and B_2. From Fig. 10.2 we infer

$$\mathcal{E}(G_3|\mathcal{G}_1,\mathcal{G}_2) = \sum_i \mathcal{E}(\alpha_\kappa^{(i)} + \alpha_\lambda^{(i)}|\mathcal{G}_1,\mathcal{G}_2)$$

$$= \sum_i [\mathcal{P}(A_j^{(i)} \to A_\kappa^{(i)})\mathcal{E}(\alpha_\kappa^{(i)}|A_j^{(i)} \to A_\kappa^{(i)})$$

$$+ \mathcal{P}(A_k^{(i)} \to A_\kappa^{(i)})\mathcal{E}(\alpha_\kappa^{(i)}|A_k^{(i)} \to A_\kappa^{(i)})$$

$$+ \mathcal{P}(A_l^{(i)} \to A_\lambda^{(i)})\mathcal{E}(\alpha_\lambda^{(i)}|A_l^{(i)} \to A_\lambda^{(i)})$$

$$+ \mathcal{P}(A_m^{(i)} \to A_\lambda^{(i)})\mathcal{E}(\alpha_\lambda^{(i)}|A_m^{(i)} \to A_\lambda^{(i)})]$$

$$= \tfrac{1}{2} \sum_i (\alpha_j^{(i)} + \alpha_k^{(i)} + \alpha_l^{(i)} + \alpha_m^{(i)})$$

$$= \tfrac{1}{2}(G_1 + G_2) , \tag{10.104}$$

where $A_j^{(i)} \to A_\kappa^{(i)}$ designates the direct descent of $A_\kappa^{(i)}$ from $A_j^{(i)}$, as in Sect. 9.2. This yields immediately

$$\mathcal{E}(G_3|G_1,G_2) = \mathcal{E}[\mathcal{E}(G_3|\mathcal{G}_1,\mathcal{G}_2)|G_1,G_2] = \tfrac{1}{2}(G_1 + G_2) . \tag{10.105}$$

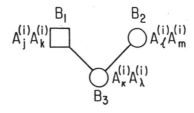

Fig. 10.2. Pedigree for parent and offspring.

The linear biparental genotypic regression (10.104) relieves us from distinguishing the sexes and referring to individual loci and alleles in the ensuing calculations. From (10.104), the stochastic independence of $\mathcal{G}_1$ and $\mathcal{G}_2$ with Y_1 fixed, (10.81), and the stochastic independence of $\mathcal{G}_1$ and E_1, we deduce the linear uniparental genotypic regression

$$\mathcal{E}(G_3|\mathcal{G}_1) = \mathcal{E}[\mathcal{E}(G_3|\mathcal{G}_1,\mathcal{G}_2)|\mathcal{G}_1]$$
$$= \tfrac{1}{2}G_1 + \tfrac{1}{2}\mathcal{E}(G_2|\mathcal{G}_1)$$
$$= \tfrac{1}{2}G_1 + \tfrac{1}{2}\mathcal{E}[\mathcal{E}(G_2|Y_1,\mathcal{G}_1)|\mathcal{G}_1]$$
$$= \tfrac{1}{2}G_1 + \tfrac{1}{2}\mathcal{E}[\mathcal{E}(G_2|Y_1)|\mathcal{G}_1]$$
$$= \tfrac{1}{2}G_1 + \tfrac{1}{2}\mathcal{E}(rh^2 Y_1|\mathcal{G}_1)$$
$$= \tfrac{1}{2}(1 + rh^2)G_1 \ . \tag{10.106}$$

Therefore,

$$\mathcal{E}(G_3|G_1) = \tfrac{1}{2}(1 + rh^2)G_1 \ , \tag{10.107}$$

which shows that the correlation between the genotypic values of parent and offspring is $\tfrac{1}{2}(1+rh^2)$. Since the *transmission* of resemblance is purely genetic, this factor will enter many of the correlations below.

Parent and Offspring

Referring to Fig. 10.2 and using (10.1), (10.104), (10.79), and (10.81), we obtain

$$\mathcal{E}(Y_3|Y_1) = \mathcal{E}(G_3|Y_1)$$
$$= \mathcal{E}[\mathcal{E}(G_3|\mathcal{G}_1,\mathcal{G}_2,Y_1)|Y_1]$$
$$= \mathcal{E}[\mathcal{E}(G_3|\mathcal{G}_1,\mathcal{G}_2)|Y_1]$$
$$= \mathcal{E}[\tfrac{1}{2}(G_1 + G_2)|Y_1]$$
$$= \tfrac{1}{2}h^2(1+r)Y_1 \ . \tag{10.108}$$

Thus, the regression of the offspring phenotype on that of a parent is linear, and the parent-offspring correlation is

$$r_{po} = \tfrac{1}{2}h^2(1+r) \ . \tag{10.109}$$

nth Parent and Offspring

We denote the phenotypic correlation between nth parent ($n = 1, 2, \ldots$, corresponding to parent, grandparent, $\ldots$) and offspring by r_n. We posit inductively that

$$\mathcal{E}(Y_1|Y_n) = r_{n-1}Y_n \tag{10.110}$$

and evaluate r_n for $n \geq 2$ ($r_1 = r_{po}$). Invoking (10.1), the stochastic independence of G_0 and Y_n with $\mathcal{G}_1$ fixed, (10.106), (10.1) again, and (10.110), we have from Fig. 10.3

$$\mathcal{E}(Y_0|Y_n) = \mathcal{E}(G_0|Y_n)$$
$$= \mathcal{E}[\mathcal{E}(G_0|\mathcal{G}_1,Y_n)|Y_n]$$
$$= \mathcal{E}[\mathcal{E}(G_0|\mathcal{G}_1)|Y_n]$$
$$= \tfrac{1}{2}(1 + rh^2)\mathcal{E}(G_1|Y_n)$$
$$= \tfrac{1}{2}(1 + rh^2)r_{n-1}Y_n \ . \tag{10.111}$$

We conclude by induction that the phenotypic regression of the offspring on the nth parent is linear and the correlation satisfies

$$r_n = \tfrac{1}{2}(1 + rh^2)r_{n-1}, \qquad n \geq 2, \qquad (10.112)$$

whence

$$r_n = r_{po}[\tfrac{1}{2}(1 + rh^2)]^{n-1}, \qquad n \geq 1. \qquad (10.113)$$

The linearity of the phenotypic regression has not been proved for the other relationships considered in this section.

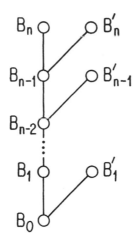

Fig. 10.3. Pedigree for nth parent and offspring.

Full Siblings

Employing (10.1), the stochastic independence of G_3 and G_4 with $\mathcal{G}_1$ and $\mathcal{G}_2$ fixed, (10.104), the stochastic independence of G_1 and G_2 with Y_1 fixed, (10.79), and (10.81), from Fig. 10.4 we deduce

$$
\begin{aligned}
\mathcal{E}(Y_3 Y_4) &= \mathcal{E}(G_3 G_4) \\
&= \mathcal{E}[\mathcal{E}(G_3 G_4 | \mathcal{G}_1, \mathcal{G}_2)] \\
&= \mathcal{E}[\mathcal{E}(G_3 | \mathcal{G}_1, \mathcal{G}_2)\mathcal{E}(G_4 | \mathcal{G}_1, \mathcal{G}_2)] \\
&= \mathcal{E}[\tfrac{1}{2}(G_1 + G_2)\tfrac{1}{2}(G_1 + G_2)] \\
&= \tfrac{1}{2}\mathcal{E}(G_1^2 + G_1 G_2) \\
&= \tfrac{1}{2}V_G + \tfrac{1}{2}\mathcal{E}[\mathcal{E}(G_1 G_2 | Y_1)] \\
&= \tfrac{1}{2}V_G + \tfrac{1}{2}\mathcal{E}[\mathcal{E}(G_1 | Y_1)\mathcal{E}(G_2 | Y_1)] \\
&= \tfrac{1}{2}h^2 V_Y + \tfrac{1}{2}\mathcal{E}(h^2 Y_1 r h^2 Y_1) \\
&= \tfrac{1}{2}h^2(1 + rh^2)V_Y.
\end{aligned} \qquad (10.114)
$$

Hence, the correlation between full siblings reads

$$r_s = \tfrac{1}{2}h^2(1 + rh^2) \ . \tag{10.115}$$

Comparing (10.109) with (10.115), we notice that $r_{po} = r_s$ if and only if $r = 0$ (random mating), $h = 0$ (no genetic variance, so all correlations are zero), or $h = 1$ (no environmental variance). If $0 < h < 1$, then $r_s < r_{po}$ for assortative mating ($r > 0$) and $r_s > r_{po}$ for disassortative mating ($r < 0$).

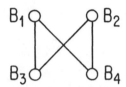

Fig. 10.4. Pedigree for full siblings.

Half-Siblings

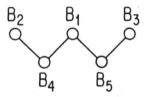

Fig. 10.5. Pedigree for half-siblings.

Glancing at Fig. 10.5, we find from (10.104)

$$\begin{aligned}
\mathcal{E}(Y_4 Y_5) &= \mathcal{E}[\mathcal{E}(G_4 G_5 | \mathcal{G}_1, \mathcal{G}_2, \mathcal{G}_3)] \\
&= \mathcal{E}[\mathcal{E}(G_4 | \mathcal{G}_1, \mathcal{G}_2)\mathcal{E}(G_5 | \mathcal{G}_1, \mathcal{G}_3)] \\
&= \tfrac{1}{4}\mathcal{E}(G_1^2 + 2G_1 G_2 + G_2 G_3) \ . \tag{10.116}
\end{aligned}$$

In deriving (10.114) we showed

$$\mathcal{E}(G_1 G_2) = rh^4 V_Y \ . \tag{10.117}$$

Now, (10.81) gives

$$\begin{aligned}
\mathcal{E}(G_2 G_3) &= \mathcal{E}[\mathcal{E}(G_2 G_3 | Y_1)] \\
&= \mathcal{E}[\mathcal{E}(G_2 | Y_1)\mathcal{E}(G_3 | Y_1)] \\
&= r^2 h^4 V_Y \ . \tag{10.118}
\end{aligned}$$

Substituting (10.117) and (10.118) into (10.116) yields the phenotypic correlation

$$r_{hs} = \tfrac{1}{4}h^2(1 + 2rh^2 + r^2h^2) \tag{10.119}$$

between half-siblings (Nagylaki, 1978, 1982; Gimelfarb, 1979, 1981). This result disagrees with Bulmer's (1980, p. 129).

Observe that (if $h > 0$) $r_{hs} = \tfrac{1}{2}r_s$ for $r = 0$ and -1, $r_{hs} > \tfrac{1}{2}r_s$ for $0 < r \le 1$, and $r_{hs} < \tfrac{1}{2}r_s$ for $-1 < r < 0$.

Uncle and Niece

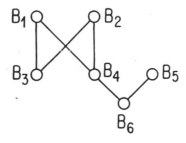

Fig. 10.6. Pedigree for uncle and niece.

From Fig. 10.6, (10.106), (10.114), and (10.115) we derive

$$
\begin{aligned}
\mathcal{E}(Y_3Y_6) &= \mathcal{E}[\mathcal{E}(G_3G_6|\mathcal{G}_4)] \\
&= \mathcal{E}[\mathcal{E}(G_3|\mathcal{G}_4)\mathcal{E}(G_6|\mathcal{G}_4)] \\
&= \mathcal{E}[\mathcal{E}(G_3|\mathcal{G}_4)\tfrac{1}{2}(1 + rh^2)G_4] \\
&= \tfrac{1}{2}(1 + rh^2)\mathcal{E}(G_3G_4) \\
&= \tfrac{1}{2}(1 + rh^2)r_sV_Y \ ,
\end{aligned}
\tag{10.120}
$$

whence

$$r_{un} = \tfrac{1}{2}(1 + rh^2)r_s = h^2[\tfrac{1}{2}(1 + rh^2)]^2 \ . \tag{10.121}$$

Note that $r_{un} = \tfrac{1}{2}r_s$ if at least one of the conditions (i) $r = 0$, (ii) $h = 0$, or (iii) $h = 1$ and $r = -1$ holds. Otherwise, $r > 0$ implies that $r_{un} > \tfrac{1}{2}r_s$ and $r < 0$ implies that $r_{un} < \tfrac{1}{2}r_s$. Furthermore, $r_{un} = r_{hs}$ if $r = 0$, $h = 0$, or $h = 1$; otherwise, $r_{un} < r_{hs}$.

First Cousins

Figure 10.7, (10.106), (10.120), and (10.121) give

$$
\begin{aligned}
\mathcal{E}(Y_7Y_8) &= \mathcal{E}[\mathcal{E}(G_7G_8|\mathcal{G}_4)] \\
&= \mathcal{E}[\mathcal{E}(G_7|\mathcal{G}_4)\mathcal{E}(G_8|\mathcal{G}_4)] \\
&= \tfrac{1}{2}(1 + rh^2)\mathcal{E}(G_4G_8) \\
&= \tfrac{1}{2}(1 + rh^2)r_{un}V_Y \ ,
\end{aligned}
\tag{10.122}
$$

from which

$$r_{fc} = \tfrac{1}{2}(1 + rh^2)r_{un} = h^2[\tfrac{1}{2}(1 + rh^2)]^3 \ . \qquad (10.123)$$

Notice that the relation between r_{fc} and r_{un} is exactly the same as between r_{un} and r_s.

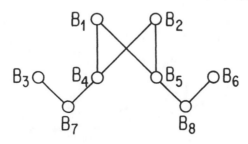

Fig. 10.7. Pedigree for first cousins.

General Remarks

Observe that all the above correlations are independent of the linkage map of the loci that influence the character. Compared with panmixia, assortative mating raises the correlations between relatives, whereas disassortative mating lowers them.

For random mating, we have seen that even with arbitrary genetic determination of the character, there is a general formula for the correlation between any relatives. In contrast, for assortative mating, despite the restrictive assumptions in this section, we have results only for certain relationships. Developing a formula even for a particular class of relatives (e.g., unilineal) would be conceptually instructive. In this connection, see the discussion of double first cousins below.

In his classic 1918 paper, Fisher used ingenious intuitive arguments to compute the correlations between relatives at equilibrium under phenotypic assortative mating. He posited stochastic independence of the environment, control of the character by many unlinked loci with contributions of the same order of magnitude, and the absence of epistasis. Because of the importance of his results, we reproduce them in our notation. As in (10.19), h^2 and D^2 denote the narrow heritability and the proportion of the phenotypic variance due to dominance; r represents the correlation between marital phenotypes. Then the descendant-ancestor, sibling-sibling, uncle-niece, and first-cousin correlations read (Fisher, 1918)

$$r_n = \tfrac{1}{2}h^2(1+r)[\tfrac{1}{2}(1+rh^2)]^{n-1} \ , \tag{10.124a}$$

$$r_s = \tfrac{1}{2}h^2(1+rh^2) + \tfrac{1}{4}D^2 \ , \tag{10.124b}$$

$$r_{un} = \tfrac{1}{4}h^2(1+rh^2)^2 + \tfrac{1}{8}rh^2 D^2 \ , \tag{10.124c}$$

$$r_{fc} = \tfrac{1}{8}h^2(1+rh^2)^3 + \tfrac{1}{16}r^2h^4 D^2 \ . \tag{10.124d}$$

In the absence of dominance, these formulae agree with the ones derived above. (Fisher did not treat half-siblings.) The dominance variance does not contribute to the correlation between nth parent and offspring; it increases the correlations between full siblings and first cousins; and it increases the uncle-niece correlation for assortative mating ($r > 0$) and decreases it for disassortative mating ($r < 0$). Only the uncle-niece correlation can ever be negative, and this can occur only for disassortative mating in the presence of dominance.

Wright (1921) obtained the parent-offspring and sib-sib correlations without dominance by applying his method of path coefficients.

Malécot (1939) proved for diallelic loci that Fisher's expressions for the parental and sibling correlations hold in the limit as the number of loci tends to infinity.

Although in the model of Wilson (1973) an individual's total probability of mating depends on his phenotype, and hence the gene frequencies are not constant, in a special case her ancestral and sibling correlations agree with Fisher's.

In the first concise derivation of the correlations, the same assumptions were made as in this section, except that instead of (10.77) and (10.79) it was supposed that the regression of the phenotype or genotype of an individual on the phenotype or genotype of any of his relatives is linear (Nagylaki, 1978). Risch (1979) extended this analysis to incorporate an X-linked component in the determination of the trait. Gimelfarb (1979, 1981) deduced the results of this section (including the linearity of the ancestral regression) under the weaker hypotheses (10.77) and (10.79). Since he introduced explicitly various probability distributions, his treatment is more computational than the one presented here, which follows Nagylaki (1982).

The results of Bulmer (1980, pp. 128–129) agree with (10.124).

Double First Cousins

We close this section with a discussion of the correlation between double first cousins (Fig. 10.8). Although this particular problem is not especially important in itself, the absence of a definitive solution suggests that we have not described or understood our model of assortative mating in sufficient detail or depth.

Fisher (1918) obtained

$$r_{dfc} = \tfrac{1}{4}h^2(1+3rh^2) + \tfrac{1}{16}D^2$$

by the same intuitive argument that gives the correct result (10.124b) for full siblings.

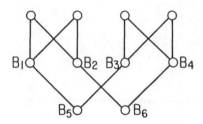

Fig. 10.8. Pedigree for double first cousins.

Bulmer's formula (Bulmer, 1980, p. 129, and personal communication),

$$r_{dfc} = \tfrac{1}{4}h^2(1+rh^2)^2 + \tfrac{1}{16}D^2 \ ,$$

which presupposes multivariate normality, does not appear to take into account the complexity of the mating structure in Fig. 10.8. For example, Bulmer assumed that the genic correlation between siblings in Fig. 10.8 is the same as for siblings chosen at random, $\tfrac{1}{2}(1+rh^2)$. Since B_1 and B_2 mate with correlated individuals (the siblings B_3 and B_4), we might expect the correlation between them to exceed this value for $r > 0$. He assumed also that the genic correlation between B_1 and B_4 is the product of the genic correlations between mates, rh^2, and siblings.

Gimelfarb (1981a) has examined this interesting problem in detail in the absence of dominance. With the assumptions in this section, he found that the results still depend on the distribution of the trait in the population. Under strong hypotheses of multivariate normality in the pedigree, he obtained the following.

Let

$$\rho = rh^2 , \qquad \rho_s = \tfrac{1}{2}(1+\rho)$$

designate the genic correlations between mates and siblings in the general population. The distribution of the character in the parents of double first cousins differs from that in the general population; its variance,

$$V_Y^* = \left[1 - h^2 + \left(\frac{1+\rho^2\rho_s^2}{1-\rho^2\rho_s^2}\right)h^2\right]V_Y \ ,$$

always exceeds the population variance V_Y. Gimelfarb could not calculate the variance in double first cousins in terms of r, h, and V_Y, and hence could not convert the phenotypic covariance between double first cousins,

$$C_{dfc} = \frac{2\rho_s^3 h^2 V_Y}{1-\rho^2\rho_s^2} \ ,$$

to a correlation that depends only on r and h. Note that both V_Y^* and C_{dfc} diverge as $\rho \to 1$. Unlike Fisher's correlation, this covariance is nonnegative for all r and h.

For the genic correlations between mates, siblings, and the affinate individuals B_1 and B_4 in Fig. 10.8, Gimelfarb's formulae lead to

$$\rho_{13} = \frac{\rho(1 + \rho_s^2)}{1 + \rho^2 \rho_s^2}, \qquad \rho_{12} = \frac{\rho_s(1 + \rho^2)}{1 + \rho^2 \rho_s^2}, \qquad \rho_{14} = \frac{2\rho\rho_s}{1 + \rho^2 \rho_s^2} .$$

Observe that, as one expects intuitively, $|\rho_{13}| \geq |\rho|$, $\rho_{12} \geq \rho_s$, and $|\rho_{14}| \geq \rho_{12}|\rho_{13}| \geq \rho_s|\rho|$, with equality throughout if and only if $\rho = 0$, 1, or -1. These relations disagree with Bulmer's assumptions.

As a rough approximation, we might suppose that the variance in double first cousins is the same as in their parents. Then

$$r_{dfc} = \frac{C_{dfc}}{V_Y^\lambda} = \frac{2h^2 \rho_s^3}{1 - \rho^2 \rho_s^2 + 2h^2 \rho^2 \rho_s^2} .$$

It is easy to show that $0 \leq r_{dfc} \leq 1$. For weak assortative mating, we recover Fisher's formula without dominance:

$$r_{dfc} = \tfrac{1}{4}h^2(1 + 3\rho) + O(\rho^2)$$

as $\rho \to 0$. This disagrees with the corresponding approximation of Bulmer's result.

10.5 Selection

Since a basic, rigorous theory of selection over evolutionary time scales is just being developed, we shall restrict ourselves to the short-term effects of selection. We shall treat these first by positing linearity of the biparental phenotypic regression. This approach will then be illuminated by a more detailed analysis through gene-frequency change. The latter is an essential ingredient of the rigorous study of evolutionary problems in quantitative genetics.

Consult Falconer (1960, Chaps. 11–13, 19–20), Bulmer (1980, Chaps. 9–12) and the references at the end of the introduction to this chapter for further discussion of selection.

1. Linear Regression

Let Y, $Y^\star$, and Y' denote the paternal, maternal, and offspring phenotypes, respectively. Although sex differences are easy to incorporate (Bulmer, 1980, pp. 144-145), for simplicity we suppose that they are absent. We assume that the biparental regression is linear. Then it must have the form

$$\mathcal{E}(Y'|Y, Y^\star) = \overline{Y}'_o + b(\tilde{Y} - \overline{Y}) , \tag{10.125a}$$

where $\overline{Y}'_o$, $\overline{Y}$, and $\tilde{Y}$ designate the offspring mean without selection, the parental mean before selection, and the mid-parent value

$$\tilde{Y} = \tfrac{1}{2}(Y + Y^\star) , \tag{10.125b}$$

respectively, and b is a constant. We shall discuss the value of the regression coefficient b and the linearity of regression below. We use S to indicate that the parents have been selected in some manner on the basis of their phenotype and evaluate the offspring mean (in juveniles) from (10.125) as

$$
\begin{aligned}
\overline{Y}' &= \mathcal{E}(Y'|S) \\
&= \mathcal{E}[\mathcal{E}(Y'|Y,Y^*,S)|S] \\
&= \mathcal{E}[\mathcal{E}(Y'|Y,Y^*)|S] \\
&= \overline{Y}'_o + b(\overline{Y}_s - \overline{Y}) \ ,
\end{aligned}
\tag{10.126a}
$$

where

$$
\overline{Y}_s = \mathcal{E}(Y|S) = \mathcal{E}(Y^*|S)
\tag{10.126b}
$$

represents the mean of the reproducing parents.

We call the deviation of the mean of the selected parents from that of the entire parental population the *selection differential*,

$$
S = \overline{Y}_s - \overline{Y} \ ;
\tag{10.127a}
$$

the *response* to selection,

$$
R = \overline{Y}' - \overline{Y}'_o \ ,
\tag{10.127b}
$$

is the deviation of the mean of the progeny from the value expected without selection. Inserting (10.127) into (10.126) produces the classic equation

$$
R = bS \ .
\tag{10.128}
$$

This statistical relation involves no genetics; indeed, it goes back to a series of papers written by Pearson in the 1890's.

The *selection intensity*,

$$
I = S/\sigma_Y \ ,
\tag{10.129}
$$

provides a convenient dimensionless measure of the strength of selection.

If the population is initially at equilibrium in the absence of selection, then

$$
\overline{Y}'_o = \overline{Y}
\tag{10.130}
$$

for the first generation, and hence (10.128) gives

$$
\Delta\overline{Y} = bS \ .
\tag{10.131}
$$

As discussed in detail by Falconer (1960, Chaps. 11-13, 19), after a generation of selection (10.130) may be violated. Furthermore, even if the linearity of the regression (10.125) is preserved under selection, b will generally change as the genotypic frequencies evolve. For at least several generations of weak selection, however, we expect (10.130) and the constancy of b to be good approximations. In that case, (10.131) implies that for a constant selection differential S, the mean of the character will differ from its initial value by about tbS after t generations of selection.

Next, we turn to the value of b and the linearity of regression. We confine ourselves to a single generation of selection and posit initial equilibrium. Then we may choose $\overline{Y}'_o = \overline{Y} = 0$ without loss of generality, and this reduces (10.125a) to

$$\mathcal{E}(Y'|Y,Y^*) = b\tilde{Y} \ . \tag{10.132}$$

If r_{po} and r represent the phenotypic parent-offspring and marital correlations, respectively, then (10.132) and (10.125b) yield

$$\begin{aligned} V_Y r_{po} &= \mathcal{E}(YY') \\ &= \mathcal{E}[Y\mathcal{E}(Y'|Y,Y^*)] \\ &= \tfrac{1}{2}b(1+r)V_Y \ , \end{aligned}$$

whence

$$b = 2r_{po}/(1+r) \ . \tag{10.133}$$

For phenotypic assortative mating with a stochastically independent environment, many comparable unlinked loci, and no epistasis, (10.124a) yields (Fisher, 1918; Malécot, 1939)

$$r_{po} = \tfrac{1}{2}h^2(1+r) \ , \tag{10.134}$$

where h^2 denotes the narrow heritability. According to (10.109), if there is also no dominance and the linear regression hypotheses (10.77) and (10.79) hold, then (10.134) applies to an arbitrary number of loci and any linkage map. In view of (10.36), the special case of (10.134) with $r = 0$ holds with panmixia for any number of loci and any linkage map, provided epistasis and genotype-environment correlation and interaction are absent. Substituting (10.134) into (10.133) establishes that, at least in these three cases, $b = h^2$ and therefore

$$\Delta\overline{Y} = h^2 S \ . \tag{10.135}$$

This basic result presupposes the linearity of the biparental regression. We can demonstrate this linearity for phenotypic assortative mating if we posit that the environment is stochastically independent, dominance and epistasis are absent, and (10.79) holds: from (10.104) and (10.79) we obtain

$$\begin{aligned} \mathcal{E}(Y'|Y,Y^*) &= \mathcal{E}[\mathcal{E}(G'|\mathcal{G},\mathcal{G}^*,Y,Y^*)|Y,Y^*] \\ &= \mathcal{E}[\mathcal{E}(G'|\mathcal{G},\mathcal{G}^*)|Y,Y^*] \\ &= \mathcal{E}[\tfrac{1}{2}(G+G^*)|Y,Y^*] \\ &= \tfrac{1}{2}\mathcal{E}(G|Y) + \tfrac{1}{2}\mathcal{E}(G^*|Y^*) \\ &= h^2\tilde{Y} \ . \end{aligned} \tag{10.136}$$

If the character has probability density $f(y)$ and is under viability selection with fitness $w(y)$, then

$$S = \frac{1}{\overline{w}}\int_{-\infty}^{\infty} yw(y)f(y)dy - \overline{Y} = \frac{\mathrm{Cov}\,(Y,w)}{\overline{w}} \ .$$

In particular, if the character is viability, then $S = V_w/\overline{w}$, and hence (10.135) becomes

$$\Delta\overline{w} = V_g/\overline{w} \ ,$$

where V_g denotes the genic variance in fitness. Notice the agreement with (8.64).

Metrical characters are often approximately normally distributed. If the number of loci is large, approximate normality will frequently follow from the Central Limit Theorem (Feller, 1971, p. 262). Note, however, that if the terms in (10.144) below are stochastically independent, *exact* normality of the phenotypic distribution would imply normality of the single-locus contributions and of the environment (Feller, 1971, p. 525). If the number of alleles at locus i is finite, $G^{(i)}$ cannot be normally distributed. In the following two examples, we suppose that the character has the normal distribution with probability density

$$f(y) = (2\pi V_Y)^{-1/2} e^{-y^2/(2V_Y)} \ . \tag{10.137}$$

(a) Truncation Selection

Truncation selection for quantitative characters is very important in plant and animal breeding, but is unlikely to occur in natural populations. Only individuals with phenotypic value at least as great as some number c are permitted to reproduce. In terms of the standard normal variable $X = Y/\sigma_Y$, the requirement $Y \geq c$ becomes $X \geq \gamma = c/\sigma_Y$. The proportion of adults saved for reproduction reads

$$\psi(\gamma) = \int_c^\infty f(y)dy = \int_\gamma^\infty \phi(x)dx \ , \tag{10.138}$$

where

$$\phi(x) = (2\pi)^{-1/2} e^{-x^2/2} \tag{10.139}$$

is the density of X. After selection, Y has density $f(y)/\psi(\gamma)$ for $y \geq c$; the selection differential S is the mean of this distribution. From (10.129), (10.137), and (10.139) we obtain the selection intensity

$$I = \phi(\gamma)/\psi(\gamma) \ . \tag{10.140}$$

Common sense suggests that I must increase as the truncation point is moved up. Indeed, successive utilization of the derivatives

$$\frac{d}{d\gamma}\phi(\gamma) = -\gamma\phi(\gamma), \qquad \frac{d}{d\gamma}\psi(\gamma) = -\phi(\gamma)$$

and the inequality $\gamma\psi(\gamma) < \phi(\gamma)$ (Feller, 1968, p. 175) proves that

$$\frac{dI}{d\gamma} > 0 \ .$$

However, the selection intensity increases rather slowly as the proportion saved decreases: the values of I corresponding to $\psi = 0.90$, 0.50, and 0.01 are approximately 0.20, 0.80, and 2.66.

For weak selection, we may insert $\psi(\gamma) \approx 1$ into (10.140). If selection is so strong that $\gamma \gg 1$, which implies that $\psi(\gamma) \ll 1$, then the asymptotic formula $\psi(\gamma) \sim \phi(\gamma)/\gamma$ (Feller, 1968, p. 175) simplifies (10.140) to $I \sim \gamma$.

(b) Stabilizing Selection

Selection for an intermediate optimum is generally considered to be common in natural populations. Assume that an individual with phenotypic value y survives to reproduce with a probability proportional to (Haldane, 1954)

$$w(y) = e^{-(y-y_o)^2/(2V_s)} \ .$$

(10.141)

Evidently, y_o designates the optimum phenotype, and increasing V_s decreases the selection intensity. The distribution of adult phenotypes is proportional to $w(y)f(y)$, and hence is Gaussian with mean

$$S = \frac{V_Y y_o}{V_Y + V_s}$$

(10.142a)

and variance

$$\beta^2 = \frac{V_Y V_s}{V_Y + V_s} \ .$$

(10.142b)

For weak selection, $V_s \gg V_Y$, so (10.142) can be approximated by

$$S \approx V_Y y_o/V_s , \qquad \beta^2 \approx V_Y \ ;$$

(10.143a)

the corresponding results for strong selection, $V_s \ll V_Y$, read

$$S \approx y_o , \qquad \beta^2 \approx V_s \ .$$

(10.143b)

2. Gene-Frequency Change

We now rederive (10.135) by evaluating the effect of selection on gene frequencies. This general approach was first employed by Kimura (1958) for truncation selection with a Gaussian phenotypic distribution. We follow Nagylaki (1984).

We assume that the population is at equilibrium under random mating, environment and genotype are stochastically independent, and there is no epistasis. Then (10.1) and (10.8) enable us to write

$$Y = \sum_i G^{(i)} + E \ ,$$

(10.144)

in which the terms on the right-hand side are mutually stochastically independent. Selection on the phenotype generally produces linkage disequilibrium, so after one generation the single-locus contributions in (10.144) will usually be correlated. We suppose also that each locus contributes only a small fraction of the phenotypic variance. This will be the case if either there are many comparable loci or the broad heritability is low with arbitrarily many loci.

From (10.144), (10.11), (10.12), and (10.13) we find

$$
\begin{aligned}
\Delta \overline{Y} &= \sum_i \Delta \overline{G}^{(i)} \\
&= \sum_{ijk} G_{jk}^{(i)} \left(p_j^{(i)} \Delta p_k^{(i)} + p_k^{(i)} \Delta p_j^{(i)} + \Delta p_j^{(i)} \Delta p_k^{(i)} \right) \\
&= 2 \sum_{ij} g_j^{(i)} \Delta p_j^{(i)} + \sum_{ijk} g_{jk}^{(i)} \Delta p_j^{(i)} \Delta p_k^{(i)} \ .
\end{aligned}
\tag{10.145}
$$

Let $w(y)$, $w_{jk}^{(i)}$, $w_j^{(i)}$, and $\overline{w}$ denote the viabilities of an individual with phenotype y, one with genotype $A_j^{(i)} A_k^{(i)}$ at locus i, one carrying $A_j^{(i)}$ at locus i, and one chosen at random, respectively. Then

$$
\Delta p_j^{(i)} = p_j^{(i)} v_j^{(i)} / \overline{w} \ ,
\tag{10.146}
$$

where

$$
v_j^{(i)} = w_j^{(i)} - \overline{w}
\tag{10.147}
$$

represents the average excess of $A_j^{(i)}$ for fitness. Substituting (10.146) into (10.145) gives

$$
\Delta \overline{Y} = 2\overline{w}^{-1} \sum_{ij} p_j^{(i)} g_j^{(i)} v_j^{(i)} + \overline{w}^{-2} \sum_{ijk} p_j^{(i)} p_k^{(i)} g_{jk}^{(i)} v_j^{(i)} v_k^{(i)} \ .
\tag{10.148}
$$

Observe that the second sum vanishes in the absence of dominance. For related expressions in continuous time, see (4.192), Kimura (1958), Crow and Nagylaki (1976), and Nagylaki (1989). Griffing (1960) treats truncation selection with a Gaussian phenotype and discrete, nonoverlapping generations.

Since (10.148) involves only deviations from single-locus means, we may conveniently take $\overline{G}^{(i)} = 0$ and $\overline{E} = 0$ in the remainder of this section. Primes will designate derivatives throughout. With an environmental contribution to the character, it is reasonable to posit that Y has a density $f(y)$ with a continuous, bounded second derivative. The mean fitness reads

$$
\overline{w} = \int_{-\infty}^{\infty} w(y) f(y) dy \ .
\tag{10.149}
$$

If $l_{jk}^{(i)}(y)$ signifies the density of Y in individuals with genotype $A_j^{(i)} A_k^{(i)}$ at locus i, we have

$$
w_{jk}^{(i)} = \int_{-\infty}^{\infty} w(y) l_{jk}^{(i)}(y) dy \ .
\tag{10.150}
$$

To approximate $l_{jk}^{(i)}$, we rely on the stochastic independence of the terms in (10.144) and the assumption that the relative contribution of each locus is small. Here, we shall require an approximation only to first order in $g_{jk}^{(i)}$. At the end of this section, by deriving Hastings' (1990) second-order approximation, we shall prove that

$$l_{jk}^{(i)}(y) \approx f(y - g_{jk}^{(i)}) \tag{10.151}$$

to first order, but only to first order (Nagylaki, 1984). Fisher (1918), Haldane (1931), Kimura (1958, 1981), Griffing (1960), Latter (1970, 1972), Bulmer (1971; 1972a; 1980, Chap. 10), and Kimura and Crow (1978) have used (10.151). The validity of (10.151) with epistasis has not been demonstrated. Problem 10.16 will reveal that (10.151) generally fails in the presence of linkage disequilibrium, although we expect first-order accuracy when linkage disequilibrium is sufficiently weak.

Expanding (10.151) in a Taylor series and substituting into (10.150) leads to

$$w_{jk}^{(i)} \approx \overline{w} + g_{jk}^{(i)} J \;, \tag{10.152}$$

where

$$J = - \int_{-\infty}^{\infty} w(y) f'(y) dy \;. \tag{10.153}$$

Inserting (10.152) into (10.147) yields (Kimura and Crow, 1978)

$$v_j^{(i)} \approx g_j^{(i)} J \;. \tag{10.154}$$

We wish to approximate (10.148) to the leading order in the single-locus effects. According to (10.154), this order is the second, and hence with a third-order error we obtain from (10.13), (10.15b), (10.17b), (10.148), and (10.154)

$$\Delta \overline{Y} \approx \overline{w}^{-1} J V_A \;, \tag{10.155}$$

where V_A denotes the additive genetic variance of the trait.

We treat two examples before investigating (10.151).

(a) Truncation Selection

The fitness has the form

$$w(y) = \begin{cases} 0, & y < c \;, \\ 1, & y \geq c \;. \end{cases} \tag{10.156}$$

From (10.149), (10.153), and (10.156) we obtain (Kimura and Crow, 1978)

$$\overline{w} = \int_c^{\infty} f(y) dy \;, \tag{10.157}$$

$$J = f(c) \;. \tag{10.158}$$

(b) Gaussian Phenotype

For the normal distribution (10.137), we have

$$f'(y) = -yf(y)/V_Y \; ; \tag{10.159}$$

substituting this into (10.153) informs us

$$J = \overline{w}S/V_Y \; , \tag{10.160}$$

in which

$$S = \overline{w}^{-1} \int_{-\infty}^{\infty} yw(y)f(y)dy \tag{10.161}$$

represents the selection differential. Inserting (10.154) and (10.160) into (10.146) reveals (Kimura and Crow, 1978)

$$\Delta p_j^{(i)} \approx p_j^{(i)} g_j^{(i)} S/V_Y \; ; \tag{10.162}$$

(10.160) immediately simplifies (10.155) to

$$\Delta \overline{Y} \approx h^2 S \; , \tag{10.163}$$

in agreement with (10.135). Since our derivation of (10.135) was crucially based on linear regressions, it is hardly surprising that we have recovered this important formula under the assumption of phenotypic normality.

For truncation selection (Haldane, 1931; Kimura, 1958; Griffing, 1960), we may calculate S from (10.129) and (10.140).

If selection is stabilizing with the Gaussian fitness function (10.141), then S is given by (10.142a). Using (10.141) and (10.149) leads to the mean relative fitness (Kimura and Crow, 1978)

$$\overline{w} = \left(\frac{V_s}{V_Y + V_s} \right)^{1/2} \exp \left[-\frac{y_o^2}{2(V_s + V_Y)} \right] \; . \tag{10.164}$$

(c) The Approximation (10.151)

Finally, we return to the critical approximation (10.151). Set $X = G^{(i)}$, for some i, and $Z = Y - X$ in (10.144), so that

$$Y = X + Z \; , \tag{10.165}$$

where X and Z have mean zero and are stochastically independent. We denote the probability distribution functions of Y, X, and Z by F, K, and Q, respectively. Then (10.165) and the mutual stochastic independence of X and Z yield

$$\mathcal{P}(Y \leq y | X = x) = \mathcal{P}(Z \leq y - x | X = x) = \mathcal{P}(Z \leq y - x) = Q(y - x) \; . \tag{10.166}$$

Therefore, we wish to examine the approximation $Q(y - x) \approx F(y - x)$, i.e.,

$$Q(y) \approx F(y) . \tag{10.167}$$

In terms of probability densities, this yields (10.151) in our simplified notation:

$$q(y) \approx f(y) . \tag{10.168}$$

Let

$$\tilde{f}(\xi) = \mathcal{E}(e^{i\xi Y}), \qquad \tilde{k}(\xi) = \mathcal{E}(e^{i\xi X}), \qquad \tilde{q}(\xi) = \mathcal{E}(e^{i\xi Z}) \tag{10.169}$$

(here and in (10.172) below, $i = \sqrt{-1}$) designate the respective characteristic functions of Y, X, and Z. By (10.165) and the mutual stochastic independence of X and Z, we obtain (Feller, 1971, p. 500)

$$\tilde{f}(\xi) = \tilde{k}(\xi)\tilde{q}(\xi) , \tag{10.170}$$

whence

$$\tilde{q}(\xi) = \tilde{f}(\xi)/\tilde{k}(\xi) . \tag{10.171}$$

Inversion of the Fourier transform in (10.171) determines Q in terms of F and K.

If the single-locus contribution X has moments of all orders, then (Feller, 1971, p. 512)

$$\tilde{k}(\xi) = \sum_{j=0}^{\infty} \frac{1}{j!} \mathcal{E}[(i\xi X)^j] = \sum_{j=0}^{\infty} \frac{(i\xi)^j}{j!} \mu_j , \tag{10.172}$$

where $\mu_j = \mathcal{E}(X^j)$. Setting $\sigma^2 = \mathrm{Var}(X)$, we have

$$\mu_0 = 1, \qquad \mu_1 = 0, \qquad \mu_2 = \sigma^2 . \tag{10.173}$$

Under the assumption that $\mu_j = O(\sigma^j)$ as $\sigma \to 0$, (10.172) and (10.173) give

$$\tilde{k}(\xi) = 1 - \tfrac{1}{2}\sigma^2 \xi^2 + O(\sigma^3) . \tag{10.174}$$

Inserting (10.174) into (10.171) produces

$$\tilde{q}(\xi) = (1 + \tfrac{1}{2}\sigma^2 \xi^2)\tilde{f}(\xi) + O(\sigma^3) , \tag{10.175}$$

which inverts to (Feller, 1971, p. 514)

$$q(y) = f(y) - \tfrac{1}{2}\sigma^2 f''(y) + O(\sigma^3) \tag{10.176}$$

if f has an integrable second derivative. Thus, (10.168) is correct to first order in σ, but not to second.

Hastings (1990) presented (10.176) as an *Ansatz* and verified that it holds to second order. Our derivation shows that more accurate approximations could be deduced by using more terms in (10.172).

Equation (10.176) enables us to evaluate the changes in the gene frequencies to second order. From (10.166) and (10.176) we find

$$\ell_{jk}^{(i)}(y) = f(y - g_{jk}^{(i)}) - \tfrac{1}{2}\sigma_i^2 f''(y - g_{jk}^{(i)}) + O(\sigma_i^3)$$

$$= f(y) - g_{jk}^{(i)} f'(y) + \tfrac{1}{2}[(g_{jk}^{(i)})^2 - \sigma_i^2]f''(y) + O(\sigma_i^3) \ , \quad (10.177)$$

where $\sigma_i^2 = \text{Var}\,(g^{(i)})$. Substituting (10.177) into (10.150) and recalling (10.149) and (10.153), we get

$$w_{jk}^{(i)} = \overline{w} + g_{jk}^{(i)} J + \tfrac{1}{2}[(g_{jk}^{(i)})^2 - \sigma_i^2]L + O(\sigma_i^3) \ , \qquad (10.178)$$

in which

$$L = \int_{-\infty}^{\infty} w(y)f''(y)dy \ . \qquad (10.179)$$

We invoke (10.178) and (10.13) to obtain (Hastings, 1990)

$$w_j^{(i)} = \sum_k w_{jk}^{(i)} p_k^{(i)} = \overline{w} + g_j^{(i)} J + \tfrac{1}{2}(V_j^{(i)} - \sigma_i^2)L + O(\sigma_i^3) \ , \qquad (10.180a)$$

where

$$V_j^{(i)} = \sum_k (g_{jk}^{(i)})^2 p_k^{(i)} \ . \qquad (10.80b)$$

Inserting (10.180a) into (10.146) and (10.147) leads to

$$\Delta p_j^{(i)} = \frac{p_j^{(i)}}{\overline{w}}[g_j^{(i)} J + \tfrac{1}{2}(V_j^{(i)} - \sigma_i^2)L + O(\sigma_i^3)] \ . \qquad (10.181)$$

This result was derived first from a tentative improvement of (10.151) (Nagylaki, 1984) and then from (10.176) (Hastings, 1990).

Since Latter's (1970, 1972), Bulmer's (1971; 1972a; 1980, Chap. 10), and Kimura's (1981) studies of stabilizing selection depend crucially on second-order terms that come from the first-order approximation (10.151), it is important to compare (10.181) with the corresponding formula from (10.151). Had we used (10.151) instead of (10.176), the terms proportional to σ_i^2 would have been missing from (10.177), (10.178), and (10.180a). Nevertheless, if $\overline{w}$ is evaluated by averaging this defective version of (10.180a), then (10.146) and (10.147) again produce (10.181) (Nagylaki, 1984). Consequently, for two alleles without dominance at locus i, (10.181) reduces to a result of Bulmer (1971; 1980, p. 167) and Kimura (1981). It must be noted, however, that this use of (10.151) is inconsistent: (10.15a) and (10.180b) demonstrate that averaging (10.180a) correctly reproduces (10.149), whereas averaging its defective version augments (10.149) by $\tfrac{1}{2}\sigma_i^2 L$, which depends even on the locus under consideration (Nagylaki, 1984). Consult Walsh (1990) and Hastings (1990) for further discussion of consistency conditions.

10.6 Mutation-Selection Balance

In this section, we examine the equilibrium distribution of a character under the joint action of mutation and stabilizing selection. Lande (1976) has argued on theoretical and empirical grounds that this is a major mechanism for the maintenance of genetic variability in quantitative traits. Our analysis relies on the lucid, rigorous work of Fleming (1979).

We posit random mating and the absence of dominance and epistasis. The environmental contribution to the character is stochastically independent of genotype and is Gaussian with mean zero and variance V_E. The number of loci, n, and the linkage map are arbitrary. Selection acts solely through viability differences. Let X_i denote the (random) contribution of an allele at locus i to the phenotype of an individual. We use a continuous model of mutation in which the allelic effects X_i have a probability density (Kimura, 1965a), and we designate the joint density of these effects in gametes by $p(\mathbf{x})$.

We translate the character Y so that its optimum value is zero: an individual with $Y = y$ survives to reproduce with probability proportional to

$$w(y) = e^{-y^2/(2V_s)} \ . \tag{10.182}$$

An individual formed by the union of gametes with allelic effects $\mathbf{x}$ and $\mathbf{x}'$ has genotypic value

$$g = \sum_{i=1}^{n}(x_i + x_i') \ ; \tag{10.183}$$

from (10.1) and (10.182) we calculate his fitness as

$$W(\mathbf{x}, \mathbf{x}') = (2\pi V_E)^{-1/2} \int_{-\infty}^{\infty} \exp\left[-\frac{(g+z)^2}{2V_s} - \frac{z^2}{2V_E}\right] dz$$

$$= (V_s/V)^{1/2} e^{-g^2/(2V)} \ , \tag{10.184}$$

where $V = V_E + V_s$. Designating the density of Y by $f(y)$, we have for the mean fitness

$$\overline{w} = \int \int W(\mathbf{x}, \mathbf{x}')p(\mathbf{x})p(\mathbf{x}')dx\,dx' \tag{10.185}$$

$$= \int_{-\infty}^{\infty} w(y)f(y)dy \ ; \tag{10.186}$$

the integration in (10.185) is over all components of $\mathbf{x}$ and $\mathbf{x}'$. The density of the allelic effects in adults has the form

$$P^\star(\mathbf{x}, \mathbf{x}') = W(\mathbf{x}, \mathbf{x}')p(\mathbf{x})p(\mathbf{x}')/\overline{w} \ . \tag{10.187}$$

Next, we introduce recombination. Let I denote a proper subset of $\{1, 2, \ldots, n\}$ that includes 1, and let J represent its complement. We designate by c_I the probability of reassociation of the genes at loci in I inherited from one parent with the genes at loci in J inherited from the other. Let $(\mathbf{x}_I, \mathbf{x}'_J)$ signify the vector with components x_i for i in I and x'_j for j in J. Then recombination produces the gametic distribution

$$p^{**}(\mathbf{x}) = \int P^{**}(\mathbf{x}, \mathbf{x}')dx' \; , \qquad (10.188\text{a})$$

where

$$P^{**}(\mathbf{x}, \mathbf{x}') = P^{*}(\mathbf{x}, \mathbf{x}') - \sum_I c_I\{P^{*}(\mathbf{x}, \mathbf{x}') - P^{*}[(\mathbf{x}_I, \mathbf{x}'_J), (\mathbf{x}'_I, \mathbf{x}_J)]\} \; .$$

$$(10.188\text{b})$$

For each i and j with $i < j$, let $\mathcal{R}_{ij}$ denote the set

$$\mathcal{R}_{ij} = \{I : \ i \text{ in } I \text{ and } j \text{ in } J, \text{ or } i \text{ in } J \text{ and } j \text{ in } I\} \; . \qquad (10.189\text{a})$$

We obtain for the recombination frequency c_{ij} between loci i and j

$$c_{ij} = \sum c_I \; , \qquad (10.189\text{b})$$

in which the sum is over all I in $\mathcal{R}_{ij}$.

We suppose that mutation acts independently at each locus and that the change in allelic effect has the same symmetric probability density m_i for every allele at locus i. Let u_i represent the mutation rate at locus i. Since we are studying the population only at equilibrium, the gametic distribution in the next generation is again $p(\mathbf{x})$, and therefore

$$p(\mathbf{x}) = \left(\prod_{i=1}^{n} \Omega_i \right) p^{**}(\mathbf{x}) \; , \qquad (10.190\text{a})$$

where the linear operator Ω_i acts only on x_i. Applying Ω_i to a probability density $\psi(x_i)$ yields

$$\Omega_i\psi(x_i) = (1 - u_i)\psi(x_i) + u_i \int_{-\infty}^{\infty} \psi(\xi)m_i(x_i - \xi)d\xi \; . \qquad (10.190\text{b})$$

Equations (10.187), (10.188), and (10.190) constitute a complicated, highly nonlinear model for $p(\mathbf{x})$. Since the exact system is apparently intractable, we approximate it by positing that mutation and selection are both weak. Then we expect the loci to be almost stochastically independent, and this assumption enables us to derive an approximation for $p(\mathbf{x})$. The existence of solutions far from stochastic independence has not been ruled out, and the stability of the solution presented below has not been established. We expect intuitively that neither great departures from stochastic independence nor instability occurs.

As (10.187) reveals, selection is directly controlled by the genotypic fitness W, rather than by the phenotypic fitness w. Therefore, (10.184) shows that we can ensure the weakness of selection by setting

$$1/V = 2a\varepsilon \; , \tag{10.191}$$

where a is fixed and has the dimensions of Y^{-2}, and the dimensionless small parameter $\varepsilon \to 0$. Because of the assumption $m_i(-x) = m_i(x)$, the odd moments of the mutational increment are zero. We define the second and fourth moments as

$$\int_{-\infty}^{\infty} x^2 m_i(x)dx = \sigma_i^2 \; , \tag{10.192a}$$

$$\int_{-\infty}^{\infty} x^4 m_i(x)dx = 3\kappa_i \sigma_i^4 \; ; \tag{10.192b}$$

κ_i is a measure of kurtosis and equals unity if m_i is Gaussian. We fix the mutation rates u_i and suppose that the variances $\sigma_i^2 \to 0$ so that

$$u_i \sigma_i^2 = 2k_i \varepsilon \; , \tag{10.193}$$

where k_i is fixed and has the dimensions of Y^2.

It is important to bear in mind that the results of this section presuppose that the mutation rates are fixed and that the mutational variances conditioned on the occurrence of a mutation, σ_i^2, are small. Thus, the dimensionless parameter combination

$$\sigma_i^2/(u_i V) = (4ak_i/u_i^2)\varepsilon^2 \to 0 \tag{10.194}$$

in our approximation. We could enforce the smallness of the absolute variances $u_i \sigma_i^2$ by assuming low mutation rates ($u_i \to 0$) with σ_i^2 fixed, but then the results in this section would generally not hold.

By a long, elaborate analysis, Fleming (1979) obtained the following approximation for the density of allelic effects. To present his solution, we define

$$v_i = (k_i/a)^{1/2} = (V u_i \sigma_i^2)^{1/2} \tag{10.195}$$

and let ϕ_i represent the normal density with mean zero and variance v_i:

$$\phi_i(x_i) = (2\pi v_i)^{-1/2} e^{-x_i^2/(2v_i)} \; . \tag{10.196a}$$

We designate by $p^0(\mathbf{x})$ the multivariate normal density with independent loci for which locus i has the distribution ϕ_i,

$$p^0(\mathbf{x}) = \prod_{i=1}^{n} \phi_i(x_i) \; , \tag{10.196b}$$

and put

$$\alpha_i = \tfrac{1}{2}v_i + \sum_{j:\, j\neq i} (c_{ij}^{-1} + 1)v_j \; , \tag{10.197a}$$

$$\eta_i = \kappa_i u_i^{-1} - 1, \qquad \rho_i = \eta_i/(16v_i) \; , \tag{10.197b}$$

$$\delta_i = \frac{\alpha_i}{v_i} + \tfrac{1}{2}(1 - \tfrac{9}{8}\eta_i) \; . \tag{10.197c}$$

Then

$$p(\mathbf{x}) = p^0(\mathbf{x})\left\{1 + \varepsilon a\left[\sum_{i=1}^n \delta_i(x_i^2 - v_i) + \sum_{i=1}^n \rho_i(x_i^4 - 3v_i^2)\right.\right.$$

$$\left.\left. - 2\sum_{i<j} c_{ij}^{-1} x_i x_j\right]\right\} + O(\varepsilon^2) \qquad (10.198)$$

as $\varepsilon \to 0$ (Fleming, 1979).

The zeroth-order density is the multivariate Gaussian (10.196). Owing to the term with x_i^4, however, $p(\mathbf{x})$ is *not* Gaussian to first order. Since $u_i \ll 1$, (10.197b) informs us that $\eta_i \gg 1$, whence we expect ρ_i to be large. Therefore, the deviations from normality may greatly exceed an estimate based only on the value of ε. Remarkably, the density depends to first order only on the re-combination frequencies c_{ij}, rather than on the complete set of recombination frequencies c_I.

Integrating (10.198) over x_k for all $k \neq i, j$ yields the joint density of allelic effects at loci i and j for $i \neq j$ (Nagylaki, 1984):

$$p_{ij}(x_i, x_j) = \phi_i(x_i)\phi_j(x_j)\{1 + \varepsilon a[\delta_i(x_i^2 - v_i) + \delta_j(x_j^2 - v_j) + \rho_i(x_i^4 - 3v_i^2)$$

$$+ \rho_j(x_j^4 - 3v_j^2) - 2c_{ij}^{-1}x_i x_j]\} + O(\varepsilon^2) . \quad (10.199)$$

The density of allelic effects at locus i reads (Nagylaki, 1984)

$$p_i(x_i) = \phi_i(x_i)\{1 + \varepsilon a[\delta_i(x_i^2 - v_i) + \rho_i(x_i^4 - 3v_i^2)]\} + O(\varepsilon^2) . \quad (10.200)$$

Even this departs from normality to first order. From (10.199) and (10.200) we calculate the regression of allelic effects at locus j on those at locus i (Nagylaki, 1984):

$$\mathcal{E}(X_j | X_i = x_i) = \int_{-\infty}^{\infty} dx_j\, x_j p_{ij}(x_i, x_j)/p_i(x_i)$$

$$= -(2\varepsilon a v_j c_{ij}^{-1})x_i + O(\varepsilon^2) . \quad (10.201)$$

Observe that this regression is linear despite the deviation from normality.

We compute the single-locus variances from (10.200) as (Fleming, 1979)

$$V_i = \mathcal{E}(X_i^2) = v_i + 2\varepsilon a v_i^2(\delta_i + \tfrac{3}{8}\eta_i) + O(\varepsilon^2) , \quad (10.202)$$

The covariances follow from (10.201) and (10.202) by conditioning:

$$V_{ij} = \mathcal{E}(X_i X_j)$$
$$= \mathcal{E}[X_i \mathcal{E}(X_j | X_i)]$$
$$= -2\varepsilon a v_i v_j c_{ij}^{-1} + O(\varepsilon^2), \qquad i \neq j , \quad (10.203)$$

in agreement with Fleming (1979). Recalling (10.191) and (10.195), we obtain from (10.203)

$$V_{ij} \approx -(u_i \sigma_i^2 u_j \sigma_j^2)^{1/2}/c_{ij} , \quad (10.204)$$

which is always negative, increases in magnitude as linkage becomes tighter, and is independent of the selection parameter V. Equations (10.202) and (10.203) lead to the correlation

$$r_{ij} = -2\varepsilon a(v_i v_j)^{1/2} c_{ij}^{-1} + O(\varepsilon^2) \tag{10.205}$$

between X_i and X_j.

Since the two gametes that form an individual fuse at random, therefore (10.197a), (10.197c), (10.202), and (10.203) enables us to evaluate the total genetic variance (Fleming, 1979):

$$V_G = 2\left(\sum_i V_i + \sum_{i \neq j} V_{ij}\right)$$

$$= V_G^0 + \varepsilon a\left[(V_G^0)^2 - \tfrac{3}{4}\sum_{i=1}^n \eta_i v_i^2\right] + O(\varepsilon^2) , \tag{10.206}$$

where

$$V_G^0 = 2\sum_{i=1}^n v_i \tag{10.207}$$

signifies the zeroth-order approximation. Notice that, to first order, the genotypic variance is independent of the linkage map.

We define the effective number of loci, n_e, in terms of the mutational variances by (Lande, 1976)

$$n_e = \left[\sum_{i=1}^n (u_i \sigma_i^2)^{1/2}\right]^2 \bigg/ \sum_{i=1}^n u_i \sigma_i^2 \tag{10.208}$$

and let

$$V_m = 2\sum_{i=1}^n u_i \sigma_i^2 \tag{10.209}$$

represent the total mutational variance. The discussion below (10.61) shows that $1 \leq n_e \leq n$, with $n_e = 1$ if and only if $n = 1$ (one locus), and $n_e = n$ if and only if $u_i \sigma_i^2$ is independent of i (exchangeable loci). Substituting (10.195), (10.208), and (10.209) into (10.207) produces

$$V_G^0 = (2n_e V V_m)^{1/2} . \tag{10.210}$$

The remainder of this section is based on Nagylaki (1984).

To obtain the distribution of the genotypic value G, from (10.198) we find its characteristic function:

$$\chi_G(s) = \mathcal{E}(e^{isG})$$

$$= e^{-\frac{1}{2}V_G^0 s^2}\left\{1 + \tfrac{1}{8}\varepsilon a s^2\left[\sum_{j=1}^n \eta_j v_j^2(3 + v_j s^2) - 4(V_G^0)^2\right]\right\} + O(\varepsilon^2) .$$

$$\tag{10.211}$$

Fourier inversion yields the density of G:

$$f_G(g) = (2\pi V_G^0)^{-1/2} e^{-g^2/(2V_G^0)} \left\{ 1 + \frac{\varepsilon a}{8 V_G^0} \left[\left(1 - \frac{g^2}{V_G^0} \right) \left(3 \sum_{j=1}^{n} \eta_j v_j^2 - 4(V_G^0)^2 \right) \right. \right.$$

$$\left. \left. + \frac{1}{V_G^0} \left(3 - \frac{6g^2}{V_G^0} + \frac{g^4}{(V_G^0)^2} \right) \sum_{j=1}^{n} \eta_j v_j^3 \right] \right\} + O(\varepsilon^2) \ . \tag{10.212}$$

This distribution is symmetric, as expected because of the symmetry of our selection and mutation schemes, but it is not Gaussian. Observe that, to first order, f_G is independent of the linkage map. As a check, a trivial computation shows that the variance of (10.212) is indeed (10.206).

To derive the distribution of the phenotypic value, note that by the stochastic independence of G and E, the characteristic function of Y is the product of those of G and E. Since E is Gaussian with mean zero and variance V_E, we have

$$\chi_Y(s) = e^{-\frac{1}{2} V_E s^2} \chi_G(s) \ , \tag{10.213}$$

which merely replaces V_G^0 in the exponent in (10.211) by

$$V_Y^0 = V_G^0 + V_E \ . \tag{10.214}$$

From (10.211), (10.213), and (10.214) we infer that the density of Y reads

$$f_Y(y) = (2\pi V_Y^0)^{-1/2} e^{-y^2/(2V_Y^0)} \left\{ 1 + \frac{\varepsilon a}{8 V_Y^0} \left[\left(1 - \frac{y^2}{V_Y^0} \right) \left(3 \sum_{j=1}^{n} \eta_j v_j^2 - 4(V_G^0)^2 \right) \right. \right.$$

$$\left. \left. + \frac{1}{V_Y^0} \left(3 - \frac{6y^2}{V_Y^0} + \frac{y^4}{(V_Y^0)^2} \right) \sum_{j=1}^{n} \eta_j v_j^3 \right] \right\} + O(\varepsilon^2) \ . \tag{10.215}$$

which is again symmetric and independent of the linkage map, but not Gaussian.

Finally, from (10.184) and (10.212) we deduce the mean fitness

$$\overline{w} = \left(\frac{V_s}{V + V_G^0} \right)^{1/2} + O(\varepsilon^2) \ . \tag{10.216}$$

Let us examine the *maintenance of genetic variability* in more detail. Since we cannot evaluate the sum in (10.206) without data on individual loci, we assume that mutation is Gaussian at each of n exchangeable loci. We have $n_e = n$, $v_i = v$, $V_G^0 = 2nv$ from (10.207), and since $\kappa_i = 1$ and $u_i = u \ll 1$, therefore $\eta_i \approx 1/u$ from (10.197b). With the aid of these relations and (10.191), we reduce (10.206) to

$$V_G \approx V_G^0 \left[1 + \frac{V_G^0}{2V} \left(1 - \frac{3}{16nu} \right) \right] \ . \tag{10.217}$$

Taking advantage of the arbitrariness of the scale of the character, we divide all the variances by the environmental variance, V_E. We require the values of three parameters, viz.,

$$\mu = nV_m/V_E, \qquad \beta = V/V_E > 1, \qquad \gamma = nu \; . \tag{10.218}$$

From (10.210), (10.216), (10.217), and (10.218) we obtain

$$\lambda = V_G^0/V = (2\mu/\beta)^{1/2} \; , \tag{10.219}$$

$$\overline{w} \approx \left[\frac{\beta - 1}{\beta(1 + \lambda)} \right]^{1/2} \; , \tag{10.220}$$

$$\theta = \frac{V_G}{V_E} \approx \beta\lambda \left[1 + \tfrac{1}{2}\lambda \left(1 - \frac{3}{16\gamma} \right) \right] \; , \tag{10.221}$$

$$h^2 = \theta/(1 + \theta) \; . \tag{10.222}$$

When is this approximation for the heritability reliable? To derive the fundamental result (10.198), it is necessary to expand the exponential in (10.184) in a Maclaurin series, which clearly requires

$$V_G/V \approx V_G^0/V = \lambda \ll 1 \; . \tag{10.223}$$

It is also necessary to expand (10.190b) in a Taylor series; on the assumption that the $2j$th moment of m_i is proportional to σ_i^{2j}, this shows that (in the exchangeable case) the lth-order contribution to (10.198) will generally include terms of order $(\varepsilon/u)^l$. This argument leads to the stringent sufficient condition

$$\lambda/u \ll 1 \; , \tag{10.224}$$

instead of the weaker criterion (10.223). Equation (10.221) suggests that (10.224) may be weakened to

$$\lambda/\gamma \ll 1 \; , \tag{10.225}$$

but this has not been proved. Perhaps the least conservative requirement is that the first-order correction in (10.221) be small. Since this will always be negative for empirical values of γ, we write

$$k = \tfrac{1}{2}\lambda \left(\frac{3}{16\lambda} - 1 \right) \ll 1 \; . \tag{10.226}$$

The empirical result $\gamma \ll 1$ implies that the third term in (10.221) is much more important than the second.

We define a standard measure of selection intensity by converting the mean relative fitness $\overline{w}$ to a phenotypic load

$$L = 1 - \overline{w} \; . \tag{10.227}$$

For strong selection, $V_s \lesssim V_E$, so β is not much greater than one, and therefore L is not much less than one. For weak selection, $V_s \gg V_E$, which implies $\beta \gg 1$, and hence $L \ll 1$. If β is sufficiently large to satisfy (10.224), the heritability will be substantial unless $\beta\lambda = (2\mu\beta)^{1/2} \ll 1$.

Writing

$$b = \overline{w}^2 = (1 - L)^2 \ , \qquad\qquad (10.228a)$$

we express β in terms of L by inverting (10.220):

$$\beta = (1 - b)^{-2}\{1 - b + \mu b^2 + b[2\mu(1 - b) + \mu^2 b^2]^{1/2}\} \ . \qquad (10.228b)$$

[The root with a minus sign before the radical does not satisfy (10.220).] Observe that for extremely weak selection, defined by $1 \gg \lambda \gg 1/\beta$, (10.220) and (10.227) yield $\lambda \approx 2L$.

Lande (1976) estimates V_m/V_E as about 0.001 or somewhat higher; for nu he uses at least 0.01 (Lande, 1980). Estimates of the effective number of loci, n_e, range from about 5 to about 25, low values being more common (Wright, 1968, Chap. 15; Lande, 1981). These estimates, however, are based on the means and variances in sundry crosses, rather than on (10.208). They presuppose unlinked loci; for linked loci, they are too small. Phenotypic loads less than about 0.01 are difficult to measure. Typical values of L range from about 0.02 to about 0.14 (Cavalli-Sforza and Bodmer, 1971, pp. 612–615; Johnson, 1976, p. 178); the high values that range up to 0.47 in Johnson may be due to directional selection or selection on correlated characters.

Table 10.2. Mutation-selection balance

$\gamma \cdot$	μ	L	β	λ	k	h^2
0.01	0.005	0.001	3420	0.00171	0.0152	0.852
0.01	0.005	0.010	99.3	0.0100	0.0890	0.476
0.01	0.010	0.001	5940	0.00183	0.0163	0.915
0.01	0.010	0.010	130	0.0124	0.110	0.589
0.01	0.050	0.001	25900	0.00196	0.0174	0.980
0.01	0.050	0.010	336	0.0173	0.153	0.831
0.01	0.100	0.001	50800	0.00198	0.0176	0.990
0.01	0.100	0.010	581	0.0185	0.165	0.900
0.10	0.005	0.001	3420	0.00171	0.000748	0.854
0.10	0.005	0.010	99.3	0.0100	0.00439	0.498
0.10	0.005	0.100	6.34	0.0397	0.0174	0.198
0.10	0.005	0.300	2.10	0.0690	0.0302	0.123
0.10	0.010	0.001	5940	0.00183	0.000803	0.916
0.10	0.010	0.010	130	0.0124	0.00544	0.615
0.10	0.010	0.100	6.84	0.0541	0.0237	0.265
0.10	0.010	0.300	2.16	0.0962	0.0421	0.166
0.10	0.050	0.001	25900	0.00196	0.000859	0.981
0.10	0.050	0.010	336	0.0173	0.00755	0.852
0.10	0.050	0.100	9.40	0.103	0.0451	0.481
0.10	0.050	0.300	2.43	0.203	0.0887	0.310
0.10	0.100	0.001	50800	0.00198	0.000868	0.990
0.10	0.100	0.010	581	0.0185	0.00812	0.914
0.10	0.100	0.100	11.8	0.130	0.0569	0.592
0.10	0.100	0.300	2.66	0.274	0.120	0.391

In Table 10.2, we present examples with various values of γ, μ, and L. In agreement with the monotone behavior in β of (10.219) and (10.220) with μ fixed, β decreases and λ increases as selection becomes more intense (L increasing). All the examples satisfy (10.226), but only those with relatively weak selection satisfy the more stringent condition (10.225). The table reveals that considerable genetic diversity is maintained even when selection is moderately strong. For weak selection, the heritabilities are quite high.

Kimura (1965a) was the first to treat the model discussed in this section. His diffusion analysis corresponds to our zeroth-order approximation. He derived the single-locus Gaussian solution (10.196a) and asserted that the genotypic value for multiple loci is normally distributed with the variance given by (10.207). This procedure yields the correct result to zeroth order because, as (10.198) shows, the loci are stochastically independent to that order.

Lande (1976) took recombination into account, but assumed multivariate normality as an approximation. This approach leads to the correct approximations (10.204) for the covariances and (10.216) for the mean fitness, but misses the sum in the expression (10.206) for the genetic variance. For weak selection, Lande's formula somewhat overestimates the genetic variance and is slightly *less* accurate than the zeroth-order approximation (10.210); if selection is strong, neither his result nor (10.206) is reliable.

Lande (1977) and Bulmer (1980, pp. 177–179) assumed, *inter alia*, the linearity of the regression $\mathcal{E}(X_i|Y)$ in their concise, heuristic derivations of Lande's (1976) formula. According to Problem 10.18, however, this linearity holds only to zeroth order in ε.

Equation (10.217) shows that Fleming's (1979) approximation fails if the mutation rate is sufficiently small. For weak selection and very low mutation rates, the approximate analysis of Turelli (1984) appears to be accurate. See also Barton and Turelli (1987) and Turelli and Barton (1990). Bürger (1986, 1988) demonstrates that Kimura's (1965a) integro-differential equation for a continuum of alleles at a single locus has a unique, globally asymptotically stable equilibrium, with a variance close to that for many alleles.

10.7 Problems

10.1 Show that for two alleles at locus i, (10.15b) and (10.15c) may be expressed in terms of genotypic contributions as

$$V_A^{(i)} = 2p_1^{(i)}p_2^{(i)}[(G_{11}^{(i)} - G_{12}^{(i)})p_1^{(i)} + (G_{12}^{(i)} - G_{22}^{(i)})p_2^{(i)}]^2 \; , \quad (10.229a)$$

$$V_D^{(i)} = [p_1^{(i)}p_2^{(i)}(G_{11}^{(i)} - 2G_{12}^{(i)} + G_{22}^{(i)})]^2 \; . \quad (10.229b)$$

10.2 For some selection problems, it is useful to extend the analysis of variance between (10.8) and (10.17) to incorporate linkage disequilibrium.

Prove that, although (10.13) still holds, the average excess, $a_j^{(i)}$, of $A_j^{(i)}$ is no longer $g_j^{(i)}$, but is now given by

$$p_j^{(i)} a_j^{(i)} = p_j^{(i)} \alpha_j^{(i)} + \sum_{k:\, k \neq i} \sum_l p_{jl}^{(ik)} \alpha_l^{(k)} , \qquad (10.230)$$

where $p_{jl}^{(ik)}$ designates the frequency of the gamete $A_j^{(i)} A_l^{(k)}$. Establish (10.17a) and the generalization of (8.57),

$$V_A = 2 \sum_{ij} p_j^{(i)} a_j^{(i)} \alpha_j^{(i)} . \qquad (10.231)$$

10.3 An infinite population consists of demes of N monoecious individuals. Posit panmixia without selfing in each deme and assume that a proportion m $(0 < m \leq 1)$ of individuals comes from the offspring of a mating pool chosen at random from the entire population. Consider two loci with recombination frequency c $(0 < c \leq \frac{1}{2})$ and denote by p_i and q_j the respective frequencies of the alleles A_i and B_j. (The use of this model for studying the stochastic dependence of genes at two loci was suggested by B. S. Weir in a personal communication.)

(a) Let P_{ij} represent the probability that a chromosome chosen at random carries A_i and B_j; Q_{ij} represent the probability that a chromosome chosen at random carries A_i and the other chromosome in the same individual carries B_j; R_{ij} represent the probability that a chromosome chosen at random carries A_i and another chromosome, chosen at random from another individual in the same deme, carries B_j; and S_{ij} represent the probability that a chromosome chosen at random carries A_i and another chromosome, chosen at random from the entire population, carries B_j. Derive the recurrence relations

$$P'_{ij} = (1 - c)P_{ij} + cQ_{ij} , \qquad (10.232a)$$
$$Q'_{ij} = (1 - m)R_{ij} + mS_{ij} , \qquad (10.232b)$$
$$R'_{ij} = (1 - m)^2 [N^{-1} \tfrac{1}{2}(P_{ij} + Q_{ij}) + (1 - N^{-1})R_{ij}]$$
$$\qquad + [1 - (1 - m)^2]S_{ij} , \qquad (10.232c)$$
$$S'_{ij} = S_{ij} . \qquad (10.232d)$$

(b) Infer that $S_{ij}(t) = p_i q_j$ and that as $t \to \infty$, $P_{ij}(t)$, $Q_{ij}(t)$, and $R_{ij}(t)$ all converge to $p_i q_j$. Note also that if

$$P_{ij}(0) = Q_{ij}(0) = R_{ij}(0) = p_i q_j , \qquad (10.233a)$$

then

$$P_{ij}(t) = Q_{ij}(t) = R_{ij}(t) = p_i q_j . \qquad (10.233b)$$

Thus, *genes* (not *genotypes*) at different loci are pairwise stochastically independent at equilibrium; if pairwise stochastic independence holds initially, then it does so in all subsequent generations.

10.4   In Problems 9.17 and 10.3, a certain proportion of the population mates at random. Although it lacks this special feature, half-sib mating by infinitely many individuals (Wright, 1921), displayed in Fig. 10.9, has qualitatively the same two-locus behavior. Define c, p_i, q_j, P_{ij}, and Q_{ij} as in Problem 10.3; let $R_{k,ij}$ denote the probability that a chromosome chosen at random carries A_i and another chromosome, chosen at random from an individual k steps away ($k \geq 1$), carries B_j.

(a) Derive the difference equations

$$P'_{ij} = (1 - c)P_{ij} + cQ_{ij} \; , \tag{10.234a}$$

$$Q'_{ij} = R_{1,ij} \; , \tag{10.234b}$$

$$R'_{1,ij} = \tfrac{1}{4}[\tfrac{1}{2}(P_{ij} + Q_{ij}) + 2R_{1,ij} + R_{2,ij}] \; , \tag{10.234c}$$

$$R'_{k,ij} = \tfrac{1}{4}(R_{k-1,ij} + 2R_{k,ij} + R_{k+1,ij}), \qquad k \geq 2 \; . \tag{10.234d}$$

(b) Notice that if

$$P_{ij}(0) = Q_{ij}(0) = R_{k,ij}(0) = p_i q_j \tag{10.235a}$$

for all k, then

$$P_{ij}(t) = Q_{ij}(t) = R_{k,ij}(t) = p_i q_j \tag{10.235b}$$

for all k.

(c) Prove that if $R_{k,ij}(0) \to p_i q_j$ as $k \to \infty$, then $P_{ij}(t)$, $Q_{ij}(t)$, and $R_{k,ij}(t)$ all converge to $p_i q_j$ as $t \to \infty$.

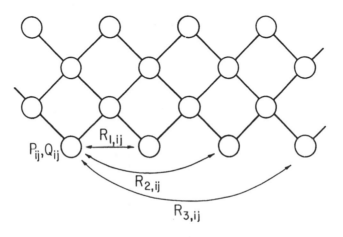

Fig. 10.9. Half-sib mating.

10.5   To generalize Problems 9.17 and 10.3, designate by $\mathbf{x}(t)$ the vector of n two-locus frequencies and write the recurrence relations in the matrix form

$$\mathbf{x}' = C\mathbf{x} \ . \tag{10.236}$$

Observe that

$$\sum_{j=1}^{n} c_{ij} = 1 \tag{10.237}$$

and consider the finite Markov chain with states $E_1, E_2, \ldots, E_n$ and transition matrix C. Suppose that $x_n(0) = p_i q_j$ and $c_{nk} = \delta_{nk}$; then $x_n(t) = p_i q_j$. Show that if the absorbing state E_n is accessible from all other states, then $x_k(t) \to p_i q_j$ as $t \to \infty$.

10.6   We can generalize Problem 10.4 along the lines of Problem 10.5. The relevant Markov chain now has a countably infinite number of states, $E_1, E_2, \ldots$. Posit that $x_k(0) \to p_i q_j$ as $k \to \infty$. Prove that if all states are transient or null recurrent, then $x_k(t) \to p_i q_j$ as $t \to \infty$.

10.7   Assume that the pedigree depicted in Fig. 10.10 occurs in an otherwise panmictic population at equilibrium. Genotype and environment are uncorrelated and do not interact; there is no epistasis; and $h^2 = 0.640$ and $D^2 = 0.064$ for the character, which has mean 2 and variance 100. Suppose that the regression of the phenotype of J (Y^*) on that of I (Y) is linear.

(a) What is the correlation between Y and Y^*?
(b) Calculate $\mathcal{E}(Y^*|Y = -8)$.
(c) Compute $\mathcal{E}[\mathrm{Var}\,(Y^*|Y)]$.
(d) Evaluate the phenotypic variance within pairs of individuals related like I and J.

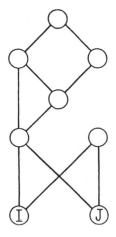

Fig. 10.10. Pedigree for relatives I and J.

10.8 Posit equilibrium, panmixia, and the absence of epistasis and genotype-environment interaction and correlation. Assume that the phenotypic regressions of offspring on one parent and on the mid-parent value are linear. Let V_1, V_2, V_3, and V_4 signify the (expected) phenotypic variances among individuals with the same trait value for one parent, half-sibs, individuals with the same mid-parent value, and full sibs. Calculate these variances in terms of V_Y, h^2, and D^2. You should find (for $h > 0$) that $V_1 > V_3$, $V_2 > V_4$, $V_1 > V_2$ for $h < 1$, and $V_3 > V_4$ for $h < 1$ or $D > 0$. Explain these inequalities intuitively.

10.9 Assume equilibrium, panmixia, the absence of epistasis, and stochastic independence of the environment. Study the regression of sundry random variables associated with individual J on those associated with his relative I. Use the notation of Sects. 10.1 and 10.2 and suppose without loss of generality that all random variables have zero expectation. Define the additive and dominance components of the genotypic value by

$$G_A = \sum_i (\alpha^{(i)} + \tilde{\alpha}^{(i)}), \qquad G_D = \sum_i d^{(i)}, \qquad (10.238)$$

so that

$$V_A = \mathcal{E}(G_A^2), \qquad V_D = \mathcal{E}(G_D^2) . \qquad (10.239)$$

(a) Show that

$$\mathcal{E}(G_A G_D) = 0 . \qquad (10.240)$$

(b) Demonstrate that

$$\mathcal{E}(G_A^*|G_A) = 2F_{IJ}G_A , \qquad (10.241a)$$
$$\mathcal{E}(G_D^*|G_D) = k_2 G_D . \qquad (10.241b)$$

(c) Deduce that

$$\mathcal{E}(Y^*|Y) = 2F_{IJ}\mathcal{E}(G_A|Y) + k_2\mathcal{E}(G_D|Y) . \qquad (10.242)$$

Thus, despite the linearity of the additive and dominance regressions exhibited in (10.241), the phenotypic (or even genotypic) regression is generally nonlinear. Clearly, (10.242) is linear if and only if $\mathcal{E}(G_A|Y)$ is linear, and $k_2 = 0$ or $\mathcal{E}(G_D|Y)$ is linear. In the absence of dominance, these conditions reduce to (10.79).

(d) Derive the correlation formula (10.35) from (10.239), (10.240), and (10.241).

10.10 Assume that (10.48) holds.

(a) If I is the parent of J, show that

$$r_{IJ} = h_0^2[\tfrac{1}{2}(1 + F_I) + F_J][(1 + F_I h_0^2)(1 + F_J h_0^2)]^{-1/2} . \qquad (10.243)$$

(b) If I and J are full siblings and their parents, A and B, have inbreeding coefficients F_A and F_B and coefficient of consanguinity F_{AB}, establish the formula

$$r_{IJ} = h_0^2[\tfrac{1}{4}(2 + F_A + F_B) + F_{AB}]/(1 + F_{AB}h_0^2) . \qquad (10.244)$$

Notice that this simplifies to

$$r_{IJ} = 9h_0^2/(16 + h_0^2) \qquad (10.245)$$

if A and B are noninbred first cousins.

10.11 A character is controlled by a very large number of additive loci without dominance. Environment and genotype are stochastically independent. Under random mating, the phenotypic variance and heritability are 1 and $\tfrac{4}{5}$. Phenotypic assortative mating with marital correlation $\tfrac{9}{16}$ commences. Assuming that (10.77) and (10.79) hold, calculate the phenotypic variance and heritability at equilibrium.

10.12 A trait is determined by two diallelic loci with alleles A and a, B and b. For each genotype, the phenotype Y is the sum of the number of A and B alleles. Individuals mate only with those who have the same phenotypic value, and mating is random within each phenotypic group, each of which contributes progeny in proportion to its frequency. Denote by c, p_A, p_a, p_B, and p_b the recombination fraction and the frequencies of A, a, B, and b; assume without loss of generality that $0 < c \le \tfrac{1}{2}$ and $0 < p_B \le p_A < 1$.

(a) Write the recursion relations for the genotypic frequencies.
(b) Show that the gene frequencies are conserved.
(c) Prove that the genotypic frequencies converge globally to the equilibrium at which $AABB$, $AAbb$, and $aabb$ have the respective frequencies p_B, $p_A - p_B$, and p_a, all other genotypic frequencies being zero (Ghai, 1973).
(d) Observe that the allelic effects at individual loci are *not* perfectly correlated with the phenotype at equilibrium. Note also that (10.77) and (10.79) hold trivially. To demonstrate consistency with the remarks below (10.97), establish that the regression of the allelic effects on Y is nonlinear. Thus, (10.83) is invalid in this model.

10.13 Assume that the regression of the mid-parent phenotype on the offspring phenotype is linear. Prove that the regression coefficient is the parent-offspring correlation, r_{po}.

10.14 Refer to Fig. 10.2 and consider assortative mating in the absence of dominance and epistasis. Suppose that the random variable R is stochastically independent of $\mathcal{G}_3$ if $\mathcal{G}_1$ and $\mathcal{G}_2$ are fixed. Show that (Nagylaki, 1978)

$$\mathcal{E}(RG_3) = \tfrac{1}{2}\mathcal{E}(RG_1) + \tfrac{1}{2}\mathcal{E}(RG_2) . \qquad (10.246)$$

This leads to a slight variant of the derivation of the correlation between relatives in Sect. 10.4. If B_3 mates with B_4 (not shown), who has a parent B_5 (also not shown), and $R = G_5$, does (10.246) hold?

10.15 Posit equilibrium, phenotypic assortative mating, purely additive gene action, stochastically independent environment, and the linear regression formulae (10.77) and (10.79).

(a) Prove that the correlation between the phenotypes of B_m and B'_n in Fig. 10.11 is

$$r_{mn} = [\tfrac{1}{2}(1 + rh^2)]^{m+n}r_s - h^2[\tfrac{1}{2}(1 \mid rh^2)]^{m+n+1} . \qquad (10.247)$$

Cloninger, Rice, and Reich (1979) used path coefficients to derive (10.247).

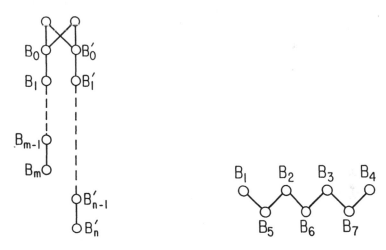

Fig. 10.11. Pedigree for distant collateral relatives.

Fig. 10.12. Pedigree for steprelatives.

(b) Show that if B_0 and B'_0 are half-sibs instead of full sibs, (10.247) must be replaced by

$$\begin{aligned} r_{mn} &= [\tfrac{1}{2}(1 + rh^2)]^{m+n}r_{hs} \\ &= \tfrac{1}{4}h^2(1 + 2rh^2 + r^2h^2)[\tfrac{1}{2}(1 + rh^2)]^{m+n} . \end{aligned} \qquad (10.248)$$

(c) In Fig. 10.12, B_5 is the stepoffspring of B_3. Establish the linear-regression formula (Gimelfarb, 1981)

$$\mathcal{E}(Y_5 \mid Y_3) = \tfrac{1}{2}rh^2(1 + r)Y_3 , \qquad (10.249)$$

which yields the correlation

$$r_{sp,so} = \tfrac{1}{2} r h^2 (1 + r) \tag{10.250}$$

between stepparent and stepoffspring.

(d) Demonstrate that the correlation between the stepsibs B_5 and B_7 reads (Gimelfarb, 1981)

$$r_{ss} = \tfrac{1}{4} r h^4 (1 + r)^2 \ . \tag{10.251}$$

Observe that $r_{sp,so} = r_{ss}$ if and only if $h = 0$, $r = 0$, $r = -1$, or $h = 1$ and $r = 1$; otherwise $r_{sp,so} > r_{ss}$.

(e) Prove that the correlations between the affinate relatives B_3 and B_1, B_3 and B_5, and B_3 and B_6 in Fig. 10.7 are rr_{po}, rr_s, and $r^2 r_s$, respectively.

10.16 By choosing X and Z in (10.165) to have a correlated bivariate normal distribution, prove that in linkage disequilibrium the phenotypic distribution with a single-locus genotype fixed *cannot* generally be approximated by shifting the unconditional phenotypic distribution (Nagylaki, 1984).

10.17 In the model of Sect. 10.6, establish that, conditioned on $X_i = x_i$, the random variable (Nagylaki, 1984)

$$Z_j = X_j - \mathcal{E}(X_j | X_i = x_i) \tag{10.252}$$

has the density $p_j(z_j)$, where p_j is given by (10.200). Thus, conditioning simply shifts the mean.

10.18 In the model of Sect. 10.6, prove that (for $V_E > 0$) $\mathcal{E}(G|Y)$ is linear only to zeroth order in ε. Infer that $\mathcal{E}(X_i|Y)$ is linear only to zeroth order. Explicit calculation demonstrates the last result even for $V_E = 0$, provided the number of loci is greater than one (Nagylaki, 1984).

10.19 If selection is weak, we may neglect deviations from linkage equilibrium and approximate (10.141) by the quadratic optimum model (Wright, 1935). Suppose that the environment does not affect the character, and dominance and epistasis are absent. Each of the n loci is diallelic; the alleles A_i and a_i at locus i have frequencies p_i and q_i, respectively. By separate translations of the allelic contributions at each locus and absorption of the total shift into the optimum y_o, we can arrange that A_i and a_i contribute to the character the constants $\tfrac{1}{2}\alpha_i$ (> 0) and $-\tfrac{1}{2}\alpha_i$, respectively (Wright uses α_i and 0). By scaling the character, we can absorb the selection intensity, so that

$$w(y) = 1 - (y - y_o)^2 \ , \tag{10.253}$$

which makes sense only if we assume that $w(y) \geq 0$ for all possible values of y. From Problem 8.14 it follows that the mean fitness is non-decreasing.

(a) Show that

$$\overline{Y} = \sum_i \alpha_i(2p_i - 1) , \qquad (10.254a)$$

$$V_Y = 2\sum_i \alpha_i^2 p_i q_i , \qquad (10.254b)$$

$$\overline{w} = 1 - V_Y - (\overline{Y} - y_o)^2 . \qquad (10.255)$$

(b) Prove that there is an equilibrium with locus i polymorphic if and only if

$$\alpha_i > 2|y_o|/(2n - 1) , \qquad (10.256)$$

in which case

$$\hat{p}_i = \tfrac{1}{2} + \frac{y_o}{(2n - 1)\alpha_i} . \qquad (10.257)$$

(c) Establish that equilibria with at least two loci segregating are unstable, and hence that (unless the population starts on the stable manifold of an unstable equilibrium) at most one locus can remain polymorphic. If the restrictive assumptions made here are relaxed, two or more loci can segregate at an asymptotically stable equilibrum (Kojima, 1959; Lewontin, 1964; Singh and Lewontin, 1966; Gale and Kearsey, 1968; Kearsey and Gale, 1968; Nagylaki, 1989a).

(d) Demonstrate that ($q_i = 1 - p_i$ in the partial derivative)

$$\overline{w}\Delta p_i = \tfrac{1}{2}p_i q_i \frac{\partial \overline{w}}{\partial p_i} = \alpha_i p_i q_i [\alpha_i(2p_i - 1) + 2(y_o - \overline{Y})] . \quad (10.258)$$

10.20 In the two-locus model of Gale and Kearsey (1968), fitness decreases linearly as the character deviates from the optimum. Suppose that the environment does not affect the character, and dominance and epistasis are absent. If A, a, B, and b contribute to the character 1, 0, $1 + k$, and 0, respectively, the optimum is at $2 + k$ (the value of $AaBb$), and we generalize the model of Gale and Kearsey (1968) by introducing the selection intensity s (> 0), then we can choose the fitnesses in Table 10.3. Without loss of generality, we take $k \geq 0$; clearly, we must require $s(2 + k) \leq 3$. Gale and Kearsey (1968) studied numerically the special case $s = 1$; since this corresponds to strong selection, their examples show considerable linkage disequilibrium. If $s(2+k) \ll 1$, then selection is weak and, by Sect. 8.2, we may posit linkage equilibrium as a good approximation. Let p_1 and p_2 denote the frequencies of A and B, respectively, and analyze the evolution of the population in the (p_1, p_2) plane as follows.

Table 10.3. Fitness pattern for Problem 10.20

	BB	Bb	bb
AA	$3 - s(2 + k)$	$3 - s$	$3 - ks$
Aa	$3 - s(1 + k)$	3	$3 - s(1 + k)$
aa	$3 - ks$	$3 - s$	$3 - s(2 + k)$

(a) Show that, in addition to the corner equilibria $(0, 0)$, $(0, 1)$, $(1, 0)$, and $(1, 1)$ and the symmetric equilibrium $(\frac{1}{2}, \frac{1}{2})$, a pair of side equilibria and a pair of unsymmetric internal equilibria exist for the ranges of k indicated:

$$\left(0, \frac{k+1}{2k}\right), \quad \left(1, \frac{k-1}{2k}\right): \qquad k > 1; \qquad (10.259)$$

$$\left(\frac{1}{2} \pm \sqrt{\alpha(\alpha - 1)}, \frac{1}{2} \mp \frac{1}{2}\sqrt{(\alpha - 1)/\alpha}\right): \quad \frac{3}{2} < k < 1 + \sqrt{2}, \quad (10.260)$$

$$\alpha = \sqrt{\frac{1}{4} + \frac{1}{2}k} . \qquad (10.261)$$

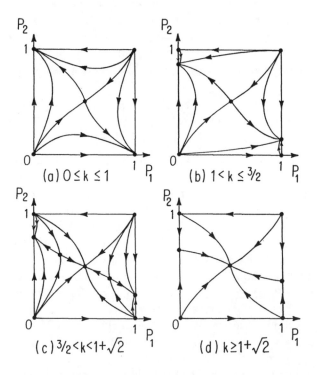

Fig. 10.13. Convergence patterns for Problem 10.20.

(b) Prove that $(0,0)$ and $(1,1)$ are always unstable; $(0,1)$ and $(1,0)$ are asymptotically stable if $k \leq 1$ and unstable if $k > 1$; (10.259) is asymptotically stable if $1 < k < 1 + \sqrt{2}$ and unstable if $k \geq 1 + \sqrt{2}$; $(\frac{1}{2}, \frac{1}{2})$ is unstable if $k \leq \frac{3}{2}$ and asymptotically stable if $k > \frac{3}{2}$; and (10.260) is unstable. The global convergence pattern sketched in Fig. 10.13 demonstrates that ultimately both loci become monomorphic if $k \leq 1$; the A-locus, which contributes less to the character, becomes monomorphic, whereas the B-locus remains polymorphic if $1 < k \leq \frac{3}{2}$; either only the B-locus remains polymorphic or both loci do if $\frac{3}{2} < k < 1 + \sqrt{2}$; and both loci remain polymorphic if $k \geq 1 + \sqrt{2}$. Thus, even with purely additive determination of the character, sufficient disparity between the contributions of the two loci leads to an asymptotically stable two-locus polymorphism under this form of centripetal selection.

Consult Nagylaki (1989a) for the analysis of the two-locus optimum model with an arbitrary fitness function.

References

van Aarde, I.M.R. 1975. The covariance of relatives derived from a random mating population. *Theor. Pop. Biol.* **8**, 166–183.

Abramowitz, M. 1964. Elementary analytical methods. Pp. 9–63 in *Handbook of Mathematical Functions* (M. Abramowitz and I.A. Stegun, eds.). National Bureau of Standards, Washington.

Abugov, R. 1985. Genetics of Darwinian fitness. II. Pleiotropic selection on fertility and survival. *J. Theor. Biol.* **114**, 109–125.

Akin, E. 1979. *The Geometry of Population Genetics.* Springer, Berlin.

Akin, E. 1982. Cycling in simple genetic systems. *J. Math. Biol.* **13**, 305–324.

Akin, E. 1983. Hopf bifurcation in the two-locus genetic model. *Mem. Am. Math. Soc.* **44**, No. 284.

Apostol, T.M. 1974. *Mathematical Analysis.* 2nd edn. Addison-Wesley, Reading, Mass.

Arzberger, P. 1985. A probabilistic and algebraic treatment of regular inbreeding systems. *J. Math. Biol.* **22**, 175–197.

Arzberger, P. 1988. Results for generalized regular inbreeding systems. *J. Math. Biol.* **26**, 535–550.

Atkinson, F.V., Watterson, G.A., and Moran, P.A.P. 1960. A matrix inequality. *Quart. J. Math.* **11**, 137–140.

Bartlett, M.S. 1955. *An Introduction to Stochastic Processes, with Special Reference to Methods and Applications.* 1st edn. Cambridge University Press, Cambridge.

Bartlett, M.S., and Haldane, J.B.S. 1934. The theory of inbreeding in autotetraploids. *J. Genet.* **29**, 175–180.

Barton, N.H. 1989. The divergence of a polygenic system subject to stabilizing selection, mutation and drift. *Genet. Res.* **54**, 59–77.

Barton, N.H., and Turelli, M. 1987. Adaptive landscapes, genetic distance and the evolution of quantitative characters. *Genet. Res.* **49**, 157–173.

Barton, N.H., and Turelli, M. 1991. Natural and sexual selection on many loci. *Genetics* **127**, 229–255.

Baum, L.E., and Eagon, J.A. 1967. An inequality with applications to statistical estimation for probabilistic functions of Markov processes and to a model for ecology. *Bull. Am. Math. Soc.* **73**, 360–363.

Bellman, R. 1949. *A Survey of the Theory of the Boundedness, Stability, and Asymptotic Behavior of Solutions of Linear and Non-linear Differential and Difference Equations.* Office of Naval Research, Washington, D.C.

Bennett, J.H. 1954. On the theory of random mating. *Ann. Eugen.* **18**, 311–317.

Bennett, J.H. 1957. Selectively balanced polymorphism at a sex-linked locus. *Nature* **180**, 1363–1364.

Bennett, J.H. 1963. Random mating and sex linkage. *J. Theor. Biol.* **4**, 28–36.

Bennett, J.H., and Binet, F.E. 1956. Association between Mendelian factors with mixed selfing and random mating. *Heredity* **10**, 51–55.

Bodmer, W.F. 1965. Differential fertility in population genetics models. *Genetics* **51**, 411–424.

Bodmer, W.F., and Cavalli-Sforza, L.L. 1968. A migration matrix model for the study of random genetic drift. *Genetics* **59**, 565–592.

Bohidar, N.R. 1964. Derivation and estimation of variance and covariance components associated with covariance between relatives under sex-linked transmission. *Biometrics* **20**, 505–521.

Borel, E. 1943. *Les probabilités et la vie*. Presses Universitaires de France, Paris. Revised translation: *Probabilities and Life*. Dover, New York (1962).

Boucher, W. 1988. Calculation of the inbreeding coefficient. *J. Math. Biol.* **26**, 57–64.

Boucher, W. 1991. Global analysis of a two-allele mating system. *J. Math. Biol.*, in press.

Boucher, W. 1991a. A deterministic analysis of self-incompatibility alleles. *J. Math. Biol.*, in press.

Boucher, W., and Cotterman, C.W. 1990. On the classification of regular systems of inbreeding. *J. Math. Biol.* **28**, 293–305.

Boucher, W., and Nagylaki, T. 1988. Regular systems of inbreeding. *J. Math. Biol.* **26**, 121–142.

Brauer, F., and Nohel, J.A. 1969. *The Qualitative Theory of Ordinary Differential Equations*. W.A. Benjamin, New York.

Bulmer, M.G. 1971. The stability of equilibria under selection. *Heredity* **27**, 157–162.

Bulmer, M.G. 1972. Multiple niche polymorphism. *Am. Nat.* **106**, 254–257.

Bulmer, M.G. 1972a. The genetic variability of polygenic characters under optimizing selection, mutation and drift. *Genet. Res.* **19**, 17–25.

Bulmer, M.G. 1980. *The Mathematical Theory of Quantitative Genetics*. Clarendon Press, Oxford.

Bürger, R. 1986. On the maintenance of genetic variation: global analysis of Kimura's continuum-of-alleles model. *J. Math. Biol.* **24**, 341–351.

Bürger, R. 1988. Mutation-selection balance and continuum-of-alleles models. *Math. Biosci.* **91**, 67–83.

Bürger, R., Wagner, G.P., and Stettinger, F. 1989. How much heritable variation can be maintained in finite populations by mutation-selection balance? *Evolution* **43**, 1748–1766.

Campbell, M.A., and Elston, R.C. 1971. Relatives of probands: models for preliminary genetic analysis. *Ann. Hum. Genet.* **35**, 225–236.

Cannings, C. 1967. Equilibrium, convergence and stability at a sex-linked locus under natural selection. *Genetics* **56**, 613–617.

Cannings, C. 1968. Fertility difference between homogamous and heterogamous matings. *Genet. Res.* **11**, 289–301.

Cannings, C. 1968a. Equilibrium under selection at a multi-allelic sex-linked locus. *Biometrics* **11**, 187–189.

Cannings, C. 1971. Natural selection at a multiallelic autosomal locus with multiple niches. *J. Genet.* **60**, 255–259.

Cavalli-Sforza, L.L., and Bodmer, W.F. 1971. *The Genetics of Human Populations*. W.H. Freeman, San Francisco.

Charlesworth, B. 1970. Selection in populations with overlapping generations. I. The use of Malthusian parameters in population genetics. *Theor. Pop. Biol.* **1**, 352–370.

Charlesworth, B. 1974. Selection in populations with overlapping generations. VI. Rates of change of gene frequency and population growth rate. *Theor. Pop. Biol.* **6**, 108–133.

Charlesworth, B. 1980. *Evolution in Age-Structured Populations*. Cambridge University Press, Cambridge.

Charlesworth, B., and Hartl, D.L. 1978. Population dynamics of the segregation distorter polymorphism of *Drosophila melanogaster*. *Genetics* **89**, 171–192.

Chia, A.B., and Pollak, E. 1974. The inbreeding effective number and the effective number of alleles in a population that varies in size. *Theor. Pop. Biol.* **6**, 149–172.

Christiansen, F.B. 1974. Sufficient condition for protected polymorphism in a subdivided population. *Am. Nat.* **108**, 157–166.

Christiansen, F.B. 1975. Hard and soft selection in a subdivided population. *Am. Nat.* **109**, 11–16.

Christiansen, F.B., and Frydenberg, O. 1977. Selection-mutation balance for two nonallelic recessives producing an inferior double homozygote. *Am. J. Hum. Genet.* **29**, 195–207.

Cloninger, C.R., Rice, J., and Reich, T. 1979. Multifactorial inheritance with cultural transmission and assortative mating. II. A general model of combined polygenic and cultural inheritance. *Am. J. Hum. Genet.* **31**, 176–198.

Cockerham, C.C. 1954. An extension of the concept of partitioning hereditary variance for analysis of covariances among relatives when epistasis is present. *Genetics* **39**, 859–882.

Cockerham, C.C. 1956. Effects of linkage on the covariances between relatives. *Genetics* **41**, 138–141.

Cockerham, C.C. 1969. Variance of gene frequencies. *Evolution* **23**, 72–84.

Cockerham, C.C. 1970. Avoidance and rate of inbreeding. Pp. 104–127 in *Mathematical Topics in Population Genetics* (K. Kojima, ed.). Springer, Berlin.

Cockerham, C.C. 1971. Higher order probability functions of identity of alleles by descent. *Genetics* **69**, 235–246.

Cockerham, C.C., and Weir, B.S. 1973. Descent measures for two loci with some applications. *Theor. Pop. Biol.* **4**, 300–330.

Conley, C.C. 1972. Unpublished University of Wisconsin lecture notes.

Cornette, J.L. 1975. Some basic elements of continuous selection models. *Theor. Pop. Biol.* **8**, 301–313.

Cornette, J.L. 1981. Deterministic genetic models in varying environments. *J. Math. Biol.* **12**, 173–186.

Cotterman, C.W. 1937. Indication of unit factor inheritance in data comprising but a single generation. *Ohio J. Sci.* **37**, 75–81.

Cotterman, C.W. 1940. A calculus for statistico-genetics. Ph.D. thesis, Ohio State University. Reprinting: Pp. 157–272 in *Genetics and Social Structure* (P. Ballonoff, ed.). Dowden, Hutchinson and Ross, Stroudsburg, Pa. (1974).

Cotterman, C.W. 1960. Relationship and probability in Mendelian populations. Unpublished ms. Printed in *Am. J. Med. Genet.* **16**, 393–440 (1983).

Crow, J.F. 1954. Breeding structure of populations. II. Effective population number. Pp. 543–556 in *Statistics and Mathematics in Biology* (O. Kempthorne, T.A. Bancroft, J.W. Gowen, and J.L. Lush, eds.). Iowa State College Press, Ames, Iowa.

Crow, J.F. 1970. Genetic loads and the cost of natural selection. Pp. 128–177 in *Mathematical Topics in Population Genetics* (K. Kojima, ed.). Springer, Berlin.

Crow, J.F. 1983. *Genetics Notes*. 8th edn. Burgess, Minneapolis.

Crow, J.F., and Denniston, C. 1988. Inbreeding and variance effective population numbers. *Evolution* **42**, 482–495.

Crow, J.F., and Felsenstein, J. 1968. The effect of assortative mating on the genetic composition of a population. *Eugen. Quart.* **15**, 85–97.

Crow, J.F., and Kimura, M. 1970. *An Introduction to Population Genetics Theory*. Harper and Row, New York.

Crow, J.F., and Morton, N.E. 1955. Measurement of gene frequency drift in small populations. *Evolution* **9**, 202–214.

Crow, J.F., and Nagylaki, T. 1976. The rate of change of a character correlated with fitness. *Am. Nat.* **110**, 207–213, 400.

Deakin, M.A.B. 1966. Sufficient conditions for genetic polymorphism. *Am. Nat.* **100**, 690–692.

de Finetti, B. 1926. Considerazioni matematiche sul l'ereditariete mendeliana. *Metron* **6**, 1–41.

Dempster, E.R. 1955. Maintenance of genetic heterogeneity. *Cold Spring Harbor Symp. Quant. Biol.* **20**, 25–32.

Denniston, C. 1967. Probability and genetic relationship. Ph.D. thesis, University of Wisconsin, Madison.

Denniston, C. 1974. An extension of the probability approach to genetic relationships: one locus. *Theor. Pop. Biol.* **6**, 58–75.

Denniston, C. 1975. Probability and genetic relationship: two loci. *Ann. Hum. Genet.* **39**, 89–104.

Donnelly, P. 1986. A genealogical approach to variable-population-size models in population genetics. *J. Appl. Prob.* **23**, 283–296.

Elston, R.C., and Campbell, M.A. 1970. Schizophrenia: evidence for the major gene hypothesis. *Behav. Genet.* **1**, 3–10, 172.

Ethier, S.N., and Nagylaki, T. 1980. Diffusion approximations of Markov chains with two time scales and applications to population genetics. *Adv. Appl. Prob.* **12**, 14–49.

Ewens, W.J. 1969. *Population Genetics.* Methuen, London.

Ewens, W.J. 1969a. A generalized Fundamental Theorem of Natural Selection. *Genetics* **63**, 531–537.

Ewens, W.J. 1973. Conditional diffusion processes in population genetics. *Theor. Pop. Biol.* **4**, 21–30.

Ewens, W.J. 1979. *Mathematical Population Genetics.* Springer, Berlin.

Ewens, W.J. 1990. Population genetics theory – the past and the future. Pp. 177–227 in *Mathematical and Statistical Developments of Evolutionary Theory* (S. Lessard, ed.). NATO Advanced Study Institutes Series C. Kluwer Academic Publishers, Dordrecht.

Falconer, D.S. 1960. *Introduction to Quantitative Genetics.* Ronald Press, New York. 2nd edn.: Longman, London (1981).

Falk, C.T., and Li, C.C. 1969. Negative assortative mating: exact solution to a simple model. *Genetics* **62**, 215–223.

Feldman, M.W. 1966. On the offspring number distribution in a genetic population. *J. Appl. Prob.* **3**, 129–141.

Feller, W. 1968. *An Introduction to Probability Theory and Its Applications.* Vol. I, 3rd edn. Wiley, New York.

Feller, W. 1971. *An Introduction to Probability Theory and Its Applications.* Vol. II, 2nd edn. Wiley, New York.

Felsenstein, J. 1981. *Bibliography of Theoretical Population Genetics.* Dowden, Hutchinson and Ross, Stroudsburg, Pa.

Finney, D.J. 1952. The equilibrium of a self-incompatible polymorphic species. *Genetica* **26**, 33–64.

Fish, H.D. 1914. On the progressive increase of homozygosis in brother-sister matings. *Am. Nat.* **48**, 759–761.

Fisher, R.A. 1918. The correlation between relatives on the supposition of Mendelian inheritance. *Trans. Roy. Soc. Edinb.* **52**, 399–433.

Fisher, R.A. 1922. On the dominance ratio. *Proc. Roy. Soc. Edinb.* **42**, 321–341.

Fisher, R.A. 1930. *The Genetical Theory of Natural Selection.* Clarendon Press, Oxford.

Fisher, R.A. 1930a. The distribution of gene ratios for rare mutations. *Proc. Roy. Soc. Edinb.* **50**, 204–219.

Fisher, R.A. 1949. *The Theory of Inbreeding*. Oliver and Boyd, Edinburgh.

Fisher, R.A. 1958. *The Genetical Theory of Natural Selection*. 2nd edn. Dover, New York.

Fleming, W.H. 1979. Equilibrium distributions of continuous polygenic traits. *SIAM J. Appl. Math.* **36**, 148–168.

Gale, J.S., and Kearsey, M.J. 1968. Stable equilibria under stabilising selection in the absence of dominance. *Heredity* **23**, 553–561.

Gantmacher, F.R. 1959. *The Theory of Matrices*. 2 vol. Chelsea, New York.

Geiringer, H. 1944. On the probability theory of linkage in Mendelian heredity. *Ann. Math. Stat.* **15**, 25–57.

Geiringer, H. 1948. On the mathematics of random mating in case of different recombination values for males and females. *Genetics* **33**, 548–564.

Ghai, G.L. 1973. Limiting distributions under assortative mating. *Genetics* **75**, 727–732.

Gillespie, J.H., and Turelli, M. 1989. Genotype-environment interactions and the maintenance of polygenic variation. *Genetics* **121**, 129–138.

Gimelfarb, A. 1979. Evolving populations: mathematical analysis of some dynamical properties. Ph.D. thesis, University of Wisconsin-Madison.

Gimelfarb, A. 1981. A general linear model for the genotypic covariance between relatives under assortative mating. *J. Math. Biol.* **13**, 209–226.

Gimelfarb, A. 1981a. Analysis of "nontraditional" relationships under assortative mating. *J. Math. Biol.* **13**, 227–240.

Gimelfarb, A. 1989. Genotypic variation for a quantitative character maintained under stabilizing selection without mutations: epistasis. *Genetics* **123**, 217–227.

Ginzburg, L.R. 1983. *Theory of Natural Selection and Population Growth*. Benjamin/Cummings, Menlo Park, Calif.

Griffing, B. 1960. Theoretical consequences of truncation selection based on the individual phenotype. *Aust. J. Biol. Sci.* **13**, 307–343.

Haldane, J.B.S. 1924. A mathematical theory of natural and artificial selection. Part I. *Trans. Camb. Phil. Soc.* **23**, 19–41.

Haldane, J.B.S. 1924a. A mathematical theory of natural and artificial selection. Part II. *Proc. Camb. Phil. Soc., Biol. Sci.* **1**, 158–163.

Haldane, J.B.S. 1926. A mathematical theory of natural and artificial selection. Part III. *Proc. Camb. Phil. Soc.* **23**, 363–372.

Haldane, J.B.S. 1927. A mathematical theory of natural and artificial selection. Part IV. *Proc. Camb. Phil. Soc.* **23**, 607–615.

Haldane, J.B.S. 1927a. A mathematical theory of natural and artificial selection. Part V. Selection and mutation. *Proc. Camb. Phil. Soc.* **23**, 838–844.

Haldane, J.B.S. 1930. A mathematical theory of natural and artificial selection. Part VI. Isolation. *Proc. Camb. Phil. Soc.* **26**, 220–230.

Haldane, J.B.S. 1931. A mathematical theory of natural and artificial selection. Part VII. Selection intensity as a function of mortality rate. *Proc. Camb. Phil. Soc.* **27**, 131–136.

Haldane, J.B.S. 1931a. A mathematical theory of natural and artificial selection. Part VIII. Metastable populations. *Proc. Camb. Phil. Soc.* **27**, 137–142.

Haldane, J.B.S. 1932. A mathematical theory of natural and artificial selection. Part IX. Rapid selection. *Proc. Camb. Phil. Soc.* **28**, 244–248.

Haldane, J.B.S. 1934. A mathematical theory of natural and artificial selection. Part X. Some theorems on artificial selection. *Genetics* **19**, 412–429.

Haldane, J.B.S. 1935. The rate of spontaneous mutation of a human gene. *J. Genet.* **31**, 317–326.

Haldane, J.B.S. 1937. Some theoretical results of continued brother-sister mating. *J. Genet.* **34**, 265–274.

Haldane, J.B.S. 1939. The equilibrium between mutation and random extinction. *Ann. Eugen.* **9**, 400–405.

Haldane, J.B.S. 1949. The association of characters as a result of inbreeding and linkage. *Ann. Eugen.* **15**, 15–23.

Haldane, J.B.S. 1949a. Some statistical problems arising in genetics. *J. Roy. Stat. Soc. B.* **11**, 1–14.

Haldane, J.B.S. 1954. The measurement of natural selection. Pp. 480–487 in Vol. 1 of *Proc. 9th Int. Cong. Genet.* (*Caryologia,* Suppl. to Vol. 9).

Haldane, J.B.S. 1955. The complete matrices for brother-sister and alternate parent-offspring mating involving one locus. *J. Genet.* **53**, 315–324.

Haldane, J.B.S., and Jayakar, S.D. 1963. Polymorphism due to selection of varying direction. *J. Genet.* **58**, 237–242.

Haldane, J.B.S., and Jayakar, S.D. 1965. Selection for a single pair of allelomorphs with complete replacement. *J. Genet.* **59**, 171–177.

Haldane, J.B.S., and Moshinsky, P. 1939. Inbreeding in Mendelian populations with special reference to human cousin marriage. *Ann. Eugen.* **9**, 321–340.

Hardy, G.H. 1908. Mendelian proportions in a mixed population. *Science* **28**, 49–50.

Harris, D.L. 1964. Genotypic covariances between inbred relatives. *Genetics* **50**, 1319–1348.

Hartl, D.L. 1970. A mathematical model for recessive lethal segregation distorters with differential viabilities in the sexes. *Genetics* **66**, 147–163.

Hartl, D.L. 1970a. Analysis of a general population genetic model of meiotic drive. *Evolution* **24**, 538–545.

Hartl, D.L. 1970b. Population consequences of non-Mendelian segregation among multiple alleles. *Evolution* **24**, 415–423.

Hartl, D.L. 1971. Some aspects of natural selection in arrhenotokous populations. *Am. Zool.* **11**, 302–325.

Hartl, D.L. 1972. A fundamental theorem of natural selection for sex linkage or arrhenotoky. *Am. Nat.* **106**, 516–524.

Hartl, D.L., and Clark, A.G. 1989. *Principles of Population Genetics.* 2nd edn. Sinauer, Sunderland, Mass.

Hastings, A. 1981. Stable cycling in discrete-time genetic models. *Proc. Natl. Acad. Sci. USA* **78**, 7224–7225.

Hastings, A. 1981a. Simultaneous stability of $D = 0$ and $D \neq 0$ for multiplicative viabilities at two loci: an analytical study. *J. Theor. Biol.* **89**, 69–81.

Hastings, A. 1990. Second-order approximations for selection coefficients at polygenic loci. *J. Math. Biol.* **28**, 475–483.

Hastings, A., and Hom, C.L. 1989. Pleiotropic stabilizing selection limits the number of polymorphic loci to at most the number of characters. *Genetics* **122**, 459–463.

Haynsworth, E.V., and Goldberg, K. 1964. Bernoulli and Euler polynomials – Riemann zeta function. Pp. 803–819 in *Handbook of Mathematical Functions* (M. Abramowitz and I.A. Stegun, eds.). National Bureau of Standards, Washington.

an der Heiden, U. 1975. On manifolds of equilibria in the selection model for multiple alleles. *J. Math. Biol.* **1**, 321–330.

Heyde, C.C., and Seneta, E. 1975. The genetic balance between random sampling and random population size. *J. Math. Biol.* **1**, 317–320.

Hoekstra, R.F. 1975. A deterministic model of cyclical selection. *Genet. Res.* **25**, 1–15.

Hughes, P.J., and Seneta, E. 1975. Selection equilibria in a multi-allele single-locus setting. *Heredity* **35**, 185–194.

Jacquard, A. 1974. *The Genetic Structure of Populations*. Springer, Berlin.

James, J.W. 1973. Covariances between relatives due to sex-linked genes. *Biometrics* **29**, 584–588.

Jennings, H.S. 1914. Formulae for the results of inbreeding. *Am. Nat.* **48**, 693–696.

Jennings, H.S. 1916. The numerical results of diverse systems of breeding. *Genetics* **1**, 53–89.

Jennings, H.S. 1917. The numerical results of diverse systems of breeding, with respect to two pairs of characters, linked or independent, with special relation to the effects of linkage. *Genetics* **2**, 97–154.

Jensen, A. 1978. Genetic and behavioral effects of nonrandom mating. Pp. 51–105 in *Human Variation: Biogenetics of Age, Race and Sex* (C.E. Nobel, R.T. Osborne, and N. Weyl, eds.). Academic Press, New York.

Johnson, C. 1976. *Introduction to Natural Selection*. University Park Press, Baltimore.

Johnson, N.L., and Kotz, S. 1969. *Distributions in Statistics: Discrete Distributions*. Houghton Mifflin, Boston.

Johnson, N.L., and Kotz, S. 1972. *Distributions in Statistics: Continuous Multivariate Distributions*. Wiley, New York.

Karlin, S. 1968. Equilibrium behavior of population genetic models with non-random mating. *J. Appl. Prob.* **5**, 231–313, 487–566. Reprinted by Gordon and Breach, New York (1969).

Karlin, S. 1975. General two-locus selection models: some objectives, results and interpretations. *Theor. Pop. Biol.* **7**, 364–398.

Karlin, S. 1976. Population subdivision and selection migration interaction. Pp. 617–657 in *Proc. Int. Conf. Pop. Genet. Ecol.* (S. Karlin and E. Nevo, eds.). Academic Press, New York.

Karlin, S. 1977. Gene frequency patterns in the Levene subdivided population model. *Theor. Pop. Biol.* **11**, 356–385.

Karlin, S. 1977a. Protection of recessive and dominant traits in a subdivided population with general migration structure. *Am. Nat.* **111**, 1145–1162.

Karlin, S. 1978. Theoretical aspects of multi-locus selection balance. I. Pp. 503–587 in *Studies in Mathematics*. Vol. 16: *Studies in Mathematical Biology* (S.A. Levin, ed.). The Mathematical Association of America, Washington.

Karlin, S. 1982. Classifications of selection-migration structures and conditions for a protected polymorphism. *Evol. Biol.* **14**, 61–204.

Karlin, S., and Feldman, M.W. 1968. Further analysis of negative assortative mating. *Genetics* **59**, 117–136.

Karlin, S., and Feldman, M.W. 1978. Simultaneous stability of $D = 0$ and $D \neq 0$ for multiplicative viabilities at two loci. *Genetics* **90**, 813–825.

Karlin, S., and Kenett, R.S. 1977. Variable spatial selection with two stages of migrations and comparisons between different timings. *Theor. Pop. Biol.* **11**, 386–409.

Karlin, S., and Lessard, S. 1986. *Sex Ratio Evolution*. Princeton University Press, Princeton, N.J.

Karlin, S., and Liberman, U. 1974. Random temporal variation in selection intensities: case of large population size. *Theor. Pop. Biol.* **6**, 355–382.

Karlin, S., and McGregor, J. 1964. Direct product branching processes and related Markov chains. *Proc. Natl. Acad. Sci. USA* **51**, 598–602.

Karlin, S., and McGregor, J. 1965. Direct product branching processes and related induced Markoff chains. I. Calculations of rates of approach to homozygosity. Pp. 111-145 in *Bernoulli (1723), Bayes (1763), Laplace (1813) Anniversary Volume* (L. LeCam and J. Neyman, eds.). Springer, Berlin.

Karlin, S., and McGregor, J. 1971.On mutation selection balance for two-locus haploid and diploid populations. *Theor. Pop. Biol.* **2**, 60–70.

Karlin, S., and McGregor, J. 1972. Application of method of small parameters to multi-niche population genetic models. *Theor. Pop. Biol.* **3**, 186–209.

Karlin, S., and McGregor, J. 1972a. Polymorphisms for genetic and ecological systems with weak coupling. *Theor. Pop. Biol.* **3**, 210–238.

Karlin, S., and Richter-Dyn, N. 1976. Some theoretical analyses of migration selection interaction in a cline: a generalized two range environment. Pp. 659–706 in *Proc. Int. Conf. Pop. Genet. Ecol.* (S. Karlin and E. Nevo, eds.). Academic Press, New York.

Karlin, S., and Taylor, H.M. 1975. *A First Course in Stochastic Processes*. Academic Press, New York.

Kearsey, M.J., and Gale, J.S. 1968. Stabilising selection in the absence of dominance: an additional note. *Heredity* **23**, 617–620.

Keightley, P.D., and Hill, W.G. 1989. Quantitative genetic variability maintained by mutation-selection balance: sampling variation and response to subsequent directional selection. *Genet. Res.* **54**, 45–57.

Kempthorne, O. 1954. The correlation between relatives in a random mating population. *Proc. Roy. Soc. B* **143**, 103–113.

Kempthorne, O. 1955. The theoretical values of the correlations between relatives in random mating populations. *Genetics* **40**, 153–167.

Kempthorne, O. 1955a. The correlations between relatives in random mating populations. *Cold Spring Harbor Symp. Quant. Biol.* **20**, 60–78.

Kempthorne, O. 1957. *An Introduction to Genetic Statistics*. Wiley, New York.

Kendall, M.G., and Stuart, A. 1973. *The Advanced Theory of Statistics*. Vol. 2, 3rd edn. Griffin, London.

Kidwell, J.F., Clegg, M.T., Stewart, F.M, and Prout, T. 1977. Regions of stable equilibria for models of differential selection in the two sexes under random mating. *Genetics* **85**, 171–183.

Kimura, M. 1955. Stochastic processes and distribution of gene frequencies under natural selection. *Cold Spring Harbor Symp. Quant. Biol.* **20**, 33–53.

Kimura, M. 1955a. Random genetic drift in multi-allelic locus. *Evolution* **9**, 419–435.

Kimura, M. 1956. Rules for testing stability of a selective polymorphism. *Proc. Natl. Acad. Sci. USA* **42**, 336–340.

Kimura, M. 1956a. A model of a genetic system which leads to closer linkage by natural selection. *Evolution* **10**, 278–287.

Kimura, M. 1956b. Random genetic drift in a tri-allelic locus; exact solution with a continuous model. *Biometrics* **12**, 57–66.

Kimura, M. 1957. Some problems of stochastic processes in genetics. *Ann. Math. Stat.* **28**, 882–901.

Kimura, M. 1958. On the change of population fitness by natural selection. *Heredity* **12**, 145–167.

Kimura, M. 1958a. Zygotic frequencies in a partially self-fertilizing population. *Ann. Rept. Natl. Inst. Genet., Japan* **8**, 104–105.

Kimura, M. 1960. *Outline of Population Genetics*. Baifukan, Tokyo (in Japanese).

Kimura, M. 1963. A probability method for treating inbreeding systems, especially with linked genes. *Biometrics* **19**, 1–17.

Kimura, M. 1965. Attainment of quasi-linkage equilibrium when gene frequencies are changing by natural selection. *Genetics* **52**, 875–890.

Kimura, M. 1965a. A stochastic model concerning the maintenance of genetic variability in quantitative characters. *Proc. Natl. Acad. Sci. USA* **54**, 731–736.

Kimura, M. 1968. Evolutionary rate at the molecular level. *Nature* **217**, 624–626.

Kimura, M. 1968a. Genetic variability maintained in a finite population due to mutational production of neutral and nearly neutral isoalleles. *Genet. Res.* **11**, 247–269.

Kimura, M. 1971. Theoretical foundation of population genetics at the molecular level. *Theor. Pop. Biol.* **2**, 174–208.

Kimura, M. 1978. Change of gene frequencies by natural selection under population number regulation. *Proc. Natl. Acad. Sci. USA* **75**, 1934–1937.

Kimura, M. 1981. Possibility of extensive neutral evolution under stabilizing selection with special reference to nonrandom usage of synonymous codons. *Proc. Natl. Acad. Sci. USA* **78**, 5773–5777.

Kimura, M. 1983. *The Neutral Theory of Molecular Evolution*. Cambridge University Press, Cambridge.

Kimura, M., and Crow, J.F. 1963. On the maximum avoidance of inbreeding. *Genet. Res.* **4**, 399–415.

Kimura, M., and Crow, J.F. 1963a. The measurement of effective population number. *Evolution* **17**, 279–288.

Kimura, M., and Crow, J.F. 1964. The number of alleles that can be maintained in a finite population. *Genetics* **49**, 725–738.

Kimura, M., and Crow, J.F. 1978. Effect of overall phenotypic selection on genetic change at individual loci. *Proc. Natl. Acad. Sci. USA* **75**, 6168–6171.

Kimura, M., and Ohta, T. 1971. *Theoretical Aspects of Population Genetics*. Princeton University Press, Princeton.

King, J.L. 1965. The effect of litter culling – of family planning – on the rate of natural selection. *Genetics* **51**, 425–429.

King, J.L., and Jukes, T.H. 1969. Non-Darwinian evolution. *Science* **164**, 788–798.

Kingman, J.F.C. 1961. On an inequality in partial averages. *Quart. J. Math.* **12**, 78–80.

Kingman, J.F.C. 1961a. A mathematical problem in population genetics. *Proc. Camb. Phil. Soc.* **57**, 574–582.

Kingman, J.F.C. 1980. *Mathematics of Genetic Diversity*. SIAM, Philadelphia.

Klebaner, F.C. 1988. Conditions for fixation of an allele in the density-dependent Wright-Fisher models. *J. Appl. Prob.* **25**, 247–256.

Kojima, K. 1959. Stable equilibria for the optimum model. *Proc. Natl. Acad. Sci. USA* **45**, 989–993.

Lande, R. 1976. The maintenance of genetic variability by mutation in a polygenic character with linked loci. *Genet. Res.* **26**, 221–235.

Lande, R. 1977. The influence of the mating system on the maintenance of genetic variability in polygenic characters. *Genetics* **86**, 485–498.

Lande, R. 1980. Genetic variation and phenotypic evolution during allopatric speciation. *Am. Nat.* **116**, 463–479.

Lande, R. 1981. The minimum number of genes contributing to quantitative variation between and within populations. *Genetics* **99**, 541–553.

Lande, R. 1991. Isolation by distance in a quantitative trait. *Genetics* **128**, 443–452.

LaSalle, J.P. 1977. *The Stability of Dynamical Systems*. SIAM, Philadelphia.

Latter, B.D.H. 1959. Genetic sampling in a random mating control population of constant size and sex-ratio. *Aust. J. Biol. Sci.* **12**, 500–505.

Latter, B.D.H. 1970. Selection in finite populations with multiple alleles. II. Centripetal selection, mutation, and isoallelic variation. *Genetics* **66**, 165–186.

Latter, B.D.H. 1972. Selection in finite populations with multiple alleles. III. Genetic divergence with centripetal selection and mutation. *Genetics* **70**, 475–490.

Léon, J.A., and Charlesworth, B. 1976. Ecological versions of Fisher's fundamental theorem of natural selection. *Adv. Appl. Prob.* **8**, 639–641.

Léon, J.A., and Charlesworth, B. 1978. Ecological versions of Fisher's fundamental theorem of natural selection. *Ecology* **59**, 457–464.

Levene, H. 1949. On a matching problem in genetics. *Ann. Math. Stat.* **20**, 91–94.

Levene, H. 1953. Genetic equilibrium when more than one ecological niche is available. *Am. Nat.* **87**, 311–313.

Levy, J. 1976. A review of evidence for a genetic component in the determination of handedness. *Behav. Genet.* **6**, 429–453.

Levy, J., and Nagylaki, T. 1972. A model for the genetics of handedness. *Genetics* **72**, 117–128.

Levy, J., and Reid, M. 1976. Variations in writing posture and cerebral organization. *Science* **194**, 337–339.

Levy, J., and Reid, M. 1978. Variations in cerebral organization as a function of handedness, hand posture in writing, and sex. *J. Exp. Psychol.: General* **107**, 119–144.

Lewontin, R.C. 1958. A general method for investigating the equilibrium of gene frequency in a population. *Genetics* **43**, 421–433.

Lewontin, R.C. 1964. The interaction of linkage and selection. II. Optimum models. *Genetics* **50**, 757–782.

Lewontin, R.C., Ginzburg, L.R., and Tuljapurkar, S.D. 1978. Heterosis as an explanation for large amounts of genic polymorphism. *Genetics* **88**, 149–169.

Lewontin, R.C., and Kojima, K. 1960. The evolutionary dynamics of complex polymorphisms. *Evolution* **14**, 458–472.

Li, C.C. 1955. *Population Genetics*. The University of Chicago Press, Chicago. 2nd edn.: Boxwood Press, Pacific Grove, Calif. (1976).

Li, C.C. 1955a. The stability of an equilibrium and the average fitness of a population. *Am. Nat.* **89**, 281–295.

Li, C.C. 1959. Notes on relative fitness of genotypes that form a geometric progression. *Evolution* **13**, 564–567.

Li, C.C. 1969. Increment of average fitness for multiple alleles. *Proc. Natl. Acad. Sci. USA* **62**, 395–398.

Li, C.C., and Sacks, L. 1954. The derivation of joint distribution and correlation between relatives by the use of stochastic matrices. *Biometrics* **10**, 347–360.

Lloyd, D.G. 1977. Genetic and phenotypic models of natural selection. *J. Theor. Biol.* **69**, 543–560.

Losert, V., and Akin, E. 1983. Dynamics of games and genes: discrete versus continuous time. *J. Math. Biol.* **17**, 241–251.

Lukacs, E. 1970. *Characteristic Functions*. 2nd edn. Griffin, London.

Lush, J.L. 1948. *The Genetics of Populations*. Mimeographed notes. Ames, Iowa.

Lyubich, Yu. I., Maistrovskii, G.D., and Ol'khovskii, Yu. G. 1980. Selection-induced convergence to equilibrium in a single-locus autosomal population. *Probs. Inf. Trans.* **16**, 66–75.

Malécot, G. 1939. Théorie mathématique de l'hérédité mendélienne généralisée. Dissertation, Faculté Sciences, University of Paris.

Malécot, G. 1941. Étude mathématique des populations "mendéliennes". *Ann. Univ. Lyon Sci. Sec. A* **4**, 45–60.

Malécot, G. 1942. Mendélisme et consanguinité. *C. R. Acad. Sci. Paris* **215**, 313–314.

Malécot, G. 1944. Sur un problème de probabilités en chaîne que pose la génétique. *C. R. Acad. Sci. Paris* **219**, 379–381.

Malécot, G. 1946. La consanguinité dans une population limitée. *C. R. Acad. Sci. Paris* **222**, 841–843.

Malécot, G. 1948. *Les mathématiques de l'hérédité*. Masson, Paris. Extended translation: *The Mathematics of Heredity*. W.H. Freeman, San Francisco (1969).

Malécot, G. 1950. Quelques schémas probabilistes sur la variabilité des populations naturelles. *Ann. Univ. Lyon Sci. Sec. A* **13**, 37–60.

Malécot, G. 1951. Un traitement stochastique des problèmes linéaires (mutation, linkage, migration) en Génétique de Population. *Ann. Univ. Lyon Sci. Sec. A* **14**, 79–117.

Mandel, S.P.H. 1959. The stability of a multiple allelic system. *Heredity* **13**, 289–302.

Mandel, S.P.H. 1970. The equivalence of different sets of stability conditions for multiple allelic systems. *Biometrics* **26**, 840–845.

Mather, K., and Jinks, J.L. 1971. *Biometrical Genetics*. 2nd edn. Chapman and Hall, London.

Maynard Smith, J. 1966. Sympatric speciation. *Am. Nat.* **100**, 637–650.

Maynard Smith, J. 1970. Genetic polymorphism in a varied environment. *Am. Nat.* **104**, 487–490.

Moody, M. 1978. A multi-locus continuous-time selection model. *J. Math. Biol.* **5**, 281–291.

Moody, M. 1979. Polymorphism with migration and selection. *J. Math. Biol.* **8**, 73–109.

Moran, P.A.P. 1958. A general theory of the distribution of gene frequencies. II. Non-overlapping generations. *Proc. Roy. Soc. B* **149**, 113–116.

Moran, P.A.P. 1958a. Random processes in genetics. *Proc. Camb. Phil. Soc.* **54**, 60–71.

Moran, P.A.P. 1958b. The effect of selection in a haploid genetic population. *Proc. Camb. Phil. Soc.* **54**, 463–467.

Moran, P.A.P. 1958c. The distribution of gene frequency in a bisexual diploid population. *Proc. Camb. Phil. Soc.* **54**, 468–474.

Moran, P.A.P. 1958d. A general theory of the distribution of gene frequencies. I. Overlapping generations. *Proc. Roy. Soc. B* **149**, 102–112.

Moran, P.A.P. 1962. *The Statistical Processes of Evolutionary Theory*. Clarendon Press, Oxford.

Moran, P.A.P. 1967. Unsolved problems in evolutionary theory. *Proc. Fifth Berkeley Symp. Math. Stat. Prob.* **4**, 457–480.

Moran, P.A.P. 1968. On the theory of selection dependent on two loci. *Ann. Hum. Genet.* **32**, 183–190.

Moran, P.A.P., and Smith, C.A.B. 1966. Commentary on R.A. Fisher's paper on "The correlation between relatives on the supposition of Mendelian inheritance". *Eugenics Laboratory Memoirs* **41**. Cambridge University Press, Cambridge.

Moran, P.A.P., and Watterson, G.A. 1959. The genetic effects of family structure in natural populations. *Aust. J. Biol. Sci.* **12**, 1–15.

Morton, N.E., Crow, J.F., and Muller, H.J. 1956. An estimate of the mutational damage in man from data on consanguineous marriages. *Proc. Natl. Acad. Sci. USA* **42**, 855–863.

Mulholland, H.P., and Smith, C.A.B. 1959. An inequality arising in genetical theory. *Am. Math. Mon.* **66**, 673–683.

Muller, H.J. 1948. Mutational prophylaxis. *Bull. N. Y. Acad. Med.* **24**, 447–469.

Murphy, E.A., and Chase, G.A. 1975. *Principles of Genetic Counseling*. Year Book Medical Publishers, Chicago.

Nagylaki, T. 1974. Continuous selective models with mutation and migration. *Theor. Pop. Biol.* **5**, 284–295.

Nagylaki, T. 1975. Polymorphisms in cyclically-varying environments. *Heredity* **35**, 67–74.

Nagylaki, T. 1975a. The deterministic behavior of self-incompatibility alleles. *Genetics* **79**, 545–550.

Nagylaki, T. 1975b. Conditions for the existence of clines. *Genetics* **80**, 595–615.

Nagylaki, T. 1975c. A continuous selective model for an X-linked locus. *Heredity* **34**, 273–278.

Nagylaki, T. 1976. The evolution of one- and two-locus systems. *Genetics* **83**, 583–600.

Nagylaki, T. 1976a. A model for the evolution of self-fertilization and vegetative reproduction. *J. Theor. Biol.* **58**, 55–58.

Nagylaki, T. 1976b. Dispersion-selection balance in localised plant populations. *Heredity* **37**, 59–67.

Nagylaki, T. 1977. The evolution of one- and two-locus systems. II. *Genetics* **85**, 347–354.

Nagylaki, T. 1977a. *Selection in One- and Two-Locus Systems*. Springer, Berlin.

Nagylaki, T. 1977b. Selection and mutation at an X-linked locus. *Ann. Hum. Genet.* **41**, 241–248.

Nagylaki, T. 1978. The correlation between relatives with assortative mating. *Ann. Hum. Genet.* **42**, 131–137.

Nagylaki, T. 1979. Dynamics of density- and frequency-dependent selection. *Proc. Natl. Acad. Sci. USA* **76**, 438–441.

Nagylaki, T. 1979a. Migration-selection polymorphism in dioecious populations. *J. Math. Biol.* **8**, 123–131.

Nagylaki, T. 1979b. Selection in dioecious populations. *Ann. Hum. Genet.* **43**, 143–150.

Nagylaki, T. 1979c. The island model with stochastic migration. *Genetics* **91**, 163–176.

Nagylaki, T. 1980. Analysis of some regular systems of inbreeding. *J. Math. Biol.* **9**, 237–244.

Nagylaki, T. 1981. The inbreeding effective population number and the expected homozygosity for an X-linked locus. *Genetics* **97**, 731–737.

Nagylaki, T. 1982. Assortative mating for a quantitative character. *J. Math. Biol.* **16**, 57–74.

Nagylaki, T. 1984. Selection on a quantitative character. Pp. 275–306 in *Human Population Genetics: The Pittsburgh Symposium* (A. Chakravarti, ed.). Van Nostrand Reinhold, Stroudsburg, Pa.

Nagylaki, T. 1986. The Gaussian approximation for random genetic drift. Pp. 629–642 in *Evolutionary Processes and Theory* (S. Karlin and E. Nevo, eds.). Academic Press, New York.

Nagylaki, T. 1987. Evolution under fertility and viability selection. *Genetics* **115**, 367–375.

Nagylaki, T. 1989. Rate of evolution of a character without epistasis. *Proc. Natl. Acad. Sci. USA* **86**, 1910–1913.

Nagylaki, T. 1989a. The maintenance of genetic variability in two-locus models of stabilizing selection. *Genetics* **122**, 235–248.

Nagylaki, T. 1989b. Gustave Malécot and the transition from classical to modern population genetics. *Genetics* **122**, 253–268.

Nagylaki, T. 1989c. The diffusion model for migration and selection. Pp. 55–75 in *Some Mathematical Questions in Biology: Models in Population Biology* (A. Hastings, ed.). American Mathematical Society, Providence, R. I.

Nagylaki, T. 1990. Models and approximations for random genetic drift. *Theor. Pop. Biol.* **37**, 192–212.

Nagylaki, T. 1991. Error bounds for the fundamental and secondary theorems of natural selection. *Proc. Natl. Acad. Sci. USA* **88**, 2402–2406.

Nagylaki, T., and Crow, J.F. 1974. Continuous selective models. *Theor. Pop. Biol.* **5**, 257–283.

Nagylaki, T., and Levy, J. 1973. "The sound of one paw clapping" isn't sound. *Behav. Genet.* **3**, 279–292.

Narain, P. 1966. Effect of linkage on homozygosity of a population under mixed selfing and random mating. *Genetics* **54**, 303–314.

Nei, M. 1968. The frequency distribution of lethal chromosomes in finite populations. *Proc. Natl. Acad. Sci. USA* **60**, 517–524.

Nei, M. 1971. Fertility excess necessary for gene substitution in regulated populations. *Genetics* **68**, 169–184.

Nei, M. 1975. *Molecular Population Genetics and Evolution*. North-Holland, Amsterdam.

Nei, M. 1987. *Molecular Evolutionary Genetics*. Columbia University Press, New York.

Norman, M.F. 1974. A central limit theorem for Markov processes that move by small steps. *Ann. Prob.* **2**, 1065–1074.

Norton, H.T.J. 1928. Natural selection and Mendelian variation. *Proc. Lond. Math. Soc.* **28**, 1–45.

O'Donald, P. 1960. Assortative mating in a population in which two alleles are segregating. *Heredity* **15**, 389–396.

O'Donald, P. 1980. *Genetic Models of Sexual Selection*. Cambridge University Press, Cambridge.

Ohta, T. 1976. Role of very slightly deleterious mutations in molecular evolution and polymorphism. *Theor. Pop. Biol.* **10**, 254–275.

O'Rourke, D.H., Gottesman, I.I., Suarez, B.K., Rice, J., and Reich, T. 1982. Refutation of the general single-locus model for the etiology of schizophrenia. *Am. J. Hum. Genet.* **34**, 630–649.

Owen, A.R.G. 1953. A genetical system admitting of two distinct stable equilibria under natural selection. *Heredity* **7**, 97–102.

Owen, R.E. 1986. Gene frequency clines at X-linked or haplodiploid loci. *Heredity* **57**, 209–219.

Palm, G. 1974. On the selection model for a sex-linked locus. *J. Math. Biol* **1**, 47–50.

Parsons, P.A. 1961. The initial progress of new genes with viability differences between the sexes and with sex linkage. *Heredity* **16**, 103–107.

Pearl, R. 1914. On the results of inbreeding a Mendelian population: a correction and extension of previous conclusions. *Am. Nat.* **48**, 57–62.

Pearl, R. 1914a. Studies on inbreeding. IV. On a general formula for the constitution of the nth generation of a Mendelian population in which all matings are of brother × sister. *Am. Nat.* **48**, 491–494.

Pearson, K. 1904. Mathematical contributions to the theory of evolution. XII. On a generalised theory of alternative inheritance, with special reference to Mendel's laws. *Phil. Trans. Roy. Soc. A* **203**, 53–86.

Penrose, L.S. 1949. The meaning of "fitness" in human populations. *Ann. Eugen.* **14**, 301–304.

Penrose, L.S., Maynard Smith, S., and Sprott, D.A. 1956. On the stability of allelic systems, with special reference to haemoglobins A, S and C. *Ann. Hum. Genet.* **21**, 90–93.

Pollak, E. 1977. Effective population numbers and their interrelations. Pp. 115–144 in *Proceedings of the Washington State University Conference on Biomathematics and Biostatistics of May 1974*. Department of Pure and Applied Mathematics, Washington State University and Pi Mu Epsilon, Washington Alpha Chapter, Pullman, Washington.

Prout, T. 1968. Sufficient conditions for multiple niche polymorphism. *Am. Nat.* **102**, 493–496.

Risch, H. 1979. The correlation between relatives under assortative mating for an X-linked and autosomal trait. *Ann. Hum. Genet.* **43**, 151–165.

Risch, N., and Baron, M. 1984. Segregation analysis of schizophrenia and related disorders. *Am. J. Hum. Genet.* **36**, 1039–1059.

Robbins, R.B. 1917. Some applications of mathematics to breeding problems. *Genetics* **2**, 489–504.

Robbins, R.B. 1918. Some applications of mathematics to breeding problems. II. *Genetics* **3**, 73–92.

Robbins, R.B. 1918a. Some applications of mathematics to breeding problems. III. *Genetics* **3**, 375–389.

Robertson, A. 1952. The effect of inbreeding on the variation due to recessive genes. *Genetics* **37**, 189–207.

Robertson, A. 1964. The effect of non-random mating within inbred lines on the rate of inbreeding. *Genet. Res.* **5**, 164–167.

Schäfer, W. 1936. Über die Zunahme der Isozygotie (Gleicherbligkeit) bei fortgesetzter Bruder-Schwester-Inzucht. *Zeit. ind. Abst. Ver.* **72**, 50–79.

Scheuer, P.A.G., and Mandel, S.P.H. 1959. An inequality in population genetics. *Heredity* **13**, 519–524.

Schnell, F.W. 1963. The covariance between relatives in the presence of linkage. Pp. 468–483 in *Statistical Genetics and Plant Breeding* (W.D. Hanson and H.F. Robinson, eds.). National Academy of Sciences-National Research Council Publication 982, Washington.

Schull, W.J., and Neel, J.V. 1965. *The Effects of Inbreeding on Japanese Children.* Harper and Row, New York.

Scudo, F.M., and Karlin, S. 1969. Assortative mating based on phenotype. I. Two alleles with dominance. *Genetics* **63**, 479–498.

Searle, A.G. 1974. Mutation induction in mice. *Adv. Rad. Biol.* 4, 131–207.

Selgrade, J.F., and Ziehe, M. 1987. Convergence to equilibrium in a genetic model with differential viability between the sexes. *J. Math. Biol.* **25**, 477–490.

Seneta, E. 1974. A note on the balance between random sampling and population size (on the 30th anniversary of G. Malécot's paper). *Genetics* **77**, 607–610.

Singh, M., and Lewontin, R.C. 1966. Stable equilibria under optimizing selection. *Proc. Natl. Acad. Sci. USA* **56**, 1345–1348.

Skellam, J.G. 1949. The probability distribution of gene differences in relation to selection, mutation, and random extinction. *Proc. Camb. Phil. Soc.* **45**, 364–367.

Slatkin, M. 1979. A note on the symmetry constraints imposed by dominance in multiple-locus genetic models. *Genet. Res.* **33**, 81–88.

Slatkin, M. 1979a. The evolutionary response to frequency- and density-dependent interactions. *Am. Nat.* **114**, 384–398.

Snyder, L.H. 1932. Studies in human inheritance. IX. The inheritance of taste deficiency in man. *Ohio J. Sci.* **32**, 436–440.

Sprott, D.A. 1957. The stability of a sex-linked allelic system. *Ann. Hum. Genet.* **22**, 1–6.

Spuhler, J.N. 1968. Assortative mating with respect to physical characteristics. *Eugen. Quart.* **15**, 128–140.

Srb, A.M., Owen, R.D., and Edgar, R.S. 1965. *General Genetics.* 2nd edn. W.H. Freeman, San Francisco.

Strobeck, C. 1974. Sufficient conditions for polymorphism with n niches and m mating groups. *Am. Nat.* **108**, 152–156.

Svirezhev, Yu. M., and Passekov, V.P. 1990. *Fundamentals of Mathematical Evolutionary Genetics.* Kluwer Academic Publishers, Dordrecht.

Szucs, J.M. 1991. Selection and mutation at a diallelic sex-linked locus. *J. Math. Biol.* **29**, 587–627.

Szucs, J.M. 1991a. Selection at a diallelic autosomal locus in a dioecious population. *J. Math. Biol.* **29**, 693–713.

Szucs, J.M. 1991b. Selection-mutation at a diallelic autosomal locus in a dioecious population. *J. Math. Biol.* **30**, 1–14.

Tachida, H., and Cockerham, C.C. 1989. Effects of identity disequilibrium and linkage on quantitative variation in finite populations. *Genet. Res.* **53**, 63–70.

Thompson, E.A. 1974. Gene identities and multiple relationships. *Biometrics* **30**, 667–680.

Thompson, E.A. 1976. A restriction on the space of genetic relationships. *Ann. Hum. Genet.* **40**, 201–204.

Thompson, E.A. 1976a. Population correlation and population kinship. *Theor. Pop. Biol.* **10**, 205–226.

Titchmarsh, E.C. 1939. *The Theory of Functions*. 2nd edn. Oxford University Press, London.

Trankell, A. 1955. Aspects of genetics in psychology. *Am. J. Hum. Genet.* **7**, 264–276.

Trustrum, G.B. 1961. The correlations between relatives in a random mating diploid population. *Proc. Camb. Phil. Soc.* **57**, 315–320.

Turelli, M. 1984. Heritable genetic variation via mutation-selection balance: Lerch's zeta meets the abdominal bristle. *Theor. Pop. Biol.* **25**, 138–193.

Turelli, M., and Barton, N.H. 1990. Dynamics of polygenic characters under selection. *Theor. Pop. Biol.* **38**, 1–57.

Vandenberg, S.G. 1972. Assortative mating, or who marries whom? *Behav. Genet.* **2**, 127–157.

Vogel, F. 1977. A probable sex difference in some mutation rates. *Am. J. Hum. Genet.* **29**, 312–319.

Wagner, G.P. 1989. Multivariate mutation-selection balance with constrained pleiotropic effects. *Genetics* **122**, 223–234.

Wahlund, S. 1928. Zusammensetzung von Populationen und Korrelationserscheinungen vom Standpunkt der Vererbungslehre aus betrachtet. *Hereditas* **11**, 65–106.

Wallace, B. 1968. *Topics in Population Genetics*. W. W. Norton, New York.

Walsh, J.B. 1990. Inconsistencies in standard approximations for selection coefficients at loci affecting a polygenic character. *J. Math. Biol.* **28**, 21–31.

Weinberg, W. 1908. Über den Nachweis der Vererbung beim Menschen. *Jahresh. Verein. f. vaterl. Naturk. Württem.* **64**, 368–382. Translations: Pp. 4–15 in *Papers on Human Genetics* (S.H. Boyer, ed.), Prentice-Hall, Englewood Cliffs, N.J. (1963) and pp. 115–125 in *Evolutionary Genetics* (D.L. Jameson, ed.), Dowden, Hutchinson and Ross, Stroudsburg, Pa. (1977).

Weinberg, W. 1909. Über Vererbungsgesetze beim Menschen. *Zeit. ind. Abst. Vererb.* **1**, 377–392, 440–460; **2**, 276–330.

Weinberg, W. 1910. Weitere Beiträge zur Theorie der Vererbung. *Arch. Rass. Gesellsch. Biol.* **7**, 35–49, 169–173.

Weir, B.S., and Cockerham, C.C. 1973. Mixed self and random mating at two loci. *Genet. Res.* **21**, 247–262.

Wilson, S.R. 1973. The correlation between relatives under the multifactorial model with assortative mating. *Ann. Hum. Genet.* **37**, 189–215.

Wright, S. 1921. Systems of mating. *Genetics* **6**, 111–178.

Wright, S. 1921a. Correlation and causation. *J. Agric. Res.* **20**, 557–585.

Wright, S. 1922. Coefficients of inbreeding and relationship. *Am. Nat.* **56**, 330–338.

Wright, S. 1922a. The effects of inbreeding and crossbreeding on guinea pigs. III. Crosses between highly inbred families. Pp. 1–60, Bull. 1121, U.S. Department of Agriculture, Washington.

Wright, S. 1931. Evolution in Mendelian populations. *Genetics* **16**, 97–159.

Wright, S. 1933. Inbreeding and recombination. *Proc. Natl. Acad. Sci. USA* **19**, 420–433.

Wright, S. 1933a. Inbreeding and homozygosis. *Proc. Natl. Acad. Sci. USA* **19**, 411–420.

Wright, S. 1934. The method of path coefficients. *Ann. Math. Stat.* **5**, 161–215.

Wright, S. 1935. Evolution in populations in approximate equilibrium. *J. Genet.* **30**, 257–266.

Wright, S. 1938. Size of population and breeding structure in relation to evolution. *Science* **87**, 430–431.

Wright, S. 1939. *Statistical Genetics in Relation to Evolution. Actual. Sci. Ind.* **802**, *Exposés de biométrie et de statistique biologique.* XIII (G. Teissier, ed.). Hermann, Paris.

Wright, S. 1942. Statistical genetics and evolution. *Bull. Am. Math. Soc.* **48**, 223–246.

Wright, S. 1949. Genetics of populations. Pp. 111–112 in *Encyclopaedia Britannica.* Vol. 10, 14th edn. Encyclopaedia Britannica, Chicago.

Wright, S. 1949a. Adaptation and selection. Pp. 365–389 in *Genetics, Paleontology and Evolution* (G.L. Jepson, G.G. Simpson, and E. Mayr, eds.). Princeton University Press, Princeton, N.J.

Wright, S. 1951. The genetical structure of populations. *Ann. Eugen.* **15**, 323–354.

Wright, S. 1955. Classification of the factors of evolution. *Cold Spring Harbor Symp. Quant. Biol.* **20**, 16–24.

Wright, S. 1965. The interpretation of population structure by F-statistics with special regard to systems of mating. *Evolution* **19**, 395–420.

Wright, S. 1968. *Evolution and the Genetics of Populations.* Vol. I. *Genetic and Biometric Foundations.* The University of Chicago Press, Chicago.

Wright, S. 1969. *Evolution and the Genetics of Populations.* Vol. II. *The Theory of Gene Frequencies.* The University of Chicago Press, Chicago.

Wright, S. 1970. Random drift and the shifting balance theory of evolution. Pp. 1–31 in *Mathematical Topics in Population Genetics* (K. Kojima, ed.). Springer, Berlin.

Wright, S. 1977. *Evolution and the Genetics of Populations.* Vol. III. *Experimental Results and Evolutionary Deductions.* The University of Chicago Press, Chicago.

Wright, S. 1978. *Evolution and the Genetics of Populations.* Vol. IV. *Variability Within and Among Natural Populations.* The University of Chicago Press, Chicago.

Yule, G.U. 1906. On the theory of inheritance of quantitative compound characters on the basis of Mendel's laws – a preliminary note. Pp. 140–142 in *Rept. 3rd Int. Conf. Genet.*

Zeng, Z.-B. 1989. A genetic model of interpopulation variation and covariation of quantitative characters. *Genet. Res.* **53**, 215–221.

Zeng, Z.-B., Tachida, H., and Cockerham, C.C. 1989. Effects of mutation on selection limits in finite populations with multiple alleles. *Genetics* **122**, 979–984.

Author Index

Page numbers in italics refer to the references.

van Aarde, I.M.R., 213, 287, 288, *340*
Abramowitz, M., 301, *340*
Abugov, R., 170, *340*
Akin, E., 60, 189, *340, 349*
Apostol, T.M., 66, *340*
Arzberger, P., 222, 227, *340*
Atkinson, F.V., 57, *340*

Baron, M., 218, *352*
Bartlett, M.S., 259, 271, *340*
Barton, N.H., 279, 280, 329, *340, 354*
Baum, L.E., 145, 199, *340*
Bellman, R., 118, *340*
Bennett, J.H., 38, 44, 46, 160, 273, *340*
Binet, F.E., 273, *340*
Bodmer, W.F., 1, 32, 50, 53, 132, 171, 174, 185, 186, 188, 219, 220, 280, 281, 283, 290, 328, *341*
Bohidar, N.R., 288, *341*
Bona, J.L., 67
Borel, E., 201, 208, *341*
Boucher, W., 114, 117, 208, 209, 222, 230, 233, 234, 275, *341*
Brauer, F., 84, 104, 192, *341*
Bulmer, M.G., 132, 135–137, 149, 279, 280, 307, 309–311, 317, 320, 329, *341*
Bürger, R., 279, 329, *341*

Campbell, M.A., 210, 217, 218, *341, 343*
Cannings, C., 124, 145, 159, 160, *341*
Cavalli-Sforza, L.L., 1, 32, 53, 132, 219, 220, 280, 281, 283, 290, 328, *341*
Charlesworth, B., 47, 75, 79, 91, 95, *341, 342, 348*
Chase, G.A., 215, *350*
Chia, A.B., 245, 247, *342*
Christiansen, F.B., 136, 138, 140–143, 149, 171, 198, *342*
Clark, A.G., 4, *345*
Clegg, M.T., 171, *347*

Cloninger, C.R., 335, *342*
Cockerham, C.C., 210, 216, 227, 230, 254, 256, 273, 274, 280, 281, 288, *342, 353–355*
Conley, C.C., 179, *342*
Cornette, J.L., 13, 65, 67, 71, 79, *342*
Cotterman, C.W., 33, 201, 202, 208, 210, 211, 213, 215, 233, 234, *341, 342*
Crow, J.F., VII, 1, 9, 31, 32, 41, 60, 76, 81, 83, 108, 115, 160, 189, 195, 200, 202, 209, 219, 221, 222, 227, 230–233, 242, 243, 245–247, 253, 256, 265–267, 271, 275, 276, 296, 316–318, *342, 343, 348, 350, 351*

Deakin, M.A.B., 136, 149, *343*
de Finetti, B., 42, 213, *343*
Dempster, E.R., 71, 134, 147, *343*
Denniston, C., 203, 210, 215, 216, 265, 267, *342, 343*
Donnelly, P., 242, *343*

Eagon, J.A., 145, 199, *340*
Edgar, R.S., 74, *353*
Elston, R.C., 210, 217, 218, *341, 343*
Ethier, S.N., 248, 249, 252, 253, 265, 267, 277, *343*
Ewens, W.J., VII, 4, 196, 200, 253, 256, 259, 276, *343*

Falconer, D.S., 279–281, 283, 289, 311, 312, *343*
Falk, C.T., 124, *343*
Feldman, M.W., 123, 124, 189, 267, *343, 346*
Feller, W., 140, 314, 315, 319, *343*
Felsenstein, J., VII, 185, 186, 188, 296, *342, 343*
Finney, D.J., 119, 120, 123, 124, *343*
Fish, H.D., 225, *343*

Fisher, R.A.,   V, 3, 4, 8, 18, 25, 47, 52, 91, 117, 169, 200, 236, 253, 256, 257, 271, 279, 283, 287, 288, 296, 308–311, 313, 317, *343*, *344*
Fleming, W.H.,   321, 323–325, 329, *344*
Frydenberg, O.,   198, *342*

Gale, J.S.,   337, *344*, *347*
Gantmacher, F.R.,   137, 140, 141, *344*
Geiringer, H.,   38, 44, *344*
Ghai, G.L.,   334, *344*
Gillespie, J.H.,   279, *344*
Gimelfarb, A.,   279, 307, 309–311, 335, 336, *344*
Ginzburg, L.R.,   62, *344*, *349*
Goldberg, K.,   241, *345*
Gottesman, I.I.,   218, *352*
Griffing, B.,   316–318, *344*

Haldane, J.B.S.,   47, 65, 66, 73, 108, 129, 132, 159, 160, 164, 201, 224, 259, 266, 271–273, 315, 317, 318, *340*, *344*, *345*
Hardy, G.H.,   29, *345*
Harris, D.L.,   216, *345*
Hartl, D.L.,   4, 35, 75, 168, 169, *342*, *345*
Hastings, A.,   189, 279, 317, 319, 320, *345*
Haynsworth, E.V.,   241, *345*
an der Heiden, U.,   60, *345*
Heyde, C.C.,   241, 242, *345*
Hill, W.G.,   279, *347*
Hoekstra, R.F.,   70, 71, *345*
Hom, C.L.,   279, *345*
Hughes, P.J.,   56, 60, *345*

Jacquard, A.,   222, *346*
James, J.W.,   288, *346*
Jayakar, S.D.,   65, 66, 70, 73, *345*
Jennings, H.S.,   36, 38, 225, 270, 271, *346*
Jensen, A.,   290, *346*
Jinks, J.L.,   279, *350*
Johnson, C.,   328, *346*
Johnson, N.L.,   297, *346*
Jukes, T.H.,   4, 200, *348*

Karlin, S.,   60, 70, 71, 102, 107, 108, 115, 123, 124, 137–139, 141, 142, 144, 148, 171, 185, 187, 189, 198, 253, 256, 271, 274, *346*, *347*, *353*
Kearsey, M.J.,   337, *344*, *347*

Keightley, P.D.,   279, *347*
Kempthorne, O.,   271, 279, 281, 287, 288, 296, *347*
Kendall, M.G.,   297, *347*
Kenett, R.S.,   144, *346*
Kidwell, J.F.,   171, *347*
Kimura, M.,   V, VII, 3, 4, 20, 22, 31, 32, 41, 60, 62, 76, 87, 90, 91, 95, 108, 115, 160, 177, 179, 182, 194, 195, 200, 202, 209, 221, 222, 224, 227, 230–233, 237, 238, 242, 245–247, 253, 256, 265, 266, 271–273, 275, 276, 315–318, 320, 321, 329, *342*, *347*, *348*
King, J.L.,   4, 73, 200, *348*
Kingman, J.F.C.,   VII, 57, 60, 200, 253, *348*
Klebaner, F.C.,   242, *348*
Kojima, K.,   177, 337, *348*, *349*
Kotz, S.,   297, *346*

Lande, R.,   280, 297, 301, 321, 325, 328, 329, *348*
LaSalle, J.P.,   19, 118, *348*
Latter, B.D.H.,   267, 317, 320, *348*
Léon, J.A.,   91, 95, *348*
Lessard, S.,   171, *346*
Levene, H.,   41, 42, 128, 136, 139, 143, 144, 146, 149, 152, 172, 173, *348*
Levy, J.,   43, 288, *348*, *349*, *351*
Lewontin, R.C.,   62, 63, 177, 337, *349*, *353*
Li, C.C.,   31, 32, 51, 59, 124, 145, 210, 215, *343*, *349*
Liberman, U.,   70, 71, *346*
Lloyd, D.G.,   99, *349*
Losert, V.,   60, *349*
Lukacs, E.,   298, *349*
Lush, J.L.,   256, *349*
Lyubich, Yu.I.,   60, *349*

Maistrovskii, G.D.,   60, *349*
Malécot, G.,   38, 132, 201, 202, 208, 222, 236–238, 240, 242, 243, 247, 256, 275, 279, 287, 290, 296, 309, 313, *349*
Mandel, S.P.H.,   56, 57, 60, 62, *349*, *353*
Mather, K.,   279, *350*
Maynard Smith, J.,   136, 149, 151, *350*
Maynard Smith, S.,   60, *352*
McGregor, J.,   137, 139, 198, 253, *346*, *347*
Mendel, G.,   29, 223
Moody, M.,   144, 189, *350*

Moran, P.A.P.,   VII, 57, 188, 195, 200,
    247, 249, 267, 277, 279, 296, *340*, *350*
Morton, N.E.,   219, 243, 245, 265, 266,
    342, *350*
Moshinsky, P.,   201, *345*
Mulholland, H.P.,   57, *350*
Muller, H.J.,   219, *350*
Murphy, E.A.,   215, *350*

Nagylaki, T.,   V, VII, 23, 43, 51, 60,
    70, 81, 83, 84, 87, 91, 95, 101, 116,
    117, 119, 125, 128, 132, 141, 142, 144,
    153, 162, 164, 170–172, 177, 184, 189,
    194, 195, 199, 200, 222, 227, 230, 248,
    249, 252, 253, 257, 261, 265, 267, 274,
    275, 277, 278, 280, 288, 297, 307, 309,
    315–317, 320, 324, 325, 334, 336, 337,
    339, *341*, *343*, *349–351*
Narain, P.,   273, *351*
Neel, J.V.,   219, *353*
Nei, M.,   4, 20, 259, *351*, *352*
Nohel, J.A.,   84, 104, 192, *341*
Norman, M.F.,   97, *352*
Norton, H.T.J.,   79, *352*

O'Donald, P.,   102, 108, *352*
Ohta, T.,   4, 177, *348*, *352*
Ol'khovskii, Yu.G.,   60, *349*
O'Rourke, D.H.,   218, *352*
Owen, A.R.G.,   171, *352*
Owen, R.D.,   74, *353*
Owen, R.E.,   167, *352*

Palm, G.,   160, *352*
Parsons, P.A.,   171, *352*
Passekov, V.P.,   VII, 200, *353*
Pearl, R.,   225, 271, 272, *352*
Pearson, K.,   287, 312, *352*
Penrose, L.S.,   50, 60, *352*
Pollak, E.,   244, 245, 247, 254, 266, 267,
    342, *352*
Prout, T.,   65, 146, 171, *347*, *352*

Reich, T.,   218, 335, *342*, *352*
Reid, M.,   43, *349*
Rice, J.,   218, 335, *342*, *352*
Richter-Dyn, N.,   138, *347*
Risch, H.,   309, *352*
Risch, N.,   218, *352*
Robbins, R.B.,   36, 38, 44, 225,
    270–272, *352*
Robertson, A.,   230, 233, 256, *353*

Sacks, L.,   210, 215, *349*
Schäfer, W.,   272, *353*
Scheuer, P.A.G.,   57, *353*
Schnell, F.W.,   288, *353*
Schull, W.J.,   219, *353*
Scudo, F.M.,   108, 115, *353*
Searle, A.G.,   154, *353*
Selgrade, J.F.,   171, *353*
Seneta, E.,   56, 60, 241, 242, *345*, *353*
Singh, M.,   337, *353*
Skellam, J.G.,   259, *353*
Slatkin, M.,   99, *353*
Smith, C.A.B.,   57, 279, 296, *350*
Snyder, L.H.,   33, *353*
Sprott, D.A.,   60, 159, *352*, *353*
Spuhler, J.N.,   290, *353*
Srb, A.M.,   74, *353*
Stettinger, F.,   279, *341*
Stewart, F.M.,   171, *347*
Strobeck, C.,   144, *353*
Stuart, A.,   297, *347*
Suarez, B.K.,   218, *352*
Svirezhev, Yu.M.,   VII, 200, *353*
Szucs, J.M.,   164, 171, *353*

Tachida, H.,   280, *353*, *355*
Taylor, H.M.,   256, *347*
Thompson, E.A.,   210, 212, 213, 235,
    353, *354*
Titchmarsh, E.C.,   67, 70, *354*
Trankell, A.,   43, *354*
Trustrum, G.B.,   287, *354*
Tuljapurkar, S.D.,   62, *349*
Turelli, M.,   279, 280, 329, *340*, *344*,
    354

Vandenberg, S.G.,   290, *354*
Vogel, F.,   154, *354*

Wagner, G.P.,   279, 280, *341*, *354*
Wahlund, S.,   40, *354*
Wallace, B.,   134, *354*
Walsh, J.B.,   320, *354*
Watterson, G.A.,   57, 247, 267, *340*,
    350
Weinberg, W.,   29, 279, 287, *354*
Weir, B.S.,   273, 330, *342*, *354*
Wilson, S.R.,   290, 309, *354*
Wright, S.,   V, 3, 4, 25, 45–47, 51, 63,
    73, 74, 98, 108, 129, 200–202, 208, 209,
    213, 221, 222, 224–227, 230, 233, 234,
    236–239, 242, 245, 253, 254, 256–259,
    261, 266, 267, 271, 275, 279–281, 283,

286, 287, 290, 296, 309, 328, 331, 336,
354, 355

Yule, G.U.,   287, 355

Zeng, Z.-B.,   280, 355
Ziehe, M.,   171, 353

Subject Index

Cross references that begin with an uppercase letter are to main entries; those that begin with a lowercase letter are to secondary entries under the same main entry.

ABO blood group, 32
Additive genetic variance, *see* Variance, genic
Age distribution, 13–17, 26, 79–81
Age independence, 16, 80–81
Age structure, 13–17, 26, 79–81
Allele, 1
— frequency, *see* Gene frequency
Allozygosity, 201, 234–235
Amino acid, 1
Analysis, 7
— complete, *see* global
— global, 7, 18–19, 21, 53, 96–99, 101, 103–105, 111–112, 122–123, 126–127
— local, 7, 103–105, 117–118, 140–141
Analysis of variance, *see* Variance
Ancestors, 201, 203, 225, 268, 269
Arrhenotoky, 35, 153, 169, 209
Asexual populations, 1, 5–27, 51, 95, 135
Assortative mating, 2, 102–103, 107–112, 125–126, 290–311, 334–336
— quantitative character, 2, 290–311, 334–336
Asymptotic equality, 54
Autosome, 35
Autozygosity, 201, 204, 209, 234–235
Average effect, 88–89, 168–169, 182–183, 282, 291
Average excess, *see* Average effect

Base, 1
Bilineal relatives, 213
Birth-and-death models, 249
Birth rate, *see* Fertility
Blood groups, 31, 32
— ABO, 32
— MN, 31
Breeding ratios, 33, 38, 42–43
Breeding value, 333

Carrying capacity, 20
Chromosome, 1, 35, 37
Circular mating, 230–234, 271
Coefficient of consanguinity, 202–204, 212, 235, 285
Coefficient of variation, 280
Common ancestor, 203
Competition, 20–23
Conditional process, 276
Conservative migration, 135–136, 141–142, 151
Continuous time, *see* Selection
Contraction mapping, 126
Convergence, 10, 11–12, 18–19, 53–55, 96–97, 101, 103–105, 117–118, 123, 126–127
— algebraic, 12, 54–55, 127
— geometric, 10, 11–12, 54, 97, 126–127
Correlation between mates, 290, 292, 294–296, 298, 310–311
Correlation between relatives, 284–290, 302–311, 333–336
— assortative mating, 302–311, 334–336
— — affinate relatives, 335–336
— — distant collateral relatives, 335–336
— — double first cousins, 309–311
— — first cousins, 307–308, 309
— — half-siblings, 306–307
— — nth parent and offspring, 304–305, 309
— — parent and offspring, 304
— — siblings, 305–306, 309
— — stepparent and stepoffspring, 335–336
— — stepsiblings, 335–336
— — uncle and niece, 307, 309
— epistasis, 287–288
— inbreeding, 289–290, 333–334

— intraclass, 287
— panmixia, 284–290
— stepsiblings, 289
— twins, 284, 288–289
— X-linkage, 288
Cotterman k-coefficients, 210–219, 269, 285–286
Coupling, 38
Covariance, 90
— conditional decomposition, 253
— genic, 90, 100–101
Crossing over, see Recombination
Cycling, 23, 189

Death rate, 14, 79
Decomposition of variance, see Variance
de Finetti diagram, 42
Deme, 134
Density dependence, 8, 18–19, 20, 60, 91–95
Deoxyribonucleic acid (DNA), 1
Detrimental equivalent, 219–220
Difference equations
— contraction, 126
— first-order
— — linear, homogeneous, 5–7, 103–105
— — linear, inhomogeneous, 10, 106
— — nonlinear, 11–13, 53, 96–97, 101, 110–112, 117–118, 122–123, 126–127
— higher-order, 106
— linear fractional, 11–13
— linearization, 117–118
— matrix, 103–105
— monotone, 96, 101
— multidimensional, 103–105
Dioecious population, 28, 31, 44, 170–173, 174–175, 238, 243, 246–247, 249, 252, 267, 268, 277; see also X-linkage
Diploids, 1
Direct descent, 203–204, 209
Disassortative mating, 2, 102–103, 116–124, 295
Discrete generations, 5
Dispersion, see Migration
DNA, 1
Dominance, 32, 51
— degree of, 51
— deviation, 88, 282

— overdominance, 3, 51, 53–54, 62, 96, 98, 188, 196
— underdominance, 51, 52, 98, 188, 196
— variance, see Variance
Dominant allele, 32, 51
Dosage compensation, 168
Double-first-cousin-mating, 227–230
Doubly stochastic migration, 136
Drift, genetic, see Random genetic drift
Duplicate genes, 38–39, 198

Effective number of loci
— assortative mating, 293–296, 298–302
— mutation, 325, 328
Effective population number, see Inbreeding and Variance
Environment, 18, 42 43, 51, 65–71, 217–219, 280
— deteriorating, 18
— periodic, 70, 71
— variable, 65–71
Environmental covariance, 284
Environmental variance, 280, 295–296
Epistasis, 4, 182, 197, 281
Epistatic deviation, 182, 197
Epistatic variance, 182–183, 281
Equilibration of fitnesses, 98–99
Equilibrium, 7; see also Analysis, Difference equations, and Polymorphism
— asymptotically stable, 7
— stable, 7
— stochastic, 242, 257–258
— unstable, 7
Events, 28
— independent, 28
— mutually exclusive, 28
Evolution, 1–4
— Fisherian, 4, 25
— neutral, 4
— non-Darwinian, 4
— Wrightian, 4, 25
Exchangeability, 261–262
Expectation, 40

Fertility, 5, 14–16, 26, 47–51, 60, 79–83, 96, 102–103, 112–116, 125–126, 153–159, 170–171, 174–176, 189–191, 249

— multiplicative, 51, 157, 170–171, 174–176, 249
— nonmultiplicative, 112–116, 125–126
Fertilization, 1
Fitness, 2–3, 5–6, 8, 13–16, 18–19, 47–51, 60, 79–83, 153–158, 164–171, 174–177, 187–191, 195–196, 313–315, 321
— absolute, 5–6, 8, 18–19, 51, 60, 95, 156–157
— continuous time, 13–16, 79–83, 189–191
— equilibration, 98-99
— function, 313–315, 321
— mean, see Mean fitness
— multiplicative, 51, 95, 162
— relative, 6, 8, 18–19, 51, 60, 95, 157
— scaling, 6, 8, 51, 60, 95, 157
— surface, 57, 60–62, 195, 198–199
— translation, 18–19
Fixation of allele, 3, 7, 276; see also Convergence
Fixation probability, 250, 256, 277
Fixed point, see Equilibrium
Frequency dependence, see Selection
Frobenius' theorem, 137
Functional iteration, see Difference equations
Fundamental Theorem of Natural Selection, 4, 8, 17–18, 57–60, 84–91, 100–101, 164–170, 177–184, 191–197, 198–199, 313–314
— continuous time, 17–18, 84–91, 100–101, 191–195
— multiple loci, 197, 198–199
— quantitative character, 313–314
— two loci, 177–184, 191–196
— X-linkage, 164-170

Gamete, 1, 37, 174, 176
Gametic dispersion, 135, 277–278
Gametic frequency, 37, 174, 176, 321
Gametic phase equilibrium, 37
Gametic selection, 51
Gene, 1
— duplicate, 38–39, 198
— frequency, 1, 6, 29, 31, 35–36, 47–51, 248
Genetic code, 1
Genetic counselling, 216–219, 269
Genetic disease and mortality, 219–220

Genetic drift, see Random genetic drift
Genetic load, 9, 18, 60, 219–220, 327–329
Genetic variance, see Variance
Genic variance, see Variance
Genotype, 1, 280–281
Genotype-environment covariance, 280–281, 284
Genotype-environment interaction, 280
Genotype-phenotype correlation, 281
Genotypic frequency, 29, 31, 35, 38, 40–42, 47–51, 153–154, 174–176, 248, 250, 254–255, 273, 321
— autosomal locus, 29, 31, 47–51, 321
— finite population, 40–42, 248, 250, 254–255
— inbred population, 63, 201–202, 234–235, 273
— multiple loci, 174–176, 321
— subdivided population, 40
— two loci, 38, 273
— X-linkage, 35, 153–154
Genotypic variance, see Variance
Geographical structure, 3, 13, 40, 99–100, 128–152, 171–173, 259–261, 277–278

Half-first-cousin mating, 226–227, 269
Half-sib mating, 225–226, 227, 271, 331
Handedness, 42–44
Haploids, 1, 5–27, 51, 95, 135, 252
Hard selection, see Selection
Hardy-Weinberg Law, 28–51, 83–87, 95, 108, 112, 126, 153–158, 170–171, 174–176, 191–194, 252, 260, 291, 321
— assortative mating, 108, 112, 126, 291
— continuous time, 83–87, 191–194
— dioecious population, 31, 44, 170–171
— monoecious population, 28–30, 42, 47–51, 83–87, 252, 321
— multiple loci, 174–176, 321
— two loci, 37–39, 44–46, 176, 191–194
— X-linked locus, 35–36, 45–46, 153–158
Heritability
— broad sense, 280–281, 288–289

— narrow sense,   283–284, 286–287, 289, 295–296, 327–329
Heterosis,   *see* Overdominance
Heterozygosity,   40, 202, 223, 234–247, 255, 274–276
Heterozygote,   1, 29
Homogeneous migration,   136
Homozygosity,   2, 40, 202, 222–223, 234–247, 274–276
Homozygote,   1, 29

Identity by descent,   201–204, 234–236, 272–273, 285–286
Identity in state,   24, 201, 222–223, 234–236
Inbreeding,   2, 63–64, 99, 103–108, 125, 200–247, 267–276, 283–284, 289–290, 333–334
— circular mating,   230–234, 271
— coefficient,   63, 201–210, 235, 267–269, 290
— coefficient of consanguinity, 202–204, 212, 235, 267–269, 285
— — X-linkage,   209–210, 269
— correlation between relatives, 289–290, 333–334
— depression,   202, 219–222
— double-first-cousin mating,   227– 230
— effective population number,   238, 239, 243–247, 274–276
— first-cousin mating,   227
— genotypic frequencies,   63, 201–202, 234–235, 254–255, 273
— half-first-cousin mating,   226–227
— half-sib mating,   225–226, 227, 271, 331
— k-coefficients,   210–219, 269, 285–286
— mating types,   271–272
— maximum avoidance of consan- guinity,   227–230, 232–233
— mean of character,   220–221
— mutation and,   275
— number of heterozygous loci, 235–236
— panmictic index,   234
— parent-offspring mating,   270–271
— partial selfing,   107–108, 125, 236–237, 272–273
— pedigrees,   203–219, 267–269
— phenotypic effects,   202, 219–222

— regular systems,   222–236, 270–272, 275
— — mutation,   275
— repeated backcrossing,   270
— selection with,   63–64, 99, 103–107
— selfing,   223–224
— selfing alternating with panmixia, 273–274
— selfing with selection,   103–107
— sib mating,   224–225, 226, 236, 271–272
— two loci,   272–273, 331
— variable population number, 239–242, 243–247
— variance of character,   221–222, 255–256, 283–284
— variance within and between lines, 255–256
— X-linkage,   209–210
Incidence,   217
Independent loci,   37
"Infinite alleles" model,   242–243, 275, 321–322
Intra-family selection,   71–73
Island model,   13, 99–100, 128–132, 171–172, 259–261, 277–278, 330–331
Iteration,   *see* Difference equations

Jensen's inequality,   57
Juvenile dispersion,   143–144, 172–173

k-coefficients,   210–219, 269, 285–286
Kronecker delta,   11

Lagrange multipliers,   57
Lethal allele,   6, 52, 96, 170, 259
Lethal equivalent,   219–220
Levene model,   136, 144–148, 152, 172–173
Life table,   13–17, 26, 79–81
Limits inferior and superior,   65–66, 99
Linear fractional transformation, 11–13
Linkage,   37–39, 44–46, 174–199, 272–273, 321–322
— disequilibrium,   37, 44–46, 176–199, 272–273, 324–325, 336
— equilibrium,   37, 44–46, 177, 272, 282–284, 330–332
Lipschitz condition,   126
Lipschitz constant,   126
Load,   *see* Genetic load

Locus, 1
Logistic equation, 20
Loss of allele, 3, 7; *see also*
    Convergence

Malthusian parameter, 16–19, 82–83,
    190–191
Mapping, *see* Difference equations
Maternal inheritance, 74
Mating, 2, 28–31, 35–36, 47–51, 79–83,
    96, 102–103, 153–158, 174–176,
    189–191, 222–242, 243–253, 261–267,
    269–275, 290
— assortative, *see* Assortative
    mating
— disassortative, 2, 102–103, 116–
    124, 295
— frequency, 30–31, 47–50, 79–83,
    102–103, 154, 174–176, 189–191, 290
— inbreeding, *see* Inbreeding
— random, *see* Panmixia
Matrix
— aperiodic, 140
— irreducible, 137, 140
— maximal eigenvalue, 137, 140
— maximal eigenvector, 137, 140
— nonnegative, 137, 140
Maximum avoidance of consanguinity,
    227–230, 232–233
Mean
— arithmetic, 66, 152
— geometric, 65, 145
— harmonic, 146, 152
Mean fitness, 6, 8–9, 17–19, 49–51,
    57–62, 82–83, 87–91, 93–95, 100–101,
    144–147, 155–158, 167–170, 176,
    181–184, 190, 194–199, 313–314,
    316–318, 326–328
— continuous time, 17–19, 82–83,
    87–91, 100–101, 190, 194–195
— density dependence, 93–95
— Levene model, 144–147
— quantitative character, 313–314,
    316–318, 326–328
— stationarity, 57, 169, 195
— two loci, 176, 181–184, 190,
    194–196
— X-linkage, 155–158, 167–170
Meiosis, 1
Meiotic drive, 75
Mendel's Law of Segregation, 3, 29–30
Migration, 3, 4, 13, 99–100, 128–152,
    171–173, 259–261, 277–278, 330–331

— conservative, 135–136, 141–142,
    151
— dioecious organism, 144, 171–173
— doubly stochastic, 136
— gametic, 135, 277–278
— haploids, 13, 135
— homogeneous, 136
— island model, 13, 99–100, 128–132,
    171–172, 259–261, 277–278, 330–331
— juvenile, 143–144, 172–173
— Levene model, 136, 144–148, 152,
    172–173
— matrix, 132–136
— mortality, 152
— plants, 144
— random drift and, 259–261,
    277–278
— random outbreeding and site
    homing, 136, 149–151
— reciprocal, 136, 142
— selection and, *see* Selection
— two niches, 148–151
— weak, 137, 139
— X-linkage, 144
— zygotic, 143–144, 172–173
MN blood group, 31
Monoecious population, 28
Monotonicity
— island model, 131
— Levene model, 152
— mapping, 96, 101
— selection, 25, 96
Moran models, 249
Mortality, 14, 79
Multinomial sampling
— gametes, 245, 247, 265, 267, 275
— genes, 248–253, 265
— genotypes, 248–253
Multiple loci, 174–176, 197, 198–199,
    279–339
Mutation, 3, 9–13, 23, 26–27, 75–79,
    96–97, 153–159, 162–164, 197–198,
    242–243, 248–253, 257–259, 275–276,
    321–329
— finite population, 242–243,
    248–253, 257–259, 275–276
— infinitely many alleles, 242–243,
    275, 321–322
— random drift and, 242–243,
    257–258, 275–276
— rate, 3, 9, 154, 322
— regular systems of inbreeding, 275
— selection and, *see* Selection

— selection, random drift, and, 248–253, 258–259

Nonoverlapping generations, 5, 28
Nucleotide, 1
Number of heterozygous loci, 235–236

O, o, 59
Offspring-number distribution, 24, 233–234, 243–247, 261–267
Optimum model, 4, 315, 318, 321, 336–339
Ordering of frequencies, 29–31, 35–36, 38, 248–250
Overdominance, 3, 51, 53–54, 62, 64, 66, 73, 74, 98, 146–147, 160–162, 185–186, 187–189
— inbreeding, 64
— intra-family selection, 73
— Levene model, 146–147
— maternal inheritance, 74
— multiallelic, 62, 98
— pairwise, 62
— total, 62
— two loci, 185–186, 187–189
— variable environment, 66
— X-linkage, 160–162
Overlapping generations, 13–17, 26, 79–81; see also Selection, continuous time

Panmictic index, 234
Panmixia, 2, 28–46, 50, 83, 157, 170–171, 174–176, 191, 233, 236–242, 243–253, 261–267, 274–275
— continuous time, 83, 191
— dioecious population, 31, 44, 170–171, 238, 246–247, 249
— effective population number, see Inbreeding and Variance
— finite population, 40–42, 233, 236–242, 243–253, 261–267, 274–275
— monoecious population, 28–30, 42–43, 50, 83, 174–176, 191, 233, 236–238, 243–246, 248–253
— multiple loci, 174–176
— two loci, 37–39, 43–46, 176, 191
— variable population number, 239–242, 243–247
— X-linkage, 35–36, 43, 45–46, 157, 274–275
Parent-offspring mating, 270–271

Parent-offspring probabilities, 33, 36, 42
Partial selfing, 107–108, 125, 223–224, 236–237, 272–273
Pedigrees, 203–219, 267–269
Penetrance, 42–43, 217–219
Perron-Frobenius theorem, 137
Phenotype, 1, 32, 280
Phenotypic variance, see Variance
Phenylketonuria (PKU), 32
Phenylthiocarbamide (PTC), 32
PKU, 32
Pleiotropy, 4
Pollen elimination, 119–124
Polymorphism, 3, 9
— assortative mating, 110–112, 125–126
— competitive, 20–22
— cyclic environment, 70, 71
— dioecious, 170–173
— fertility selection, 113–114
— finite population, 239–242
— frequency-dependent, 19–20, 25, 26, 98–99
— inbreeding, 226–227
— inbreeding and mutation, 275
— inbreeding and selection, 64, 105, 107
— intra-family selection, 73
— island model, 99–100, 130–131, 171, 172, 259–261, 277–278
— Levene model, 144–148, 152, 172–173
— maternal inheritance, 74
— migration, 151
— migration and random drift, 259–261, 277–278
— migration-selection, 99–100, 128–152, 171–173
— multiallelic, 56–62, 98, 162–164, 195–197
— multiple niches, 132–152, 171–173
— mutation, 9–10, 26, 159, 197
— mutation and random drift, 242–243, 257–258, 275–276
— mutation-selection, 10–13, 27, 75–79, 96–97, 162–164, 197–198, 321–329
— mutation, selection, and random drift, 258–259
— overdominant, 53–54, 62, 98
— partial selfing, 107–108
— periodic environment, 70, 71

— pollen elimination,   119–124
— protected, *see* Protected polymorphism
— random genetic drift,   239–242
— random outbreeding and site homing,   149–151
— self-incompatibility,   117
— selfing with selection,   105, 107
— stabilizing selection,   336–339
— two-locus,   187–189, 195–197, 337–339
— two niches,   148-151
— underdominant,   52, 54
— variable environment,   65–71, 99–100
— X-linked,   159–164, 170
— zygote elimination,   119–124
Population regulation,   8, 18–19, 20, 60, 91–95, 132–136, 249
Population subdivision,   40
Position effect,   176
Principal minor,   56
Probability,   28
— conditional,   28
— generating function,   250–252, 262–264
— — means and covariances,   251
— joint,   28
Progeny-number distribution,   24, 233–234, 243–247, 261–267
Protected polymorphism,   65, 136–138, 140–141
— dioecious population,   171–173
— frequency-dependent,   98
— intra-family selection,   73
— island model,   99–100, 129, 171–172
— Levene model,   144–148, 152, 172–173
— maternal inheritance,   74
— migration,   151
— migration-selection,   99–100, 128–152, 171–173
— — hard selection,   138–140, 142–143, 146–148, 172
— — juvenile dispersion,   143–144, 172–173
— — recessive allele,   140–144, 146, 150–151, 172–173
— — soft selection,   135–138, 140–142, 144–146, 148–151, 171–173
— multiple niches,   132–152, 171–173
— self-incompatibility alleles,   119
— two niches,   148–151

— variable environment,   65–71, 99–100
— X-linkage,   160
Protection of rare alleles,   61–62, 65–71, 136–144, 171–173, 185–187; *see also* Protected polymorphism
Protein,   1
PTC,   32

Quantitative character
— distribution,   2, 279, 296–298, 313–314, 318–320, 323–326
— variance,   *see* Variance
Quantitative genetics,   2, 87–90, 255–256, 279–339

Random genetic drift,   3–4, 24–25, 236–267, 273–278
— basic model,   248–253
— — conditional means and covariances,   250, 252
— — probability-generating functions,   250–252
— conditional process,   276
— dioecious population,   238, 243, 246–247, 249, 252, 277
— effective population number,   *see* Inbreeding *and* Variance
— fixation probability,   250, 256, 277
— genotypic frequencies,   254–255
— heterozygosity,   24–25, 236–242, 255, 274–275
— island model,   259–261, 277–278
— means and covariances of gene frequencies,   253–254, 276, 277
— migration and,   259–261, 277–278
— mutation and,   242–243, 257–258, 275–276
— mutation, selection, and,   248–253, 258–259
— variable population number,   239–242, 243–247
— variance within and between lines,   255–256
— X-linkage,   274–275, 277
Random mating,   *see* Panmixia
Rate of convergence,   *see* Convergence
Recessive allele,   1, 32–34, 36, 38–39, 43–44, 51, 54–55, 70–71, 77, 96–99, 108–116, 120, 140–144, 146, 150–151, 161, 172–173, 198, 202, 215–217, 222, 269
Reciprocal migration,   136, 142

Recombination,   2, 37, 44–46, 174–176, 321–322
— frequency,   37, 44–46
Recursion relations,   see Difference equations
Regression
— additive effects,   333
— allelic effects,   298, 324, 329, 334, 336
— biparental,   302–303, 311–314
— dominance deviations,   333
— genotype on phenotype,   281, 295, 297–299, 334, 336
— marital,   295, 297, 334
— midparent on offspring,   334
— offspring on nth parent,   304–305
— offspring on parent,   304
— phenotypic,   333
— stepoffspring on stepparent,   335
Regular systems of inbreeding, 222–236
Regulation of population,   see Population regulation
Relatives,   201
— bilineal,   213
— conditional probabilities,   33, 36, 38, 42–43, 215–216, 269
— correlation between,   see Correlation between relatives
— genetic counselling,   216–219, 269
— joint probabilities,   33–34, 38–39, 42–44, 215–216, 269
— k-coefficients,   210–219, 269
— regular,   210
— unilineal,   213
Repeated backcrossing,   270
Replication,   203–204, 209
Reproductive variance,   243–247, 261–267
Repulsion,   38

Sampling with replacement,   24, 236, 243, 248–253
Scaling of fitnesses,   6, 8, 51, 60, 95, 157
Schizophrenia,   218, 269
Segregation distortion,   75
Selection,   2–23, 25–27, 47–101, 103–107, 108–110, 112–116, 125, 126, 128–199, 248–253, 258–259, 311–329
— age structure with,   see continuous time

— assortative mating and,   108–110, 125, 126
— autosomal locus,   47–101, 103–107, 108–110, 112–116, 125, 126, 170–173
— coefficient,   3, 6, 51, 59
— competitive,   20–23
— continuous time,   13–23, 26–27, 79–91, 99, 100–101, 153, 189–195, 199
— cyclic,   70, 71
— density-dependent,   8, 18–19, 20, 60, 91–95
— differential,   312–315, 318
— dioecious population,   170–173, 174–175, 249, 252
— fertility,   112–116, 125–126
— finite population,   248–253, 258–259
— fluctuating,   3, 65–71, 99–100
— frequency-dependent,   3, 19–23, 25, 26, 91–95, 98–99
— gametic,   51
— haploid,   5–27, 51, 95, 135
— hard,   134, 135, 138–140, 142–143, 146–148, 172, 173
— inbreeding with,   63–64, 99, 103–107
— intensity,   3, 6, 51, 59, 312–315
— intra-family,   71–73
— island model,   13, 99–100, 128–132, 171–172
— Levene model,   136, 144–148, 152, 172–173
— maternal inheritance,   74
— meiotic drive,   75
— migration with,   13, 99–100, 128–152, 171–173
— — island model,   13, 99–100, 128–132, 171–172
— — Levene model,   136, 144–148, 152, 172–173
— — random outbreeding and site homing,   136, 149–151
— — two niches,   148–151
— monotonicity,   25, 96
— multiple loci,   174–176, 197, 198–199, 311–329, 336–337
— multiple niches,   128–152, 171–173
— mutation and random drift, 248–253, 258–259
— mutation with
— — diploids,   75–79, 96–97
— — haploids,   9–13, 23, 26–27

—— multiple loci,   321–329
—— quantitative character,   321–329
—— two loci,   197–198
—— X-linkage,   153–159, 162–164
— nonmultiplicative fertilities,
   112–116, 125–126
— optimum model,   see stabilizing
— overdominant,   see Overdominance
— overlapping generations,   see
   continuous time
— periodic,   70, 71
— quantitative character,   87–90,
   311–329, 336–339
— random outbreeding and site
   homing,   136, 149–151
— recessive allele,   see Recessive allele
— response to,   312, 315–320
— selfing with,   103–107
— sex linkage,   see X-linkage and
   Y-linkage
— soft,   134, 135–138, 140–142,
   144–146, 148–151, 171–173
— stabilizing,   4, 315, 318, 321,
   336–339
— time-dependent,   65–71, 99–100
— truncation,   314–315, 317
— two loci,   174–199, 337–339
—— additive loci,   187–188, 196–197
—— continuous time,   189–195, 199
—— cycling,   189
—— multiple alleles,   174–184,
   189–197, 198–199
—— multiplicative loci,   187,
   188–189, 195–196
—— mutation-selection balance,
   197–198
—— rare alleles,   185–187
—— stabilizing,   337–339
—— two alleles,   185–189, 197–198
— two niches,   148–151
— underdominant,   see Underdominance
— weak,   9, 27, 59–60, 72–73, 74, 84–
   95, 99, 164–170, 177–184, 191–195,
   198–199, 252–253, 315, 321–329,
   336–339
— X-linkage,   153–170
—— multiple alleles,   153–159,
   162–170
—— mutation-selection balance,
   162–164
—— two alleles,   159–162, 170
— Y-linkage,   5, 35

Self-fertilization,   103–108, 125,
   223–224
Self-fertilization alternating with
   panmixia,   273–274
Self-incompatibility,   116–124
— alleles,   116–119
— pollen elimination,   119–124
— zygote elimination,   119–124
Self-sterility,   see Self-incompatibility
Sex chromosome,   35
Sex linkage,   see X-linkage and
   Y-linkage
Sex ratio,   31, 36, 156–159, 170–171,
   174
Shifting-balance theory,   4, 25
Sib mating,   224–225, 226, 236,
   271–272
Sibling distributions,   33–34, 38–39,
   42–44
Sickle-cell anemia,   53–54, 66
Simplex,   57
Snyder's ratios,   33
Soft selection,   see Selection
Solution, complete,   7
Stability,   see Equilibrium
Stabilizing selection,   see Selection
Stationary state,   see Equilibrium
Subdivision of population,   40

Translation of fitnesses,   18–19
Truncation selection,   314–315, 317
Twins,   284, 288–289
Two loci,   37–39, 43–46, 174–199,
   272–273, 330–332, 334, 337–339
— additive loci,   187–188, 196–197
— assortative mating,   334
— continuous time,   189–195, 199
— cycling,   189
— dioecious population,   44
— duplicate loci,   38–39, 198
— half-sib mating,   331
— inbreeding,   272–273
— island model,   330–331
— multiple alleles,   37–38, 44–46,
   174–184, 189–197, 198–199
— multiplicative loci, 187, 188–189,
   195–196
— rare alleles,   185–187
— stabilizing selection,   337–339
— two alleles,   38, 43–44, 185–189,
   197–198, 337–339
— X-linkage,   45–46

Underdominance,   51, 52, 54, 64, 73,
    74, 98, 160, 187–189
Unilineal relatives,   213
Unlinked loci,   37

Variance,   8, 40
— additive genetic,   *see* genic
— additivity of,   88–89, 168–169,
    182–183, 280–284
— assortative mating,   290–302
— between relatives,   287
— conditional decomposition,   253
— dominance,   59, 88–89, 168–169,
    282–284, 285, 296, 329, 333
— effective population number,
    261–267
— environmental,   280, 295–296
— epistatic,   182–183, 281
— gametic,   182–183
— genic,   8–9, 18, 59, 88–89, 168–169,
    182–183, 255–256, 282–284, 285,
    290–302, 314, 325–330, 333
— genotypic,   59, 88–89, 168–169,
    182–183, 221–222, 280–284, 285, 296
— haploid,   8–9, 18
— inbreeding,   283–284
— multiple loci,   281–284
— mutational,   325
— phenotypic,   280, 283–284, 295,
    326, 333
— total genetic,   *see* genotypic
— twins,   288–289
— two loci,   182–183
— within and between lines,   255–256
— within relatives,   287, 310, 333
— X-linkage,   168–169, 288
Variation of parameters,   84, 165, 192
Viability,   5, 47–48, 95, 153, 170, 175,
    248, 252
— multiplicative,   48, 95, 252

Wahlund's principle,   40
Wright-Fisher model,   252–259

X chromosome,   35
X-linkage,   35–36, 43, 45–46, 144,
    153–170, 209–210, 249, 267, 269,
    270–271, 274–275, 277

Y chromosome,   35
Y-linkage,   5, 35

Zygote,   1, 47
— elimination,   119–124
— migration,   143–144, 172–173

Volume 20

J. E. Cohen, F. Briand, C. M. Newman

Community Food Webs

Data and Theory

With a Contribution by Z. J. Palka

1990. XI, 308 pp. 46 figs. Hardcover DM 148,– ISBN 3-540-51129-6

Food webs hold a central place in ecology. They describe which organisms feed on which others in natural habitats. This book describes recently discovered empirical regularities in real food webs: it proposes a novel theory unifying many of these regularities, as well as extensive empirical data.

After a general introduction, reviewing the empirical and theoretical discoveries about food webs, the second portion of the book shows that community food webs obey several striking phenomenological regularities. Some of these unify, regardless of habitat. Others differentiate, showing that habitat significantly influences structure. The third portion of the book presents a theoretical analysis of some of the unifying empirical regularities. The fourth portion of the book presents 13 community food webs. Collected from scattered sources and carefully edited, they are the empirical basis for the results in the volume. The largest available set of data on community food webs provides a valuable foundation for future studies of community food webs.

Volume 19

J. D. Murray

Mathematical Biology

1989. XIV, 767 pp. 262 figs. Hardcover DM 98,– ISBN 3-540-19460-6

This textbook gives an in-depth account of the practical use of mathematical modelling in several important and diverse areas in the biomedical sciences.

The emphasis is on what is required to solve the real biological problem.

The subject matter is drawn, for example, from population biology, reaction kinetics, biological oscillators and switches, Belousov-Zhabotinskii reaction, neural models, spread of epidemics.

Volume 18

S. A. Levin, T. G. Hallam, L. J. Gross (Eds.)

Applied Mathematical Ecology

1989. XIV, 491 pp. 114 figs. Hardcover DM 98,– ISBN 3-540-19465-7

Contents: Introduction. – Resource Management. – Epidemiology: Fundamental Aspects of Epidemiology Case Studies. – Ecotoxicology. – Demography and Population Biology. – Author Index. – Subject Index.

Springer-Verlag
Berlin
Heidelberg
New York
London
Paris
Tokyo
Hong Kong
Barcelona
Budapest

Volume 17
T. G. Hallam, S. A. Levin (Eds.)

Mathematical Ecology
An Introduction
1986. XII, 457 pp. 84 figs. Hardcover DM 178,–
ISBN 3-540-13631-2

Volume 16
J. L. Casti, A. Karlqvist (Eds.)

Complexity, Language, and Life: Mathematical Approaches
1986. XIII, 281 pp. Hardcover DM 158,–
ISBN 3-540-16180-5

Volume 15
D. L. DeAngelis, W. M. Post, C. C. Travis

Positive Feedback in Natural Systems
1986. XII, 290 pp. 90 figs. Hardcover DM 168,–
ISBN 3-540-15942-8

Volume 14
C. J. Mode

Stochastic Processes in Demography and Their Computer Implementation
1985. XVII, 389 pp. 49 figs. 80 tabs.
Hardcover DM 258,– ISBN 3-540-13622-3

Volume 13
J. Impagliazzo

Deterministic Aspects of Mathematical Demography
An Investigation of the Stable Theory of Population Including an Analysis of the Population Statistics of Denmark
1985. XI, 186 pp. 52 figs. Hardcover DM 130,–
ISBN 3-540-13616-9

Volume 12
R. Gittins

Canonical Analysis
A Review with Applications in Ecology
1985. XVI, 351 pp. 16 figs. Hardcover DM 148,–
ISBN 3-540-13617-7

Volume 11
B. G. Mirkin, S. N. Rodin

Graphs and Genes
Translated from the Russian by H. L. Beus
1984. XIV, 197 pp. 46 figs. Hardcover DM 75,–
ISBN 3-540-12657-0

Volume 10
A. Okubo

Diffusion and Ecological Problems: Mathematical Models
1980. XIII, 254 pp. 114 figs. 6 tabs.
Hardcover DM 114,– ISBN 3-540-09620-5

Volume 9
W. J. Ewens

Mathematical Population Genetics
1979. XII, 325 pp. 4 figs. 17 tabs.
Hardcover DM 96,– ISBN 3-540-09577-2

Volume 7
E. R. Lewis

Network Models in Population Biology
1977. XII, 402 pp. 187 figs.
Hardcover DM 78,– ISBN 3-540-08214-X

Volume 6
D. Smith, N. Keyfitz

Mathematical Demography
Selected Papers
1977. XI, 514 pp. 31 figs.
Hardcover DM 96,– ISBN 3-540-07899-1

Volume 5
A. Jacquard

The Genetic Structure of Populations
Translated by D. Charlesworth, B. Charlesworth
1974. XVIII, 569 pp. 92 figs.
Hardcover DM 128,– 3-540-06329-3

Volume 4
M. Iosifescu, P. Tautu

Stochastic Processes and Applications in Biology and Medicine II
Models
1973. 337 pp.
Hardcover DM 84,–
ISBN 3-540-06271-8

Springer-Verlag
Berlin
Heidelberg
New York
London
Paris
Tokyo
Hong Kong
Barcelona
Budapest

Printed by Publishers' Graphics LLC USA